London Mathematical Society Lecture Note Series. 95

Low Dimensional Topology

Edited by

ROGER FENN
University of Sussex

The right of the University of Cambridge to print and sell all manner of books was granted by Henry VIII in 1534. The University has printed and published continuously since 1584.

CAMBRIDGE UNIVERSITY PRESS
Cambridge
London New York New Rochelle
Melbourne Sydney

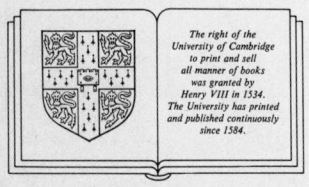

Published by the Press Syndicate of the University of Cambridge
The Pitt Building, Trumpington Street, Cambridge CB2 1RP
32 East 57th Street, New York, NY 10022, USA
10, Stamford Road, Oakleigh, Melbourne 3166, Australia

First published 1985

Printed in Great Britain at the University Press, Cambridge

Library of Congress catalogue card number: 85-47811

British Library cataloguing in publication data

Low dimensional topology. - (London Mathematical
Society lecture note series, ISSN 0076-0052; 95)
1. Topological spaces
I. Fenn, Roger II. Series
514'.2 QA611.3

ISBN 0 521 26982 2

LOW DIMENSIONAL TOPOLOGY

Sussex, 1982

This volume is a record of the talks given at the third topology seminar of the University of Sussex held at The White House, Chelwood Gate, from the 2nd to the 6th August, 1982. I would like to thank Jill Foster and Sheila Collier for their immaculate typing of the transcript and all the staff at The White House for their efforts in making us feel so welcome. On the suggestion of Laurie Siebenmann, I have dedicated this volume to Professor Seifert.

Roger Fenn

DEDICATION

This volume is dedicated to Professor Herbert Seifert, consummate craftsman, on his 78th birthday.

LONDON MATHEMATICAL SOCIETY LECTURE NOTE SERIES

4 Algebraic topology, J.F.ADAMS
5 Commutative algebra, J.T.KNIGHT
8 Integration and harmonic analysis on compact groups, R.E.EDWARDS
11 New developments in topology, G.SEGAL (ed)
12 Symposium on complex analysis, J.CLUNIE & W.K.HAYMAN (eds)
13 Combinatorics, T.P.McDONOUGH & V.C.MAVRON (eds)
15 An introduction to topological groups, P.J.HIGGINS
16 Topics in finite groups, T.M.GAGEN
17 Differential germs and catastrophes, Th.BROCKER & L.LANDER
18 A geometric approach to homology theory, S.BUONCRISTIANO, C.P.ROURKE & B.J.SANDERSON
20 Sheaf theory, B.R.TENNISON
21 Automatic continuity of linear operators, A.M.SINCLAIR
23 Parallelisms of complete designs, P.J.CAMERON
24 The topology of Stiefel manifolds, I.M.JAMES
25 Lie groups and compact groups, J.F.PRICE
26 Transformation groups, C.KOSNIOWSKI (ed)
27 Skew field constructions, P.M.COHN
29 Pontryagin duality and the structure of LCA groups, S.A.MORRIS
30 Interaction models, N.L.BIGGS
31 Continuous crossed products and type III von Neumann algebras, A.VAN DAELE
32 Uniform algebras and Jensen measures, T.W.GAMELIN
34 Representation theory of Lie groups, M.F. ATIYAH et al.
35 Trace ideals and their applications, B.SIMON
36 Homological group theory, C.T.C.WALL (ed)
37 Partially ordered rings and semi-algebraic geometry, G.W.BRUMFIEL
38 Surveys in combinatorics, B.BOLLOBAS (ed)
39 Affine sets and affine groups, D.G.NORTHCOTT
40 Introduction to Hp spaces, P.J.KOOSIS
41 Theory and applications of Hopf bifurcation, B.D.HASSARD, N.D.KAZARINOFF & Y-H.WAN
42 Topics in the theory of group presentations, D.L.JOHNSON
43 Graphs, codes and designs, P.J.CAMERON & J.H.VAN LINT
44 Z/2-homotopy theory, M.C.CRABB
45 Recursion theory: its generalisations and applications, F.R.DRAKE & S.S.WAINER (eds)
46 p-adic analysis: a short course on recent work, N.KOBLITZ
47 Coding the Universe, A.BELLER, R.JENSEN & P.WELCH
48 Low-dimensional topology, R.BROWN & T.L.THICKSTUN (eds)
49 Finite geometries and designs, P.CAMERON, J.W.P.HIRSCHFELD & D.R.HUGHES (eds)
50 Commutator calculus and groups of homotopy classes, H.J.BAUES
51 Synthetic differential geometry, A.KOCK
52 Combinatorics, H.N.V.TEMPERLEY (ed)
53 Singularity theory, V.I.ARNOLD
54 Markov process and related problems of analysis, E.B.DYNKIN
55 Ordered permutation groups, A.M.W.GLASS
56 Journêes arithmêtiques, J.V.ARMITAGE (ed)
57 Techniques of geometric topology, R.A.FENN
58 Singularities of smooth functions and maps, J.A.MARTINET
59 Applicable differential geometry, M.CRAMPIN & F.A.E.PIRANI
60 Integrable systems, S.P.NOVIKOV et al
61 The core model, A.DODD

62 Economics for mathematicians, J.W.S.CASSELS
63 Continuous semigroups in Banach algebras, A.M.SINCLAIR
64 Basic concepts of enriched category theory, G.M.KELLY
65 Several complex variables and complex manifolds I, M.J.FIELD
66 Several complex variables and complex manifolds II, M.J.FIELD
67 Classification problems in ergodic theory, W.PARRY & S.TUNCEL
68 Complex algebraic surfaces, A.BEAUVILLE
69 Representation theory, I.M.GELFAND et al.
70 Stochastic differential equations on manifolds, K.D.ELWORTHY
71 Groups - St Andrews 1981, C.M.CAMPBELL & E.F.ROBERTSON (eds)
72 Commutative algebra: Durham 1981, R.Y.SHARP (ed)
73 Riemann surfaces: a view towards several complex variables,A.T.HUCKLEBERRY
74 Symmetric designs: an algebraic approach, E.S.LANDER
75 New geometric splittings of classical knots, L.SIEBENMANN & F.BONAHON
76 Linear differential operators, H.O.CORDES
77 Isolated singular points on complete intersections, E.J.N.LOOIJENGA
78 A primer on Riemann surfaces, A.F.BEARDON
79 Probability, statistics and analysis, J.F.C.KINGMAN & G.E.H.REUTER (eds)
80 Introduction to the representation theory of compact and locally compact groups, A.ROBERT
81 Skew fields, P.K.DRAXL
82 Surveys in combinatorics, E.K.LLOYD (ed)
83 Homogeneous structures on Riemannian manifolds, F.TRICERRI & L.VANHECKE
84 Finite group algebras and their modules, P.LANDROCK
85 Solitons, P.G.DRAZIN
86 Topological topics, I.M.JAMES (ed)
87 Surveys in set theory, A.R.D.MATHIAS (ed)
88 FPF ring theory, C.FAITH & S.PAGE
89 An F-space sampler, N.J.KALTON, N.T.PECK & J.W.ROBERTS
90 Polytopes and symmetry, S.A.ROBERTSON
91 Classgroups of group rings, M.J.TAYLOR
92 Representation of rings over skew fields, A.H.SCHOFIELD
93 Aspects of topology, I.M.JAMES & E.H.KRONHEIMER (eds)
94 Representations of general linear groups, G.D.JAMES
95 Low-dimensional topology 1982, R.A.FENN (ed)
96 Diophantine equations over function fields, R.C.MASON
97 Varieties of constructive mathematics, D.S.BRIDGES & F.RICHMAN
98 Localization in Noetherian rings, A.V.JATEGAONKAR
99 Methods of differential geometry in algebraic topology, M.KAROUBI & C.LERUSTE
100 Stopping time techniques for analysts and probabilists, L.EGGHE
101 Groups and geometry, ROGER C.LYNDON
102 Topology of the automorphism group of a free group, S.M.GERSTEN
103 Surveys in combinatorics 1985, I.ANDERSEN (ed)
104 Elliptical structures on 3-manifolds, C.B.THOMAS
105 A local spectral theory for closed operators, I.ERDELYI & WANG SHENGWANG
106 Syzygies, E.G.EVANS & P.GRIFFITH

CONTENTS

GEOMETRIC TOPOLOGY CONFERENCE

2 - 6 August 1982

University of Sussex

LIST OF PARTICIPANTS

BIRMAN, Joan	Department of Mathematics, Columbia University, New York, NY 10027, U.S.A.
BOILEAU, Michel	Institut de Mathematiques, 2-4 rue du Lievre, CH 1211, CP 124 Geneve, Switzerland.
BONAHON, Francis	Mathematiques (Batiment 425), Universite de Paris-Sud, 91405 Orsay, Cedex, France.
BRYANT, John	Department of Mathematics and Computing Science, Florida State University, Tallahassee 32306, U.S.A.
BRYANT, Virginia	Department of Mathematics and Computing Science, Florida State University, Tallahassee 32306, U.S.A.
CAUDRON, Alain	13 Rue Khaourizmi, 2070 Lamarsa, Tunisia.
CESAR DE SA, Eugenia	Seccad de Matematica, Faculdade de Ciencias, da Universidade do Porto, 4000 Porto, Portugal.

CLARK, Doug	Department of Mathematics, Heriot Watt University, Edinburgh EH14 4AS.
DUNWOODY, Martin	Mathematics Division, University of Sussex, Falmer, Brighton BN1 9QH.
FENN, Roger	Mathematics Division, University of Sussex, Falmer, Brighton BN1 9QH.
GORDON, Cameron	Department of Mathematics, University of Texas, Austin, TX 78712, U.S.A.
HENDRIKS, Harrie	Mathematisch Institut, Katholieke Universiteit, Nijmegen, Netherlands.
HOUGHTON, Chris H.	Department of Pure Mathematics, University College, P.O. Box 73, Cardiff CF1 1XL.
KEARTON, Cherry	Mathematics Department, Durham University, Durham DH0 3LE.
LICKORISH, Raymond	Department of Pure Mathematics and Mathematical Statistics, University of Cambridge, 16, Mill Lane, Cambridge CB2 1SB.
LINES, Daniel	Institut de Mathematiques, Universite de Neuchatel, 2000 Neuchatel, Switzerland.
LITHERLAND, Rick	Department of Pure Mathematics and Mathematical Statistics, University of Cambridge, 16, Mill Lane, Cambridge CB2 1SB.
MAEDA, T.	Department of Mathematics, University of Toronto, Toronto, M5S 1A1, Canada.

MAGAJNA, Zlatan	Department of Mathematics, University of Ljubljana, PP14, 6111 Ljubljana, Yugoslavia.
MARTIN, Nigel	Department of Mathematics, Science Laboratories, University of Durham, Durham DH1 3LE.
MORTON, Hugh	Department of Pure Mathematics, The University, Liverpool L69 3BX.
MURASUGI, Kunio	Department of Mathematics, University of Toronto, Toronto, M5S 1A1, Canada.
OTAL, Jean Pierre	Mathematiques (Batiment 425), University de Paris Sud, 91405 Orsay, Cedex, France.
PRIDE, Steven J.	Department of Mathematics, University of Glasgow, University Gardens, Glasgow G12 8QW.
REEVE, John	Department of Mathematics, University of East Anglia, Norwich.
REGO, Eduardo	Mathematics Institute, University of Warwick, Coventry CV4 7AL.
ROLFSON, Dale	Mathematics Department, University of British Columbia, Vancouver, Canada B.C.
ROURKE, Colin	Mathematics Institute, University of Warwick, Coventry CV4 7AL.
SCOTT, Peter	Department of Pure Mathematics, University of Liverpool, P.O. Box 147, Liverpool.

SHORT, Hamish
Department of Pure Mathematics,
University of Liverpool,
P.O. Box 147,
Liverpool.

SIEBENMANN, Larry
Department of Mathematics,
Universite de Paris-Sud,
91405 Orsay,
France.

STRICKLAND, Paul
Department of Mathematics,
Science Laboratories,
University of Durham,
Durham DH1 3LH.

THISTLETHWAITE, Morwen
26, Queen's Road,
Loughton,
Essex.

TROTMAN, David
c/o D. Lodge,
Trillgate,
The Slad,
Stroud,
Gloucs.

VRABEC, Jose
Department of Mathematics,
University of Ljubljana,
PP14, 6111 Ljubljana,
Yugoslavia.

WALTON, Thomas
33, Gwydr Crescent,
Uplands,
Swansea,
South Wales.

WICHA-KRAUSE, Barbel
Department of Mathematics,
Ruhr-Universitat,
Universitat str. 150,
463 Bochum,
West Germany.

YAHIA, M.
School of Mathematics,
University of Khartoum,
Khartoum,
Sudan.

ZIESCHANG, Heiner
Department of Mathematics,
Ruhr-Universitat,
Universitat str. 150,
463 Bochum,
West Germany.

ZIESCHANG, Mrs.

Department of Mathematics,
Ruhr-Universitat,
Universitat str. 150,
463 Bochum,
West Germany.

ZIMMERMAN, Bruno

Department of Mathematics,
Ruhr-Universitat,
Universitat str. 150,
463 Bochum,
West Germany.

NOEUDS RIGIDEMENT INVERSIBLES

C. Michel Boileau

Abstract. A knot is called *invertible* if there is an orientation preserving homeomorphism of space which reverses the orientation of the knot. It is called *rigidly invertible* if the homeomorphism is an involution.

There are knots known to be invertible but not rigidly so (the Montesinos conjecture). In this paper, the author shows that they all have companions of a specific type. In particular, the invertible fibred knots are rigidly invertible.

Un noeud K est une sous-variété lisse, connexe, close de dimension 1, plongée dans la sphère orientée S^3.

Un noeud K est dit inversible s'il existe un homéomorphisme du couple (S^3, K) qui est de degré $+1$ dans S^3 et de degré -1 sur K (arbitrairement orienté). Le noeud K est dit *"rigidement inversible"* s'il est inversible et s'il peut être inversé par une involution de S^3; cette involution admet alors un cercle de points fixes non noué qui rencontre K en deux points (cf. Montesinos, J.M., 1975).

J.M. Montesinos a conjecturé (cf. Montesinos, J.M., 1975, et Kirby, R., 1978, pb.1-6) que: "tout noeud inversible est rigidement inversible".

Des contre-exemples à cette conjecture ont été exhibés indépendamment par R. Hartley (1980) et W. Whitten (1981)(1980). Ces contre-exemples K ont tous la propriété suivante : ils admettent tous un compagnon non inversible K_0, pour lequel K a un *"nombre de tours"* (ou "winding number") nul. C'est-à-dire qu'il existe dans $S^3 - K$ un tore plongé T_0 incompressible, non périphérique, tel que le nombre d'enlacement d'un méridien de T_0 avec K est nul; l'âme K_0 du tore plein bordé par T_0 dans S^3 est appelé compagnon de K et la valeur absolue du nombre d'enlacement du méridien de T_0 avec K est par définition le nombre de tours de K par rapport à son compagnon K_0. Un contre-exemple typique à la conjecture de Montesinos est n'importe quel double au sens de Whitehead d'un noeud non inversible (voir (Whitten, W., 1981), Hartley, R., 1980)).

Le but de cet article est de démontrer le théorème suivant:

Théorème.

Soit K un noeud inversible, n'ayant aucun compagnon pour lequel K a un nombre de tours nul. Alors, le noeud K est rigidement inversible.

En particulier, si on considère la classe des noeuds fibrés K dans S (c'est-à-dire les noeuds dont le complément S - K est fibré sur le cercle S avec pour fibre une surface de Seifert du noeud K ; voir par exemple (Kervaire, M. et Weber, C., 1977), on obtient le corollaire suivant :

Corollaire.

Un noeud fibré K dans S est inversible si et seulement s'il est rigidement inversible.

Au § 4, nous donnerons une condition nécessaire et suffisante pour qu'un noeud soit rigidement inversible, en utilisant la notion d'arbre caractéristique de compagnonnage (cf.(Bonahon, F., et Siebenmann, L.); voir § 1). Si la caractérisation, donnée au § 4, ne permet pas de décrire explicitement les noeuds rigidement inversibles, elle généralise les résultats obtenus dans cette direction par R. R. Hartley (1980) et W. Whitten (1981),(1980). En particulier elle permet de donner des exemples de noeuds inversibles qui ne sont pas rigidement inversibles, mais dont tous les compagnons au sens de Schubert sont isotopes à un même noeud rigidement inversible, contrairement aux exemples donnés par Hartley et Whitten.

Je tiens à remercier Daniel Lines et Claude Weber pour de nombreuses et utiles conversations durant la rédaction de ce travail.

1. *Préliminaires.*

Dans tout ce qui suit on suppose qu'on est dans la catégorie P L. ou C^∞.

Pour un noeud K dans S^3, on note $N(K)$ un voisinage tubulaire de K et $X = S^3 - \mathring{N}(K)$ l'extérieur de K.

La démonstration du théorème utilise l'existence, donnée par Johannson (Johannson, K., 1979) et Jaco-Shalen (Jaco, W. et Shalen, P., 1979), d'une famille finie $\mathcal{T}$ de tores disjoints, incompressibles, non parallèles entre eux, plongés dans X (l'un des tores correspond au bord ∂X), et vérifiant les propriétés suivantes :

(1) Les composantes connexes fermées de $X - \mathcal{T}$ sont soit des fibrés de Seifert (c'est-à-dire admettent une action différentiable du cercle S^1), soit des variétés atoroïdales et anannulaires (c'est-à-dire ne contiennent aucun tore incompressible, plongé, non périphérique, ni aucun anneau incompressible, ∂-incompressible, proprement plongé et non parralèle au bord); ces dernières variétés portent sur leur intérieur une structure hyperbolique complète à volume fini, d'après le théorème d'hyperbolisation de Thurston (Thurston, W.P., (1982), (1980)).

(2) La famille $\mathcal{T}$ est minimale, et donc unique à isotopie près.

On considère le graphe dont les sommets sont les composantes connexes fermées de $X - \mathcal{T}$ et les arêtes correspondent aux tores de la famille $\mathcal{T}$ (au tore ∂X correspond une arête de valence libre dans le graphe; voir (Bonahon, F., et Siebenmann, L., Chapter 3). Ce graphe est en fait un arbre car $H_1(X) \cong Z$ est engendré par un méridien de $\partial X = \partial N(K)$. Comme la famille $\mathcal{T}$ est unique à isotopie près, cet arbre est appelé l'arbre caractéristique de compagnonnage de K (cf. Bonahon, F., et Siebenmann, L.).

Chaque tore T_j de $\mathcal{T}$ correspond à un compagnon K_j de K, qui est l'âme du tore plein V_j bordé par T_j dans S^3. Ce tore plein contient K dans son intérieur car T_j est incompressible dans X.

Par définition, la complexité $c(K)$ du noeud K est égal au nombre d'éléments de la famille $\mathcal{T}$. Si $c(K)$ est nulle, K est le noeud trivial. Si $c(K) = 1$, K est un noeud simple au sens de Schubert (Schubert, H., 1953), c'est-à-dire un noeud torique ou hyperbolique d'après le théorème d'hyperbolisation de Thurston (Thurston, W.P., (1982), (1980)).

La démonstration du théorème s'effectuera par récurrence sur la complexité $c(K)$ du noeud K.

Dans la démonstration du théorème nous serons particulièrement intéressés par la composante fermée X_0 de $X - \mathcal{T}$, dont le bord contient ∂X. La sous-famille $\mathcal{T}_0 = \{\partial X, T_1, \dots, T_n\}$ de $\mathcal{T}$, telle que $\partial X \cup T_1 \cup \dots \cup T_n = \partial X_0$, est elle-même unique à isotopie près. De plus, $X = X_0 \cup Y_1 \cup \dots \cup Y_n$, où chaque variété Y_i, $1 \leq i \leq n$, est une composante connexe fermée de $X - X_0$ et s'identifie à l'extérieur du compagnon K_i de K, associé au tore incompressible non périphérique T_i plongé dans X. Ces tores T_i (ou ces compagnons K_i), $1 \leq i \leq n$, peuvent être considérés comme les tores incompressibles non périphériques (ou les compagnons) "les plus proches" au sens de Schubert (Schubert, H., 1953) de ∂X dans X (ou de K dans $S^3 - K$).

En particulier, on a la relation $\sum_1^n c(K_i) + 1 = c(K)$, d'où la complexité de chacun des compagnons K_i, les plus proches de K, est inférieure strictement à celle de K dès que $n \geq 1$. De plus, si l'un des noeuds K_i, $1 \leq i \leq n$, admet un compagnon pour lequel il a un nombre de tours nul, c'est aussi un compagnon de K pour lequel K a un nombre de tours nul, car tout compagnon de K_i est un compagnon de K et le nombre de tours se comporte multiplicativement par compagnonnage (cf. (Schubert, H., 1953) ou (Weber, C., 1980)).

Si on suppose que K n'admet pas de compagnon pour lequel il a un nombre de tours nul, cette propriété reste vraie pour les compagnons K_i de K. En particulier, si le noeud K n'est pas simple au sens de Schubert ($c(K) \geq 2$), l'hypothèse de récurrence pourra s'appliquer aux compagnons inversibles K_i les plus proches de K.

Dans le § 2 nous allons effectuer le premier pas de la démonstration par récurrence, en considérant le cas des noeuds simples au sens de Schubert, c'est-à-dire de complexité 1.

Dans la suite, nous utiliserons souvent le lemme suivant (cf. (Hartley, R., 1980)) :

Lemme.

Soit K un noeud tel qu'il existe un homéomorphisme h de degré +1 de l'extérieur X de K, dont la restriction à ∂X est donnée par la matrice $\begin{pmatrix} -1 & 0 \\ 0 & -1 \end{pmatrix}$ *dans la base de* $H_1(\partial X; \mathbb{Z})$ *formée d'un méridien et d'une longitude préférée de K. Alors K est inversible. De plus si h est une involution, K est rigidement inversible.*

Rappel.

Un méridien de K est une courbe simple fermée m sur X qui est le bord d'un disque rencontrant K transversalement en un seul point. Une telle courbe m est unique à isotopie près sur ∂X.

Une longitude préférée de K est une courbe simple fermée ℓ sur ∂X qui rencontre un méridien de K transversalement en un seul point et qui est homologue à zéro dans X. Une telle courbe ℓ est unique à isotopie près sur ∂X.

Dans la suite m et ℓ sont supposés orientés de telle facon que l'intersection $m \cdot \ell = +1$.

Preuve du lemme.

La restriction de l'homéomorphisme h à ∂X est isotope à l'involution standard h_0 du tore ∂X, qui admet 4 points fixes, comme sur la figure :

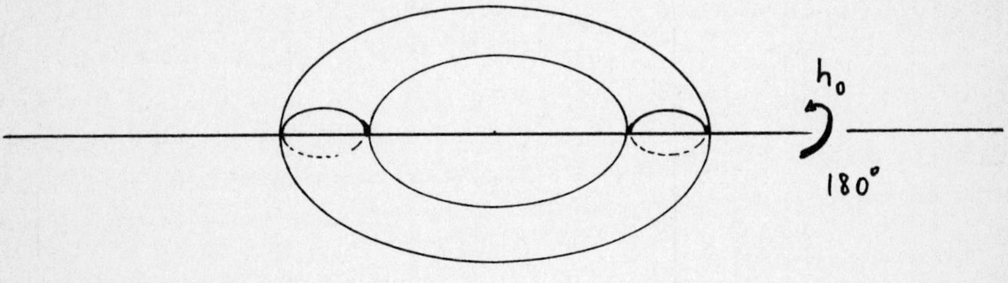

Par une isotopie sur un collier $\partial X \times [0, \varepsilon]$ de ∂X, on se ramène au cas où la restriction de h à ∂X est h_0. On peut alors prolonger h au tore plein N(K) par un homéomorphisme qui inverse K, car h_0 préserve un méridien de ∂X en renversant son orientation.

Si h est une involution, la classification des involutions du tore qui préservent homologiquement, au signe près, un méridien et une longitude préférée (cf. (Hartley, R., 1980)) montre que h est "fortement équivalente" à h_0 : il existe un homéomorphisme f de ∂X, isotope à l'identité, tel que $fhf^{-1} = h_0$. En utilisant cette isotopie on peut prolonger h en une involution de S^3, inversant K.

2. *Inversibilité des noeuds simples*

Nous allons montrer dans ce paragraphe que tout noeud simple est inversible si et seulement s'il est rigidement inversible.

D'après le théorème d'hyperbolisation de W.P. Thurston (1982), (1980) un noeud simple est torique (son complément porte une fibration de Seifert) ou hyperbolique (son complément porte une métrique hyperbolique, complète, à volume fini).

Il est bien connu que les noeuds toriques sont toujours rigidement inversibles : l'involution, inversant le noeud, est donnée par la conjugaison complexe, lorsqu'on définit le noeud torique de type (p, q) comme la trace sur la sphère unité de la singularité $z_1^p + z_2^q = 0$ dans C^2, p et q étant premiers entre eux.

On sait aussi (cf. Kawauchi, A., 1979) que les noeuds hyperboliques inversibles sont rigidement inversibles. Cependant nous aurons besoin, pour la suite de la démonstration, d'un résultat plus précis et plus général, qui puisse s'appliquer en particulier à la sous-variété X_0 de l'extérieur X de K lorsque K n'est plus un noeud simple (si K est simple $X = X_0$). C'est pourquoi nous donnons de ce résultat une démonstration géométrique, ne faisant pas appel à la conjecture de Smith comme dans (Kawauchi, A., 1979).

Proposition.

Soit K un noeud hyperbolique inversible. Tout homéomorphisme h du couple (S^3, K) de degré $+1$ dans S^3 et -1 sur K, est isotope à une involution de S^3, par une isotopie respectant K. En particulier K est rigidement inversible.

Demonstration.

Soit h un homéomorphisme du couple (S^3, K), de degré $+1$

dans S^3 et -1 sur K. On peut isotoper h dans $S^3 - K$ pour que h respecte un voisinage tubulaire $N(K)$ de K et fixe au moins un point x_0 sur $\partial N(K)$. Alors h induit sur le groupe $\pi_1(S^3 - K, x_0)$ un isomorphisme h_* tel que : $h_*(m) = m^{-1}$ et $h_*(\ell) = \ell^{-1}$, où le couple (m, ℓ) est un couple méridien, longitude préférée, du noeud K sur $N(K)$, passant par le point base x_0. (On a orienté m et ℓ pour que le nombre d'intersection $m \cdot \ell = +1$).

Puisque K est un noeud hyperbolique, le revêtement universel de $S^3 - K$ est l'espace hyperbolique $\mathbb{H}^3$ et le groupe $\pi_1(S^3 - K, x_0)$ s'identifie à un sous-groupe discret, à volume fini, de $PSL_2(C)$. En particulier, au sous-groupe périphérique π, abélien de rang 2, engendré par m et ℓ, correspond des transformations paraboliques de $\mathbb{H}^3$ qui admettent un point fixe unique, commun, p_0 sur $\partial \bar{\mathbb{H}}^3$. ($\mathbb{H}^3$ est identifié à l'intérieur de la boule unité dans R^3.)

D'après le théorème de rigidité de Mostow (Mostow, G., 1968), l'isomorphisme h_* de $\pi_1(S^3 - K)$ peut être réalise géométriquement par conjugaison par une isométrie A de $\mathbb{H}^3$ telle que : $A m A^{-1} = m^{-1}$ et $A \ell A^{-1} = \ell^{-1}$.

Puisque $A(p_0)$ est fixé par les éléments du sous-groupe périphérique π, $A(p_0) = p_0$. L'isométrie A admet un autre point fixe p_1 sur $\partial \bar{\mathbb{H}}^3$, sinon A commuterait avec les éléments de π.

Par contre A^2 commute avec les éléments de π et fixe p_0 et p_1. Donc A^2 fixe tous les transformés de p_1 par les éléments de π et c'est l'identité, car A^2 fixe plus de deux points distincts sur $\partial \bar{H}^3$.

L'involution h_0 induite sur $S^3 - \mathring{N}(K)$ par A est isotope à h sur $S^3 - N(K)$ d'après Waldhausen (Waldhausen, F., 1968). La restriction de h_0 à $\partial N(K)$ qui envoie m sur m^{-1} et ℓ sur ℓ^{-1}, peut être prolongée en une involution h_0 de S^3 inversant K, d'après le lemme du § 1. Les homéomorphismes h et h_0 sont alors isotopes dans S^3 par une isotopie respectant K.

3. *Cas général.*

Dans ce paragraphe nous achevons par recurrence la démonstration du théorème :

Théorème 1.

Soit K un noeud inversible, n'ayant aucun compagnon pour lequel il a un nombre de tours nul. Alors, le noeud K est rigidement inversible.

Démonstration du théorème 1.

Dans la proposition du § 2, on a démontré le théorème dans le cas où $c(K) = 1$. On suppose, par recurrence, avoir démontré le théorème dans le cas où $c(K) = r$.

Soit K un noeud de complexité $c(K) = r+1$, vérifiant les hypothèses du théorème. D'après le § 1, l'extérieur X de K est décomposé par une famille de tores incompressibles, non parallèles au bord, $T_1, \ldots, T_n$, en $X = X_0 \cup Y_1 \cup \ldots \cup Y_n$, où X_0 est une variété atoroidale et chaque Y_i $1 \leq i \leq n$, correspond à l'extérieur d'un compagnon K_i de K, parmi les plus proches de K au sens de Schubert (cf. figure).

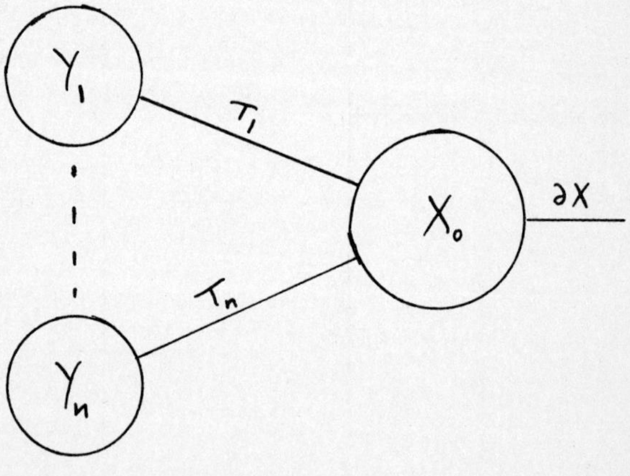

On distingue alors deux cas :

1er cas. La sous-variété X_0 est anannulaire, c'est-à-dire porte une métrique hyperbolique, complète, à volume fini, d'après le théorème de Thurston (Thurston, W.P., (1982),(1980)).

Soit h un homéomorphisme du couple (S^3, K), qui est de degré +1 dans S^3 et -1 sur K. Puisque la famille des tores ∂X, $T_1, \dots, T_n$ est unique à isotopie près dans $S^3 - K$, on peut supposer, après une isotopie respectant K, que h respecte la décomposition $X = X_0 \cup Y_i \cup \dots \cup Y_n$ en permutant peut-être les composantes Y_i, $1 \leq i \leq n$.

Comme dans la démonstration de la proposition du § 2 on peut isotoper h dans S^3, de telle facon que la restriction de h à X_0 soit une involution. En effet, la démonstration ne tient pas compte du nombre de composantes de ∂X_0, mais du fait que la composante ∂X est respectée par h et que la restriction de h à ce tore est donnée par la matrice $\begin{pmatrix} -1 & 0 \\ 0 & -1 \end{pmatrix}$.

L'homéomorphisme h induit alors une permutation d'ordre 2 de la famille des tores $T_1, \dots, T_n$; on distingue deux types de tores :

(a) les tores T_i tels que $h(T_i) \cap T_i = \emptyset$. Alors, il existe un unique tore T_j, $j \neq i$, tel que $h(T_i) = T_j$, et h échange les deux sous-variétés Y_i et Y_j. On remplace h sur $Y_i \amalg Y_j$ par h' qui est défini come suit :

sur Y_i, $h' = h$
sur Y_j, $h' = h^{-1}$.

Alors, h' est d'ordre 2 sur $Y_i \amalg Y_j$ et la restriction de h' à $\partial(Y_i \amalg Y_j)$ coïncide avec celle de h puisque h est déjà d'ordre 2 sur X_0.

(b) les tores T_i tels que $h(T_i) = T_i$. Alors h respecte la sous-variété Y_i. Le nombre de tours de K vis-à-vis de T_i étant *non nul*, h change homologiquement le signe d'un méridien m_i du tore plein V bordé par T_i. Comme h préserve l'orientation de S^3, h change homologiquement le signe d'une longitude préférée ℓ_i de T_i et la matrice de la restriction de h à $T_i = Y_i$ est $\begin{pmatrix} -1 & 0 \\ 0 & -1 \end{pmatrix}$; d'après le lemme du § 1 le compagnon K_i de K, associè au tore T_i, est inversible.

Ainsi, les compagnons K_i de K, associés a ces tores T_i sont inversibles et vérifient les hypothèses du théorème. Comme $c(K_i) \leq r$, par hypothèse de récurrence K_i est rigidement inversible. Il existe donc une involution h' sur Y_i que l'on peut identifier sur ∂Y_i avec l'involution donnée par h (définie dans X_0), car les deux involutions sont fortement équivalentes sur T_i.

En utilisant (a) et (b), on peut construire, par recollement une involution h' sur X (coincidant avec h sur X_0) et dont la restriction à ∂X est donnée par la matrice $\begin{pmatrix} -1 & 0 \\ 0 & -1 \end{pmatrix}$. Le noeud K est donc rigidement inversible d'après le lemme du § 1.

2e cas. La variété X_0 contient un anneau essentiel. X_0 est alors un espace fibré de Seifert. On en deduit aisément (cf. Jaco, W. et Shalen, P., 1979 ou Swarup, G.A., 1980) que K est soit un câble autour d'un noeud K_1 $(n = 1)$, soit la somme connexe de n noeuds, $K = K_1 \# K_2 \# \ldots \# K_n$.

(c) K est un câble autour d'un noeud K_1. On a alors $0 < c(K_1) \leq r$, puisque K n'est pas un noeud simple. Comme dans la démonstration précédente, si K est inversible, K_1 est inversible, puisque le nombre de tours de K par rapport à K_1 est non nul. Donc, par hypothèse de récurrence, K_1 est rigidement inversible.

Or, tout câble d'un noeud rigidement inversible est lui-même rigidement inversible. Cela vient du fait qu'un noeud torique est rigidement inversible par une involution respectant le tore sur lequel on a placé le noeud torique (la conjugaison complexe respecte le tore $|z_1| = a$, $|z_2| = b$, $a > 0$ et $b > 0$ vérifiant $a^2 + b^2 = 1$ et $a^p = b^q$, sur lequel se trouve le noeud torique (p, q)).

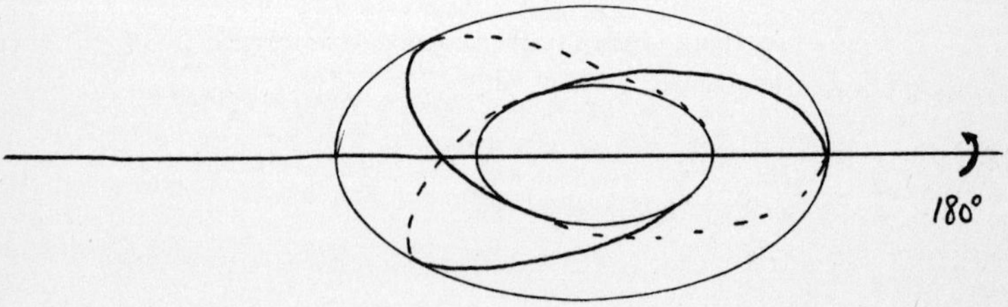

Lorsqu'on considère le câble K d'un noeud rigidement inversible K_1, on peut recoller les deux involutions, celle sur l'extérieur Y_1 de K_1 et celle sur le voisinage tubulaire de K_1 qui inverse la câble K dans $N(K_1)$, pour obtenir une involution dans S^3 inversant le noeud K.

(d) K est la somme connexe $K_1 \# K_2 \# \dots \# K_n$ de n noeuds premiers, de complexité $c(K_i)$ telle que : $0 < c(K_i) < r$.

Puisque la famille des tores incompressibles $T_1, \dots, T_n$, associés dans $S^3 - K$ aux facteurs $K_1, \dots, K_n$ de K, est unique à isotopie près, on peut supposer que l'homéomorphisme h, de degré $+1$ dans S^3 et -1 sur K, respecte la décomposition en somme connexe et induit une permutation des facteurs de K. Par unicité de cette décomposition pour *les noeuds orientés,* on distingue deux types de facteurs K_i, $1 \leq i \leq n$:

(1) les facteurs K_i dont l'orbite par h, $\{K_i, h(K_i), \dots, h^{2n_i-1}(K_i)\}$, a un cardinal pair. Alors le noeuds $h^{2q}(K_i) \# h^{2q+1}(K_i)$, $0 \leq q \leq n_i - 1$, sont rigidement inversibles.

(2) les facteurs K_i dont l'orbite par h a un cardinal impair. Alors K_i est tous ses transformés par h sont des noeuds inversibles. Par hypothèse de récurrence, l'orbite de K_i est formé de noeuds rigidement inversibles.

Donc K est une somme connexe de noeuds rigidement inversibles; il est facile de voir que le noeud K est rigidement inversible.

Ceci achève la preuve du théorème 1.

Dans le cas particulier des noeuds fibrés, on obtient le corollaire suivant :

Corollaire.

Un noeud fibré est inversible si et seulement s'il est rigidement inversible.

Démonstration du corollaire.

Un noeud fibré n'admet aucun compagnon pour lequel il a un nombre de tours nul. En effet, un tel compagnon correspond dans le groupe $\pi_1(S^3 - K)$ à un sous-groupe abélien libre de rang 2 , $\mathbb{Z} \oplus \mathbb{Z}$, contenu dans le groupe des commutateurs de $\pi_1(S^3 - K)$: le méridien et la longitude préférée du tore incompressible, associé à un tel compagnon, sont des commutateurs par définition même du nombre de tours. Le sous-groupe des commutateurs du groupe d'un noeud fibré est un groupe libre, il ne contient donc aucun sous-groupe libre abélien de rang 2 .

Remarque 1. Le corollaire précédent peut s'étendre au cas des noeuds K dont toutes les surfaces de Seifert S , de genre minimal, sont libres (c'est-à-dire que $\pi_1(S^3 - S)$ est un groupe libre). En effet Schubert (cf. Schubert, H., 1953, § 12-1) montre que si K admet un compagnon pour lequel il a un nombre de tours nul, on peut construire une surface de Seifert S pour K , de genre minimal, qui ne rencontre pas le tore incompressible associé à ce compagnon. En particulier ce tore est incompressible dans $S^3 - S$, et $\pi_1(S^3 - S)$ ne peut pas être libre.

Remarque 2. Dans le cas des noeuds fibrés, il existe une preuve directe du corollaire, utilisant la classification (donnée par W.P. Thurston, voir Fathi, A., et al, 1979) des homéomorphismes d'une surface de dimension 2 . On se ramène à ce cas en utilisant l'unicité à isotopie près de la surface fibre d'un noeud fibré (cf. Waldhausen, F., 1968).

Remarque 3. L'involution que l'on construit au cours de la démonstration du théorème, n'est pas isotope à l'application h de départ, par une isotopie respectant K . En particulier l'application h de départ peut être d'ordre infini dans le groupe $\pi_0(\mathrm{Diff}_+(S^3, K))$, alors qu'une involution inversant K est toujours d'ordre 2 dans ce groupe.

4. *Noeuds inversibles qui ne sont pas rigidement inversibles.*

Pour donner une condition nécessaire et suffisante pour qu'un noeud inversible soit rigidement inversible, nous avons besoin de la notion d'arbre caractéristique de compagnonnage (cf. Bonahon, F. et Siebenmann, L., ch. 3) que nous avons défini au § 1 .

Soit $\mathcal{T}$ la famille de tores disjoints, incompressibles, non parallèles entre eux, donnée par Jaco-Shalen (Jaco, W., et Shalen, P., 1979) et Johannson (Johannson, K., 1979) dans l'extérieur X du noeud K .

On associe à cette famille $\mathcal{T}$ un arbre $\Gamma(K)$ dont les sommets sont les composantes connexes fermées de $X - \mathcal{T}$ et les arêtes sont les tores de la famille $\mathcal{T}$ (l'arête de valence libre correspond à ∂X ; cf. Bonahon, F., et Siebenmann, L., Chapter 3).

Par unicité à isotopie près de la famille $\mathcal{T}$, tout homéomorphisme h du couple (S^3, K), de degré +1 dans S^3, induit après une isotopie convenable, respectant K, un automorphisme de l'arbre $\Gamma(K)$ noté σ_h. De plus à un tel homéomorphisme h on associe aussi une pondération des arêtes de $\Gamma(K)$ de la facon suivante :

(i) le poids d'une arête qui est invariante par l'automorphisme σ_h est 1 ou -1, suivant que la restriction de h au tore incompressible représenté par cette arête est isotope à l'identité ou à l'involution standard ayant 4 points fixes sur le tore (cf. figure au § 1 ; ce sont là les deux seules possibilités puisque h est de degré +1 dans S^3).

(ii) dans le cas où l'arête n'est pas invariante par σ_h, le poids est 0.

Ainsi, à un homéomorphisme h du couple (S^3, K), de degré +1 dans S^3, on associe un arbre pondéré, noté $\Gamma(K, h)$, et un automorphisme σ_h de cet arbre respectant les pondérations.

Par une démonstration analogue à celle du § 3, on peut démontrer le théorème suivant, qui est une généralisation du théorème 1 :

Théorème 2.

Un noeud K dans S^3 est rigidement inversible si et seulement s'il existe un homéomorphisme h du couple (S^3, K) de degré +1 dans S^3, pour lequel l'arbre pondéré $\Gamma(K, h^{2q+1})$ n'admet aucune arête de poids 1, quel que soit $q \geq 0$.

Remarque 1. Si on oublie la notion d'arbre caractéristique de compagnonnage, le théorème 2 dit que le noeud K est rigidement inversible si et seulement s'il est inversible et s'il existe un homéomorphisme h du couple (S^3, K), inversant K, respectant la famille caractéristique de tores incompressibles $\mathcal{T}$ et dont la restriction à chaque tore de la famille $\mathcal{T}$ qu'il laisse invariant n'est jamais isotope à l'identité, et de même pour h^{2q+1}, $q > 0$, (la condition sur h^{2q+1}

est dictée par le cas (d) du § 3).

Remarque 2. Le théorème 2 implique le théorème 1. En effet, si l'homéomorphisme h du couple (S^3, K) est de degré +1 dans S^3 et -1 sur K, les arêtes de poids 1 du graphe pondéré $\Gamma(K, h)$ sont associées à des tores pour lesquels le nombre de tours est nul (cf. la démonstration du théorème 1, cas (b)). Donc, si K est inversible et n'admet aucun compagnon pour lequel le nombre de tours est nul, l'arbre pondéré, associé a n'importe quel homéomorphisme de S^3 inversant K, n'admet aucune arête de poids 1 et le théorème 2 s'applique.

Demonstration du théorème 2.

(A) La condition imposée aux arbres pondérés $\Gamma(K, h^{2q+1})$ est suffisante pour construire une involution dans S^3, de degré +1 et inversant K. En effet, la démonstration est la même que pour le théorème 1 : on peut raisonner par récurrence sur la complexité de K, puisque tout compagnon K' de K, associé à un tore de la famille $\mathcal{T}$ invariant par h (ou par une puissance impaire h^{2q+1} de h) vérifie l'hypothèse de récurrence, car $\Gamma(K', h)$ (resp. $\Gamma(K', h^{2q+1})$) est un sous-arbre pondéré de $\Gamma(K, h)$ (resp. $\Gamma(K, h^{2q+1})$).

(B) La condition imposée est nécessaire, cela résulte par exemple de la variété caractéristique équivariante de F. Bonahon (1979), (cf. Bonahon, F. et Siebenmann, L.) : si h est une involution de degré +1 dans S^3, inversant K, alors on peut déformer la famille caractéristique de tores incompressibles $\mathcal{T}$ par une isotopie de $S^3 - K$, pour la rendre invariante par l'involution h. On peut aussi démontrer directement ce résultat, en utilisant les résultats de M. Freedman, J. Hass et P. Scott (cf. Freedman, M., et al.) sur le plongement des surfaces incompressibles d'aire minimale (voir par exemple M. Freedman et S.T. Yau, 1983).

Soit alors T un tore de la famille $\mathcal{T}$ invariant par l'involution h. Pour montrer que la restriction de h à T n'est pas isotope à l'identité, on reproduit l'argument donné par W. Whitten dans (Whitten, W., 1981) :

Si la restriction de h à T est isotope à l'identité, les points fixes A de h dans S^3 ne rencontrent pas T. Comme A rencontre K, A est contenu dans le tore plein V, bordé par T dans S^3

et contenant K. D'après F. Waldhausen (1969), A est un cercle non noué dans S^3. Comme le tore plein V est noué dans S^3, A est inclus dans une boule B^3 contenue dans l'intérieur de V, sinon $S^3 - A$ contiendrait un tore incompressible. La variété $V - A$ contient une sphère $S = \partial B^3$ qui ne borde aucune boule dans $V - A$. D'après le théorème de la sphère "équivariant" de W. Meeks et S.T. Yau (1980) (voir aussi Kim-Tollefson (Kim, P. et Tollefson, J., 1980)), il existe une sphère S' invariante par l'involution h et qui ne borde aucune boule dans $V - A$ (on ne peut pas avoir $h(S') \cap S' = \emptyset$, car la boule bordée par S' dans V contient A). On considère alors le boule bordée par S' dans S^3 et ne contenant pas A; cette boule est invariante par h et par le théorème de Brouwer h admet un point fixe dans cette boule ce qui est impossible. Ceci achève la preuve du théorème 2.

Remarque 3. Le theoreme 2 ne permet pas de decrire explicitement la classe des noeuds pour lesquels la conjecture de Montesinos est vraie. Tout au plus, il permet de ramener le probleme de la determination des noeuds rigidement inversibles a la realisation geometrique de certains automorphismes de l'arbre de compagnonnage et de certaines symetries d'entrelacs simples (c'est-a-dire dont le complement ne contient aucun tore incompressible non peripherique).

Cependant, le théorème 2 est très utile pour construire de nouveaux contre-exemples à la conjecture de Montesinos :

Soit $L = K_0 \cup C_1 \cup \ldots \cup C_n$ l'entrelacs de la figure ci-dessous, où K_0 est le noeud de pretzel $(2, 3, 7)$ et les composantes $C_1, \ldots, C_n$ sont des cercles non noués et non enlacés, entourant la bande centrale du pretzel comme sur la figure. L'entrelacs L est un entrelacs de Montesinos généralisé (cf. Montesinos, J.M. 1973 et 1979).

Soit K' un noeud non trivial, on construit un noeud K de la facon suivante (c'est un cas particulier de la "splicing construction", voir (Bonahon, F. et Siebenmann, L., Chapter 3) : soit $f_1 : S^3 - N(C_1) \to N(K')$ un homéomorphisme qui préserve les longitudes préférées. On note $L_1 = f_1(L - C_1)$; c'est un entrelacs dans S^3 dont les composantes $f_1(C_i)$, $2 \le i \le n$, sont toujours non nouées et non enlacées, car les composantes C_i, $2 \le i \le n$, sont inessentielles dans le tore plein $S^3 - N(C_1)$. On répète alors ce processus de

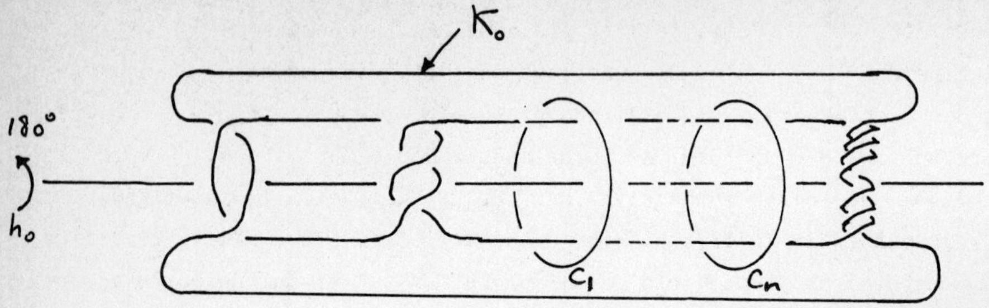

satellisation autour du noeud K' avec la composante $f_1(C_2)$ de l'entrelacs L_1. Au bout de n satellisations on obtient un noeud K qui est l'image du noeud K_0 après ces satellisations.

Puisque l'entrelacs L est simple d'après Bonahon et Siebenmann (cf. Bonahon, F. et Siebenmann, L.), le noeud K admet pour compagnons les plus proches au sens de Schubert n noeuds isotopes au noeud K'. On obtient alors la décomposition suivante de l'extérieur X de K : $X = X_0 \cup X' \cup \ldots \cup X'$, où X_0 est l'extérieur de L et X', qui apparaît n fois, est l'extérieur de K' (voir figure).

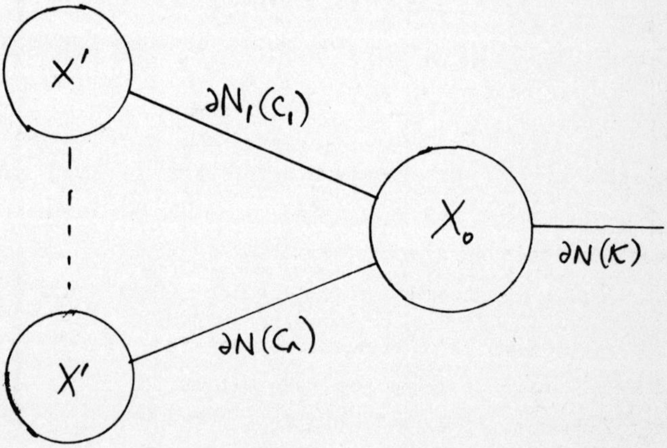

L'entrelacs L est invariant par une involution h_0 de degré +1 dans S^3, respectant chaque composante de L, ayant deux points fixes sur K_0 et aucun point fixe sur les C_i, $1 \le i \le n$; cette involution est la rotation de 180° autour d'un axe horizontal, décrite sur la projection donnée de L. L'involution h_0 permet d'inverser K dans $S^3 - \coprod_1^n X'$, et comme la restriction de h_0 au bord $\coprod_1^n \partial X' = \coprod_1^n \partial N(C_i)$ est isotope à l'identité, on peut prolonger h_0 en un homéomorphisme de S^3, de degré +1 et inversant K. Le noeud K est donc inversible.

D'après le calcul du groupe de symétrie, π_0 (Homéo (S^3, L)), de l'entrelacs L, effectué par Bonahon et Siebenmann (cf. Bonahon, F. et Siebenmann, L., Chapter 16), tout homéomorphisme du couple (S^3, L), de degré +1 dans S^3 et -1 sur K_0, est isotope à h_0 par une isotopie respectant L. En particulier, tout homéomorphisme h du couple (S^3, K), de degré +1 dans S^3 et -1 sur K, induit, après une isotopie convenable respectant K, un homéomorphisme de l'extérieur X_0 de L isotope à la restriction de h_0 ; donc le graphe pondéré associé, $\Gamma(K, h)$, admet toujours au moins $n \geq 1$ poids égaux à +1. Le théorème 2 montre que le noeud K n'est pas rigidement inversible.

Dans ce qui précède, le noeud non trivial K' est arbitraire. On peut le choisir simple au sens de Schubert et inversible. On obtient alors un noeud K inversible, non rigidement inversible, n'ayant pas de compagnon maximal unique (si $n \geq 2$), et dont tous les compagnons sont isotopes à un même noeud rigidement inversible, avec même ordre 2 et même nombre de tours.

REFERENCES

Bonahon, F., (1979). "Involutions et fibrés de Siefert dans les variétés de dimension 3", Thèse 3e cycle, Orsay.

Bonahon, F., et Siebenmann, L. "Algebraic Knots", à paraître.

Fathi, A., Laudenbach, F., et Poenaru, V., (1979). "Travaux de Thurston sur les surfaces", Astérisque 66-67.

Freedman, M., Hass, J., et Scott, P. "Imbedding least area incompressible surfaces", à paraître dans Invent. Math.

Freedman, M., et Yau, S.T., (1983). "Homotopically trivial symmetries of Haken manifolds are toral", Topology 22, 179-189.

Hartley, R., (1980). "Knots and Involutions", Math. Z. 171, 175-185.

Johannson, K., (1979). "Homotopy equivalence of 3-manifolds with boundaries", Springer Lecture Notes Math. 761.

Jaco, W., et Shalen, P., (1979). "Seifert fibred spaces in 3-manifolds", Memoirs A.M.S. 220.

Kawauchi, Q., (1979). "The invertibility problem on amphicheiral excellent knots", Proc. Japan Acad. 55, 399-402.

Kervaire, M., et Weber, C., (1977). "A survey of multidimensional knots", Proceedings des Plans-sur-Bex, Springer Lecture Notes Math. 685, 61-134.

Kirby, R., (1978). "Problems in low dimensional topology", Proceedings of Symposia in Pure Math., volume 32, 272-312.

Kim, P., et Tollefson, J., (1980). "Splitting PL involutions on non prime 3-manifolds", Michigan Math. J. 27, 259-274.

Montesinos, J.M., (1975). "Surgery on links and double branched covers of S^3", dans Knots, Groups and 3-Manifolds, Ann. Math. Studies 84, 227-259.

Montesinos, J.M., (1973). "Variedades de Seifert que son recubridadores ciclicos ramificados de dos hojas", Bol. Soc. Mat. Mexicana 18, 1-32.

Montesinos, J.M., (1979). "Revêtements doubles ramifiés de noeuds, variétés de Seifert et diagramme de Heegaard", Prépublication Orsay.

Mostow, G., (1968). "Quasiconformal mappings in n-space and the rigidity of hyperbolic space forms", Publ. Math. I.H.E.S. (Paris) 34, 53-104.

Meeks, W. III, et Yau, S.T., (1980). "Topology of three dimensional manifolds and the embedding problems in minimal surface theory", Ann. of Math. 112, 481-484.

Schubert, H., (1953). "Knoten und Vollringe", Acta Math. 90, 131-286.

Swarup, G.A., (1980). "Cable knots in homotopy 3-spheres", Quart. J. Math. Oxford (2) 31, 97-104.

Thurston, W.P., (1982). "Three dimensional manifolds, Kleinian groups and Hyperbolic geometry", Bull. A.M.S. 6, 357-381.

Thurston, W.P., (1980). "Hyperbolic structures on 3-manifolds", Preprints, Princeton University.

Waldhausen, F., (1969). "Über Involutionen der 3-Sphäre", Topology 8, 81-91.

Waldhausen, F., (1968). "On irreducible 3-manifolds which are sufficiently large", Ann. of Math. 87, 56-88.

Weber, C., (1980). "Sur le module d'Alexander des noeuds satellites", Preprint.

Whitten, W., (1981). "Inverting double knots", Pacific J. Math. 97, 209-216.

Whitten, W., (1980). "Switching and inverting knots", Preprint.

THE CLASSIFICATION OF SEIFERT FIBRED 3-ORBIFOLDS

F. Bonahon and L. Siebenmann

With surprising frequency there occurs an object of the following type: a metrizable 3-manifold (with possibly some isolated interior singularities) equipped with suitably well-behaved or 'tame' decomposition into circles and (compact) intervals. Those tame decompositions involving circles only, where in addition the manifold is orientable near each circle, are precisely the well-studied Seifert-fibrations of 3-manifolds, (Seifert, H., 1932; Orlik, P., 1972; Siebenmann, L.C., 1978). Tame decompositions involving intervals have been studied far less, but they are by no means rare. They occur in many quotient spaces of classical crystallographic groups (139 out of 219)*, and also for many non-classical crystallographic groups; see Bonahon, F. & Siebenmann, L.C. (to appear); Dunbar, W.D. (1983); Seifert, H. & Threlfall, W. (1930 & 1932). We have encountered them in our study of classical knots (Bonahon, F. & Siebenmann, L.C., to appear), through the so-called Montesinos and arborescent knots (that include respectively 174 and 208 of the 250 ten crossing knots). We will mention these examples further when we have clear definitions in hand.

The purpose of this article is to give an elementary and complete classification of all 'tame' decompositions in dimension 3 into circles and intervals, thus extending Herbert Seifert's 1932 classification of what have since been called Seifert fibrations.

The classification of tame decompositions for which the ambient space is oriented is widely known and has been touched on by several authors following Montesinos' stimulating descriptions in the early 1970's, cf. Montesinos, J.-M. (1980); Thurston, W.P. (1976-79); Dunbar, W.D. (1981); Zieschang, H. & Zimmerman, B. (1982). The classification in the non-orientable case is not well known. However, Zieschang &

*45 more of the 219 have tame decompositions involving circles only.

Zimmerman have in fact discussed it using the language of combinatorial presentations of extensions of co-compact hyperbolic planar groups by $\mathbb{Z}$ or $(\mathbb{Z}/2) * (\mathbb{Z}/2)$; indeed such groups act on $\mathbb{R} \times \mathbb{H}^2$ respecting projection to $\mathbb{R}$ so as to yield any given tame decomposition as quotient. (Beware that for trivial reasons many different groups give the same tame decomposition; cf. remarks on actions below.) Setting aside differences of language (which are enormous and worth sorting out), readers may find that we have made the classification more explicit than in Zieschang, H. & Zimmerman, B. (1982), notably by using our so-called weak data functions in §6. Also, we have completed the classification by dealing with two cases that Zieschang and Zimmermann had to exclude, cf. (Zieschang and Zimmermann (1982), §3.7).

In later sections, following W. Thurston, we will systematically use the language of orbifolds (or V-manifolds) in place of decompositions. To make amends for thus demanding of our readers sophistication that is not really necessary, we use this introduction to present the more down-to-earth decomposition viewpoint, which in spirit is Seifert's.

The first and most succinct defintion of *tameness* of a decomposition of a 3-dimensional metrizable space X^3 into (disjoint) circles and intervals is as follows: near each element (interval or circle) the decomposition is either as in the compact solid torus $S^1 \times D^2$ decomposed into the circles $S^1 \times x$, or as in some quotient $(S^1 \times D^2)/G$ of $S^1 \times D^2$ by a finite group G acting isometrically, respecting these circles, and decomposed by the quotients of these circles; here $D^2 = \{(x_1,x_2) \in \mathbb{R}^2 \,|\, x_1^2 + x_2^2 \leq 1\}$ is the standard 2-disc and S^1 is its boundary circle. (Seifert in effect allowed only those model decompositions $(S^1 \times D^2)/G$ where G acts freely preserving orientation.)

The models $(S^1 \times D^2)/G$ are too various to describe geometrically in a brief introduction. To remedy this we present a more local and more visible defintion of tameness analogous to the definition of a Seifert fibration as a foliation by circles$^{(\dagger)}$. By considering the possible isotropy subgroups of the finite isometry groups of $S^1 \times D^2$,

---

$^{(\dagger)}$For non-compact X^3 , upper semicontinuity of the decomposition must be supposed (see remarks below on work of Reeb and Epstein).

one readily finds that near any point x of X^3 the decomposition looks as near *some* point of one of the following compact model decompositions into intervals:

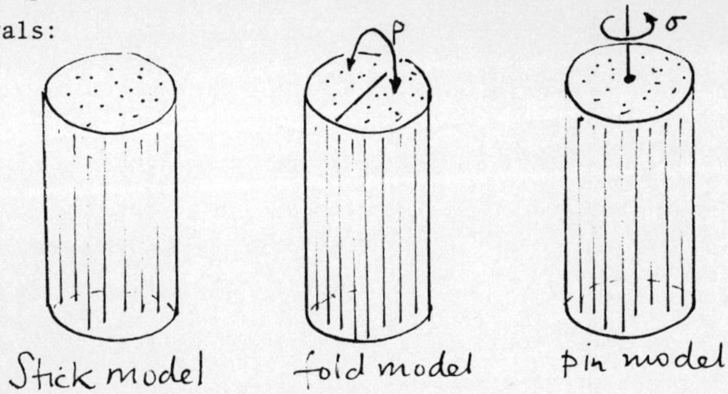

The *stick model* (a) is the solid cylinder $D^2 \times [0,1]$ trivially decomposed into the intervals $x \times [0,1]$. The *fold model* (b) and the *pin model* (c) are derived from it by identifying the end 2-disc $D^2 \times 0$ to itself under a reflection ρ and the 180° rotation σ respectively; each interval of the decomposition is the image of one or two intervals of the stick model. The fold model is easily visualized  ; it is a product of an interval with an analogous 2-dimensional model  ; the underlying space is a 3-ball and interval ends form an unknotted arc or 'fold line' crossing the ball. The pin model is a decomposition of the cone on the real projective plane, and the cone point is an isolated interval extremity, called the 'pinhead'.

The correspondence between the isotropy groups $G_x \subset G$ and the local models describing $(S^1 \times D^2)/G$ near the image (orbit) of points $x \in S^1 \times D^2$ is as follows. The pin model is required precisely if G_x is cyclic acting by rotation of even order on D^2 and by reflection on S^1 . The fold model is required precisely if G_x is dihedral, of order 2p say, acting standardly on D^2 and by reflection on S^1 . (This case splits into two subcases: either none of the two standard order 2 generators of G_x acts by the identity on S^1 and G_x respects the orientation of $S^1 \times D^2$, or one of these generators fixes S^1 and p necessarily is even while x necessarily corresponds to a point in the topological boundary of the fold model. Otherwise the stick model suffices: indeed G_x is of the form $H' \times H'' \subset O_2 \times O_2$.

From these models we see that near a given point x of X the space X is a topological manifold, unless the behaviour is as at the pinhead in (c), in which case a neighbourhood of x is a cone on the real projective plane $\mathbb{R}P(2)$. As X is a manifold except for isolated singularities, we can speak of its boundary. Near the given point x in X the behaviour of the decomposition is of one of the few local types seen in the models, seven in fact, of which four occur at boundary points of X , three in the interior.

These local models provide a 'local' definition of tameness as follows: a decomposition of a metrizable 3-dimensional space X into circles and intervals is *tame* if it is upper semi-continuous (see below) and on some open neighbourhood of each point of X the induced decomposition into connected 1-manifolds is as on some open subset of the fold model or the pin model. (Upper-semi-continuity of a decomposition of a space X means that each decomposition element has arbitrarily small open neighbourhoods that are made up of whole decomposition elements, i.e. are saturated.)

Why are the two definitions of tameness equivalent? We have already seen that the first definition of tameness implies this local definition. Conversely, granting the local definition consider any decomposition element A . First suppose for simplicity that A is a circle in the interior of X and choose a small 2-disc D pierced by A at its centre p only and transverse to the decomposition (which is a foliation near A) ; next choose a saturated open neighbourhood U of A so small that $U \cap \partial D = \phi$. Leaving one side of D and sliding along elements of the decomposition until a first return to D , is a procedure that defines a monodromy homeomorphism $\mu : D_0 \to D_0$ where D_0 is the component of $D \cap U$ containing p . Now D. Montgomery (1937) (cf. Edwards, R.D., Millett, K.C. & Sullivan, D., 1977) proved the remarkable theorem that any self-homeomorphism of a topological manifold is of finite order if its orbits are finite (as are those of μ!) . But a more pedestrian result asserts that a finite order self-homeomorphism of a 2-manifold is a diffeomorphism for a suitably constructed smooth structure, cf. Epstein, D.B.A. (1981) . It follows that, near the given circle A , the decomposition has a form allowed by the first definition; namely, it is a mapping torus of a finite order rotation of the 2-disc.

Near an arbitrary decomposition element A (which may be an interval), a similar argument applies (see Epstein, D.B.A. (1976), §7 for an easy but convenient extension of Montgomery's theorem); alternatively, by application near A of doubling tricks and branched covering tricks one can reduce to the situation just treated.

The reader should be warned that upper semi-continuity cannot be dispensed with in the local definition of tameness, as certain smooth circle decompositions due to G. Reeb (1952) show. On the other hand there is a famous difficult theorem of D. Epstein (Epstein, D.B.A., 1972 & 1976) (cf. Edwards, R.D., Millett, K.C. & Sullivan, D., 1977) asserting that a topological foliation by circles of a compact 3-manifold is always upper semi-continuous as a decomposition. Very likely, the same arguments prove upper semi-continuity is redundant in the local definition of tame decomposition by circles and intervals, in case the space X^3 being decomposed is compact.

There is a reason for our using the word 'tame', a term with simplicial connotations. Indeed, it seems to us that there is a third definition of *tame* decomposition of X into arcs and circles as follows: X is a compact 3-dimensional pseudo-manifold with $c(\mathbb{R}P(2))$ singularities at worst; this decomposition into arcs and circles is such that none departs from the boundary of X in its own interior, and that for some triangulation of X each 3-simplex inherits a decomposition that is in a suitable sense linear. Admittedly, a more topological characterization of tameness would have more chance of being useful. We leave these ideas aside.

The first and perhaps still the most significant sort of example of these decompositions was already encountered by H. Seifert and W. Threllfall in 1930; it occurs when a discrete group Γ of isometries of a complete simply connected homogeneous Riemannian 3-manifold $\tilde{X}$ acts with compact quotient respecting a given family of parallel lines (= geodesies). Granting that (as often happens) each line has compact image in $X = \tilde{X}/\Gamma$, a tame decomposition into arcs and circles arises.

Here the quotient X is more than just a topological space; it is an orbifold (see §1), from which one can retrive the symmetric space $\tilde{X}$ as the universal covering orbifold and Γ as the covering translation group. Further, the decomposition gives what we call a locally trivial orbifold bundle structure to X with 1-dimensional fibre, or an S-*fibration* (where of course S recalls Seifert's contributions as well as S^1 . The (topological) orbifold structure of X is fortunately almost determined by the tame decomposition: indeed the following will become clear on reading §1.

Fact

Given a tame decomposition of a compact metrizable space X^3 into intervals and circles, a topological closed 3-orbifold structure on X making the decomposition an S-fibration amounts to choosing arbitrarily a finite number of decomposition elements (circles or intervals), and assigning to each an integer ≥ 1 that specifies the order of the rotational part of the orbifold isotropy group of X along the element (endpoints excepted).

The foregoing suggests an approach to the study and classification of the actions Γ on symmetric 3-spaces $\tilde{X}$ introduced above: given $\tilde{X}$ examine the classification of S-fibred orbifolds X to see which might arise as $\tilde{X}/\Gamma$; then impose all possible orbifold Riemannian metrics on X that (away from singularities) are locally like that of $\tilde{X}$ and make the fibres geodesic and parallel; finally, pass to the universal covering to get $\tilde{X}$ and Γ . This approach works well with six of the eight homogeneous spaces that W. Thurston needs in his geometrization program for 3-manifolds - all but solvable geometry dealt with by Dunbar, W.D. (1983), and the difficult hyperbolic geometry currently studied by Thurston, W.P. (1976-79). This procedure is illustrated in the following article (Bonahon, F. and Siebenmann, L.C., to appear) of this volume for the euclidean geometry E^3, where the 219 types of classical crystallographic groups have to be found. For other geometries we refer the reader to Dunbar, W.D. (1983), cf. Bonahon, F. & Siebenmann, L.C. (to appear).

Another occurrence of these decompositions is in the study of classical knots. Many knots and links in the classical tables, namely the 2-bridge or rational knots and the stellar or Montesinos knots arise as a pair (X^3, K^1) where $X^3 \approx S^3$ is a 3-sphere decomposed as above and the knot string K^1 is formed by all the end points of all the intervals of the decomposition. Many more knots still (the algebraic knots of Conway = the arborescent knots) break up naturally into such pieces. We have studied all these knots and their symmetries in Bonahon, F. and Siebenmann, L.C. (to appear).

Here is a recipe to construct a typical stellar knot with a corresponding tame decomposition of S^3. Begin with S^3 tamely decomposed so that the intervals form a standard annulus 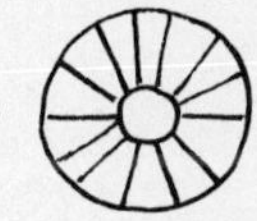 , the complement of the annulus being trivially fibred by circles. Remove a thin saturated (fibred) regular neighbourhood of one of these intervals. Cut it in two like a worm and note that it breaks into two copies of the model:

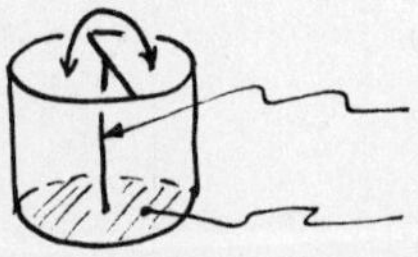

half the interval selected

the cut

Glue these two fold models back together at the cut using a rigid twist of rational angle; and note that by a minor miracle the decomposition of the new worm's skin by intervals and circles is equivalent by a homeomorphism to that for the original worm. Use any such homeomorphism to replace the original worm by the new one. Perform such a replacement or 'surgery' successively and disjointly for several different intervals to get the typical decomposition of S^3. (As a matter of definition, the typical Montesinos link in Montesinos, J.-M. (1980) may include as pieces of the string K^1 finitely many of the circle fibres as well as all the interval ends.) The intervals of the decomposition snarl up in a curious way near each core interval of a new worm; indeed in each of the two fold models they appear as  , i.e. the quotient of

I × (p diameters of D^2) . Here p is the order of the twist rotation that created the new worm; the fold axis is one of the diameters; and the angle between adjacent diameters is π/p . Note that this snarl is a self-transverse immersed 2-manifold only for $p \leq 2$. For another more geometric view of this snarl, see the tehedral figure in §2 .

Getting a global view of the intervals of the tame decomposition for a given Montesinos knot is not so easy. The reader should try first on simple examples like:

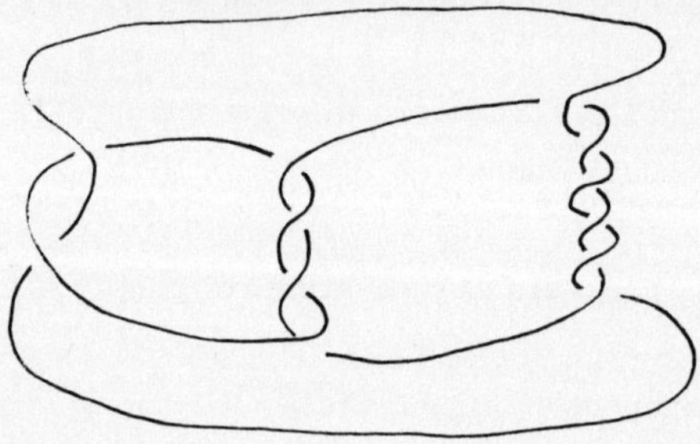

What is easy is to spot a collection consisting of one interval in each isotopy class (respecting the decomposition). For the illustrated knot there are six, dotted in the next figure:

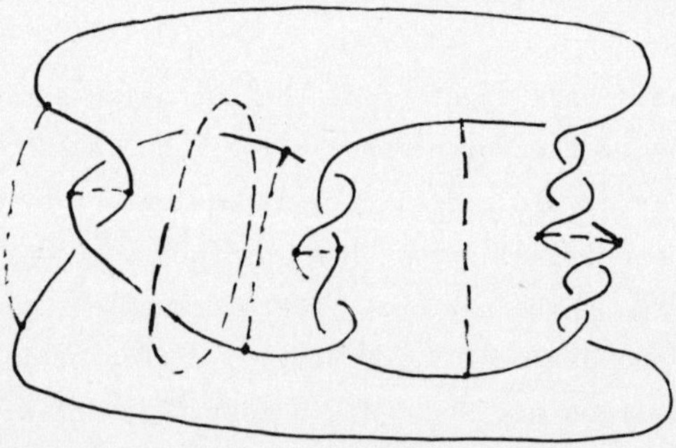

In addition, there is always just one isotopy class of circle fibres (also dotted).

Leaving details to the computational sections §§2-7, here are a few remarks on the classification obtained. We have adopted notations and orientation conventions that agree well (any exceptions are noted) with those of Waldhausen, F. (1967); Montesinos, J.-M. (1980); Bonahon, F. & Siebenmann, L.C. (to appear). The main deviation from Seifert, H. (1932) is our use of e where Seifert uses $-b$, as e corresponds to classical Euler number. We give our classification in a C^∞ smooth context (which makes good sense because of our use of orbifolds); but it will become clear that the classification is the same for the topological context of this introduction or for a piecewise linear one.

The first invariant of the tame decomposition of M to grasp is the decomposition space B together with the function assigning to each point of it the local type of the decomposition near the corrresponding interval or circle. When M is oriented, one also has a rational Euler number $e_0 \in Q$ for the projection $M \to B$. These basic invariants always determine the decomposition up to a *finite* number of possibilities. These must be sorted out by a finite combinatorial computation, which in turn we try to give in a very practical form. Sometimes we arrive in effect at tidy normalized closed forms (as Seifert did in every case), but sometimes not.

We have found the non-orientable classification comparatively complex and subtle, not so much because of the numerous new singularities one meets, but rather because three successively more intrinsic approaches, that conveniently all gave identical views in the oriented case, must be carefully distinguished, namely: (i) (primary) obstruction theory, (ii) fibred classification over a fixed base, (iii) fibred classification ignoring the base.

This considerable complexity has on occasion led us into error; we can only hope that we have by now managed to eliminate all serious blunders.

This complexity in the non-orientable case is to be contrasted with an appearance of simplicity: namely every such decomposition can be mapped to the product decomposition $B \times [0,1]$ by a map that, locally in M, is division by action of a finite group. Again, it arises from a product decomposition $B' \times S^1$ on dividing by a (global) finite group action, respecting the product structure. (For easy proofs, see §8.)

In the orientable case, what one can say in this vein is that the decomposition arises from a classical Seifert fibration on dividing by a finite group action respecting the fibration.

Such observations are of little help in the classification. Rather, the classification tells a good deal about the group actions just mentioned.

In conclusion, we mention that there is an extensive literature on general properties of orbifold fibrations (clothed, however, in the language of group actions). It was spurred by work of Conner, P.E. & Raymond F. (1977), cf. Holmann, H. (1964); Lee, K.B. & Raymond, F. (1983). The emphasis has been on the relation between fibrations, the fundamental group extensions they determine, and cohomology. On the other hand, W. Neumann has recently (see Jankins, M. & Neumann, W., 1983) indicated how these ideas lead quite directly to Seifert's classification in the orientable case. Thus in time they should lead to a third classification as self-contained as ours or that in Zieschang, H. & Zimmerman, B. (1982).

1 BASIC FACTS ABOUT ORBIFOLDS

When one studies a manifold M^n with a C^∞ action of a finite group G it often proves convenient to work in the quotient topological space M/G (the orbit space). This quotient M/G becomes more useful when equipped with information of a local character intended to capture just the right amount of detail concerning the action. A time-honoured way to attempt this is to stratify M/G by isotropy type, assigning to each point y in M/G the conjugacy class of the isotropy group $G_x = \{g \in G | g(x) = x\}$ for the action at (any) one point x in M with quotient orbit y. For C^∞ actions one can assume $G_x \subset O(n)$ and without loss understand conjugacy in $O(n)$, cf. Bredon, G.E. (1972). Wanting to do on M/G the sort of differential analysis one does on a manifold, I. Satake (1956) was led to strengthen the above 'traditional' structure. The resulting notion of (C^∞) orbifold (or V-manifold in Satake's terminology) has gained wide currency only recently when W. Thurston revealed the richness and power of the concept in his notes (Thurston, W.P., circa 1978, §13) (which we advise the reader to consult); in particular Thurston introduced the 3-orbifold S-fibrations whose classification we propose to complete in this article.

Since the appearance of (Thurston, W.P., 1976-79, §13) we have advertised a slightly simplified definition of orbifold designed to emphasize parallelism with the notion of manifold while pushing the traditional isotropy data into the background. It runs as follows.

A *pseudo-group* CAT of homeomorphisms is given (Kobayashi, S. & Nomizu, K., 1963), as is traditional in the definition of manifolds (but the spaces involved need not be subspaces of a single space). The basic pseudo-group will be that of diffeomorphisms of smooth manifolds possibly with boundary. When no pseudo-group is mentioned, this one will be understood. (The use of manifolds instead of half-space models $\mathbb{R}^n_+$ is not an essential point, but it often makes for brevity.)

A *folding map* is a continuous map $f : \tilde{X} \to X$ from an object $\tilde{X}$ of CAT onto a topological space X such that the *folding group* G, defined as the group of CAT automorphisms g of $\tilde{X}$ with $fg = f$, be finite and the natural induced continuous map $|\tilde{X}/G| \to X$ be a homeomorphism, $|\tilde{X}/G|$ denoting the topological orbit space.

A CAT orbifold M consists of a metrizable topological space $|M|$ and an *'atlas'* of folding maps $f_i : \tilde{U}_i \to U_i$ called *charts*, where the U_i's form an open covering of $|M|$ and where the f_i's are compatible in the following sense: for every $x \in \tilde{U}_i$ and $y \in \tilde{U}_j$ such that $f_i(x) = f_j(y) \in U_i \cap U_j$, there exists a CAT isomorphism $\psi : \tilde{V}_x \to \tilde{V}_y$ from an open neighbourhood of x in $\tilde{U}_i$ to an open neighbourhood of y in $\tilde{U}_j$ such that $f_j\psi(x') = f_i(x')$ for all x' in $\tilde{V}_x$.

Two such atlases give the same CAT orbifold structure if their union is again a compatible atlas.

An *isomorphism* $M \to M'$ of CAT orbifolds is a homeomorphism $|M| \to |M'|$ making the atlases compatible.

A typical example of orbifold is provided by a properly discontinuous (e.g. finite) action of a group G by CAT isomorphisms on a CAT manifold $\tilde{M}$: one considers the topological orbit space $|\tilde{M}/G|$ equipped with an atlas of folding maps obtained by suitable restrictions of the quotient map $\tilde{M} \to |\tilde{M}/G|$. This orbifold is denoted by $\tilde{M}/G$ (although its structure is much richer than that of the underlying topological orbit space that is traditionally so denoted).

The reader allergic to the general definition of orbifolds should always keep this fundamental example in mind, as it covers most interesting examples.

As a subcase of the above example (when $G = 1$), CAT manifolds can naturally be regarded as orbifolds and they are isomorphic as orbifolds if and only if they are CAT isomorphic. The one sign $\cong$ can therefore indicate isomorphism of manifolds *or* orbifolds.

To each point x of the topological space $|M|$ underlying an n-dimensional smooth orbifold M is naturally associated a finite subgroup G_x of $GL_n(\mathbb{R})$, well-defined up to conjugacy. Indeed, given a folding chart $f_i = \tilde{U}_i \to U_i \cong \tilde{U}_i/G_i$ of the atlas of the orbifold and given a point x_i with $f_i(x_i) = x$, it readily follows from definitions that the action of the stabilizer G_x of x_i in G_i on the tangent space of $\tilde{U}_i$ at x_i is independent of i up to linear conjugacy. This group $G_x \subset GL(n,\mathbb{R})$ is called the *isotropy group* of x. An open neighbourhood of x in M is isomorphic to $\mathbb{R}^n/G_x$ as an orbifold (see the equivariant tubular neighbourhood theorem (Bredon, G.E., 1972). A point x is *regular* or *generic* when its isotropy group is trivial, and *singular* otherwise. Observe that the set of regular points is a dense *connected* open subset of the topological space underlying a connected orbifold. The set of singular points is often denoted ΣM, or Σ for short; this ΣM is empty precisely if M is a manifold.

Compact 1- and 2-orbifolds are particularly simple. Indeed, up to isomorphism, connected compact 1-orbifolds are the manifolds S^1 or I, and the two orbifolds $[[0,\pi]] = S^1/(\mathbb{Z}/2)$ and $[0,\frac{1}{2}]] = I/(\mathbb{Z}/2)$, where $\mathbb{Z}/2$ acts on S^1 and I by reflection. The notation $[[a,b]]$ or $[a,b]]$ is meant to suggest that the underlying topological space is the interval $[a,b]$ while the symbols $[[$ and $]]$ mark points with reflection isotropy group $\mathbb{Z}/2$. (See a similar convention for 2-orbifolds below.)

A 2-orbifold M is determined up to isomorphism by $|M|$, which is a topological 2-manifold with boundary $\partial|M| \subset |\partial M| \cup \Sigma M$, together with the function assigning to each point x of $|M|$ its isotropy group G_x (up to conjugacy). This function will frequently be indicated diagrammatically on a picture of $|M|$ by:

(a) Labelling by the integer $\alpha > 1$ each *conical point* x in $\operatorname{int}|M|$, where the isotropy group G_x is the cyclic rotation group C_α of order α, and likewise labelling by $\alpha > 1$ each *dihedral point* x in $\partial|M| - |\partial M|$ where G_x is the dihedral group $\Delta_{2\alpha}$ of order 2α.

(b) Marking by double lines ══ the circles and open intervals of $\partial|B|$ consisting of *mirror points* x , where the isotropy group G_x is Δ_2 acting by reflection. (This contrasts with the points of $|\partial M|$ that one draws with a single line.)

Here is an example:

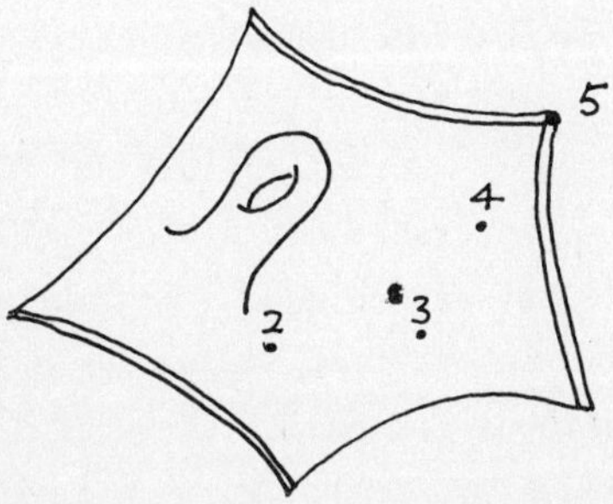

Note that the topological manifold boundary $\partial|M|$ is a disjoint union of $|\partial M|$ (which is a compact 1-manifold), of circles and open intervals of mirror points, and of isolated dihedral points.

Most connected compact 2-orbifolds are *good* in Thurston's terminology, namely obtained as quotient of a 2-manifold by a properly discontinuous group action. The only connected bad 2-orbifolds are the following (see Thurston, W.P., 1976-79, §13):

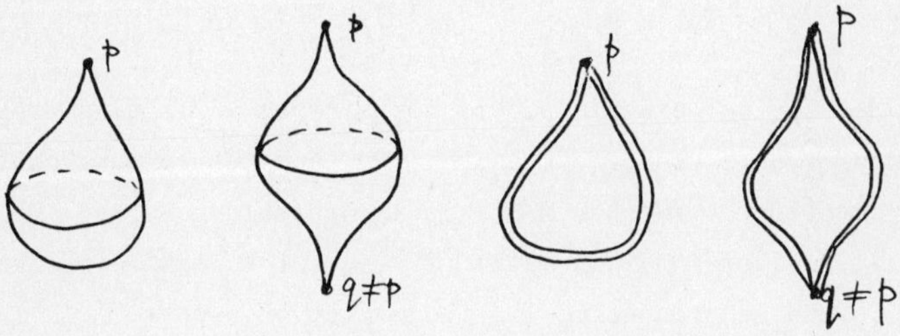

All other compact connected 2-orbifolds are quotients of the 2-sphere S^2 , the euclidean plane E^2 or the hyperbolic plane $\mathbb{H}^2$ by a properly discontinuous group of isometries (see Thurston, W.P., 1976-79, §13).

Many notions involving manifolds extend readily to orbifolds. Often, this extension is naive or obvious, as for: *compact*, *connected* (these two topological terms referring to the underlying topological space), *automorphism*, *isotopy*, *product*, *boundary*, *interior*, *closed orbifold*, *suborbifold*, *Riemmannian metric*, *geodesic*, C^∞-*triangulation*.

Here is a rough criterion that usually suggests the right extension: suppose the orbifold M is $\widetilde{M}/G$ for a finite group G acting on a manifold $\widetilde{M}$. The wanted orbifold notion in M should naturally lift to a G-equivariant version of the corresponding manifold notion in $\widetilde{M}$.

Here are now a few notions whose extension to orbifolds is a bit surprising, or at least requires some decision:

A *regular neighbourhood* $N(X)$ of a suborbifold or other 'sufficiently nice' closed subset X of an orbifold M is one that for a Riemannian metric on M can be expressed as a closed ε-neighbourhood for a sufficiently small $\varepsilon > 0$. They are unique up to ambient isotopic fixing the closed subset. For a suborbifold, a regular neighbourhood is just a tubular neighbourhood without its bundle structure. The singularity set ΣM is 'sufficiently nice'; so is a finite collection of transversally intersecting suborbifolds or more generally subcomplexes of any C^∞ Whitehead triangulation of M, cf. Illman, S., (1978). We will use only very rudimentary examples of regular neighbourhoods and C^∞ triangulations.

An *orbifold covering map* $h : N \to M$ is a continuous map $h : |N| \to |M|$ for which $|M|$ admits an atlas of folding charts $f : \widetilde{U} \to U \subset |M|$ such that, for *each* component V of $h^{-1}(U) \subset |N|$, there is a folding chart $g : \widetilde{U} \to V$ in the maximal atlas of N with $f = hg$.

A typical example occurs for a group G acting properly and discontinuously on a connected manifold $\widetilde{M}$ and G' a subgroup of G: the natural projection $|\widetilde{M}/G'| \to |\widetilde{M}/G|$ is an orbifold covering $\widetilde{M}/G' \to \widetilde{M}/G$.

The *geometric degree* in $\mathbb{N} \cup \infty$ of an orbifold covering $h : N \to M$, with M connected, is the number of points of $h^{-1}(y)$ where y is any regular point of M; this makes sense because h induces an ordinary covering of manifolds $N - \Sigma N \to M - \Sigma M$ and $M - \Sigma M$ is connected.

An *orientation* for an orbifold M is defined by a subatlas of folding charts $f_i : U_i \to U_i$ where all manifolds U_i are oriented and all folding groups and changes of chart are orientation-preserving. Any orbifold M has a unique *oriented 2-fold covering* orbifold $\hat{M}$: its existence and uniqueness is readily established when $M = W/G$ is the quotient orbifold of a manifold W by a finite group G (take $\hat{M} = \hat{W}/G$ with $\hat{W}$ the orientation 2-fold covering of W as manifold and with the natural action of G on W). This extends straightforwardly to any orbifold by a direct limit argument, as any orbifold is locally of this form.

During calculations, the reader must keep in mind the following rules concerning orientation of boundaries and products.

1) The preferred orientation of $\mathbb{R}$ is the positive one represented by vectors pointing to $+\infty$; this orients $\mathbb{R}^n$ and the unit disc $D^n \subset \mathbb{R}^n$. The boundary $S^1 = \partial D^2$ of D^2 gets the Gauss 'counterclockwise' orientation of increasing angle.

2) $M^m \times N^n \cong (-1)^{mn} N^n \times M^m$ by the map $(x, y) \to (y, x)$.

3) $\partial(M^m \times N^n) = (\partial M) \times U \cup (-1)^m M \times \partial N$.

These standard rules are in part a matter of convention; as they were not respected in the 1930's extra care is required in reading (Seifert, H., 1932).

An *orbifold bundle map* is a continuous map $p : M \to B$ of orbifolds such that for each point x in B there is an open neighbourhood U and a commutative square

$$(*) \qquad \begin{array}{ccc} \tilde{U} \times F & \xrightarrow{\text{projection}} & \tilde{U} \\ \phi \downarrow & & \downarrow f \\ p^{-1}U & \xrightarrow{\quad p \quad} & U \end{array}$$

in which f and ϕ are regular orbifold covering projections and F is some orbifold. (Without loss of generality, f can be a folding chart for B; likewise ϕ for M, provided F has a finite covering that is a manifold.) We call B the *base*, M the *total space* and p the *projection*. The fibre above $x \in B$ is the suborbifold $p^{-1}(x) \subset M$.

A typical example of an orbifold bundle is provided by a (manifold) locally trivial bundle $\tilde{p} : \tilde{M} \to \tilde{B}$ together with a group G acting properly discontinuously on $\tilde{M}$, sending fibre to fibre, and such that the action induced on $\tilde{B}$ is also properly discontinuous. Indeed $\tilde{p}$ naturally induces an orbifold bundle $p : \tilde{M}/G \to \tilde{B}/G$.

In the commutative square (*), the covering translation group G_ϕ for ϕ respects the projection $\tilde{U} \times F \to \tilde{U}$. It will not in general respect projection to F. When (*) can be chosen so that G_ϕ does respect projection to F and hence G_ϕ gives a product action on $\tilde{U} \times F$, we say that the bundle $p : M \to B$ is *locally trivial* (near $x \in B$). One readily proves that any orbifold bundle is locally trivial near $x \in B$ *if* $p^{-1}(x)$ *is compact* (is compactness perhaps unnecessary?). Indeed, given any square (*) one modifies the product structure of $\tilde{U} \times F$ near $\tilde{x} \times F$ (where $p(\tilde{x}) = x$) using a new (local) projection to $\tilde{x} \times F$, namely any G_ϕ-equivariant smooth neighbourhood retraction to $\tilde{x} \times F$ in $\tilde{U} \times F$ constructed for example using a G_ϕ-equivariant Riemannian metric.

This local triviality lemma can also be proved (under the same compactness hypothesis) for the piecewise linear pseudo-group; likewise for the topological one, provided all group actions involved are assumed to be locally orthogonal. For the proof, which is more difficult, see Lashof, R.

Local triviality of an orbifold bundle map $p : M \to B$ has some very basic consequences. Dividing F by the subgroup of G_ϕ fixing pointwise the factor $\tilde{U}$ we see that F can always be chosen to be a copy of a *generic fibre*, i.e. one over a nonsingular point of $U \subset B$. Thus, when B is connected (and hence $B - \Sigma B$), the generic fibres are all isomorphic and form a locally product bundle, while all other fibres are quotients of the generic fibre by a finite group action.

The present paper extensively studies a special kind of orbifold bundle $p : M \to B$, namely where M is 3-dimensional and the fibre above each point is a closed 1-orbifold (topologically a circle or an interval). In particular B is 2-dimensional. In this case, the fibres of p define a 'tame decomposition' or 'fibration' of $|M|$, such that each fibre admits a neighbourhood that is orbifold isomorphic to $(D^2 \times S^1)/G$, the finite group G respecting each factor, so that nearby fibres are exactly the images of the circles $* \times S$. The Seifert fibrations of 3-manifolds are celebrated examples of these. By analogy, we will call such a decomposition of M an S-*fibration*.

Note that, conversely, any S-fibration of the 3-orbifold M defines an orbifold bundle $p : M \to B$. Indeed, the decomposition space $|B|$ of the fibration inherits a natural orbifold structure B called the *decomposition orbifold* or *base orbifold*, so that the natural projection $|M| \to |B|$ is an orbifold bundle map. Therefore, giving an orbifold bundle $p' : M^3 \to B'$ with closed 1-dimensional fibres is equivalent to giving

1) an S-fibration of the 3-orbifold M, and

2) an orbifold isomorphism between the base orbifold B' and the decomposition orbifold of the S-fibration.

In view of this tight relationship, we will often allow ourselves to call such an orbifold bundle an S-fibration.

S-fibrations of course make sense in all dimensions. In particular, we will encounter S-fibrations in dimension 2 as 2-suborbifolds of an S-fibred 3-orbifold that are union of fibres.

Up to isomorphism, the only two connected orientable closed 2-orbifolds admitting such a fibration are the *standard torus* $\mathbb{R}^2/\mathbb{Z}^2$ and the *standard* 4-*corner pillow orbifold* $\mathbb{R}^2/\underline{\underline{R}}$. Here, $\underline{\underline{R}}$ is the group of isometries of the euclidean plane $\mathbb{R}^2$ generated by all π-rotations around the points of $(\frac{1}{2}\mathbb{Z})^2 \subset \mathbb{R}^2$. The pillow orbifold $\mathbb{R}^2/\underline{\underline{R}}$ has underlying space homeomorphic to S^2, and its singular set consists of four conical points with isotropy group $\mathbb{Z}/2$ (acting by rotation). As $\underline{\underline{R}}$ contains the translation group $\mathbb{Z}^2$ as a normal subgroup of index 2, the orbifold $\mathbb{R}^2/\underline{\underline{R}}$ is also the quotient of the torus $\mathbb{R}^2/\mathbb{Z}^2$ by the elliptic involution. This $\mathbb{R}^2/\underline{\underline{R}}$ is also the orientation covering orbifold of the square orbifold $[[0,1]] \times [[0,1]]$.

These two 2-orbifolds share a fundamental property which we will frequently use: say that a 1-suborbifold of a 2-orbifold F is *essential* when it does not bound a 2-suborbifold $\Delta \subset F$ whose underlying topological space is a disc and whose singular set consists of at most one (conical) point. Then, *any essential connected* 1-*suborbifold* K *of* $\mathbb{R}^2/\mathbb{Z}$ *or* $\mathbb{R}^2/\underline{\underline{R}}$ *is isotopic to the image of a straight line of rational slope in* $\mathbb{R}^2$. Moreover, the slope of this line depends only on K, and is by definition the *slope* of K (see Hatcher, A. & Thurston, W.P., 1978). These two facts are easily proved 'by hand', splitting the 2-orbifolds along the images of suitable straight lines in $\mathbb{R}^2$.

For both of these orbifolds every S-fibration is up to isotopy covered by a linear rational foliation of $\mathbb{R}^2$, whose *slope* characterizes the S-fibration up to isotopy. Up to orbifold automorphism the S-fibration is unique:

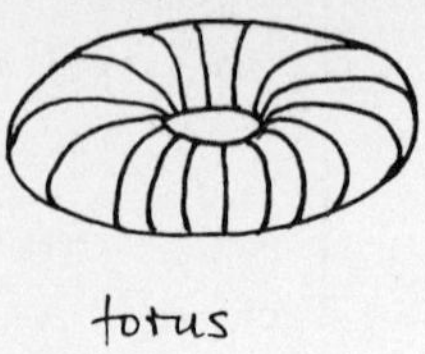

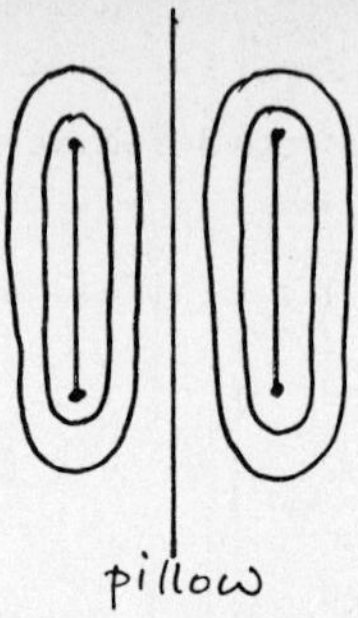

2 *THE LOCAL ORIENTED CLASSIFICATION OF* S-*FIBRATIONS*

In this section and in the next two, we restrict attention to oriented 3-orbifolds. Note that the base 2-orbifold of an S-fibration of such an oriented 3-orbifold may very well be non-orientable.

The local classification amounts to the fibred oriented classification in the case where the base B is (abstractly) a regular neighbourhood of a possibly singular point x. That the following list is complete is a routine exercise.

We distinguish cases according to the type of $x \in B$:

(0) x *is non-singular in* $\operatorname{int} B$:  x

(0*) x *is non-singular in* ∂B :  x .

In both cases, M is a manifold with a product fibration $M = B \times S^1$.

(1) x *is a conical point with angle* $2\pi/\alpha$ $(\alpha > 1)$:  x .

S-fibrations with this base corresponding to the integer $\alpha > 1$ are classified by a fraction β/α modulo 1 , with β not necessarily coprime to α . This $\beta/\alpha \in Q/\mathbb{Z}$, called the *slope modulo 1* of the fibration $p : M \to B$, is defined by the property that M is isomorphic as an S-fibred orbifold to $(D^2 \times S^1)/(\mathbb{Z}/\alpha)$ where the generator of $\mathbb{Z}/\alpha$ rotates D^2 by angle $2\pi/\alpha$ and rotates S^1 by an angle $-2\pi\beta/\alpha$ (beware of the sign!). In particular, M is a manifold precisely when α and β are coprime; otherwise, its singular set consists of the fibre $p^{-1}(x)$ with isotropy group $\mathbb{Z}/\delta$ where δ is the greatest common divisor of α and β .

Looking at a fundamental domain, we see that the underlying topological space $|M|$ is a solid torus with core the fibre $|p^{-1}(x)|$, while the other fibres are drawn on concentric tori and wrap α/δ times around in meridianal direction and β/δ times around in longitudinal direction. In other words, the fibration foliates $|M|$ as a Seifert fibred solid torus (Seifert, H., 1932). The relation of the fraction $\beta/\alpha \in \mathbb{Q}/\mathbb{Z}$ to the Seifert invariants of this solid torus (Seifert, H., 1932 and Seifert, H. & Threlfall, W., 1945) is as our notation suggests. Namely, if ∂M is degree 1 parametrized by the standard torus $\mathbb{R}^2/\mathbb{Z}^2$ so that the fibres have infinite slope, then the boundary of a meridian disc, of $|M|$ has slope $\equiv \beta/\alpha$ mod 1 .

The fraction $\beta/\alpha \in \mathbb{Q}/\mathbb{Z}$ has an interpretation as *monodromy* in the sense of differential geometry: the standard product metric on $D^2 \times S^1$ produces a connexion on the (manifold) S -bundle $p| : (M - p^{-1}(x)) \to (B - x)$. Parallel transport of fibre over a loop λ based at $y \in B - x$ defines a monodromy map on $p^{-1}(y)$ which turns out to be the isometric translation t_λ of the fibre $p^{-1}(y)$ over itself which, viewed as a rotation, has angle

$$i_x(\lambda)(\beta/\alpha)2\pi$$

where $i_x(\lambda)$ is the index of the loop λ about x . When M is equipped with other Riemannian metrics making the fibres geodesic (for instance arising from Thurston's geometries (Thurston, W.P., 1982; Bonahon, F. & Siebenmann, L.C., (to appear) and Scott, G.P., 1983), the monodromy m_λ of the loop λ will in general differ from t_λ ; but, as λ varies making the area it encloses tend to 0 , m_λ tend to t_λ . For this reason β/α is also called the *monodromy fraction* (mod 1) of the S-fibration at x .

The monodromy fraction $\beta/\alpha \in \mathbb{Q}/\mathbb{Z}$ must not be confused with the closely related *holonomy fraction* $\beta^*/\alpha \in \mathbb{Q}/\mathbb{Z}$ associated to the circle foliation of M induced by p . This last fraction is defined by the property that the first return map for nearby fibres to an (oriented) disc transverse to the (coherently oriented) singular fibre is a rotation of angle $2\pi\beta^*/\alpha$; it is related to β/α by

$$\begin{cases} (\beta/\delta)(\beta^*/\delta) \equiv 1 \text{ modulo } \alpha/\delta \\ \text{where } \delta = \text{g.c.d.}(\alpha,\beta) = \text{g.c.d.}(\alpha,\beta^*) \end{cases}$$

(2) x *is a mirror point in* int(B) : 

x

There is just one such fibration. Namely M is isomorphic to $(D^2 \times S^1)/(\mathbb{Z}/2)$ where $\mathbb{Z}/2$ acts by complex conjugation (i.e. reflection) on each of D^2 and S^1. Then $|M|$ is a 3-ball in which the singular set consists of two unknotted arcs K^1 as in Figure 1. The annulus $[-1, 1] \times S^1$ in $D^2 \times S^1$ gets folded onto a square Q of arc fibres in M,

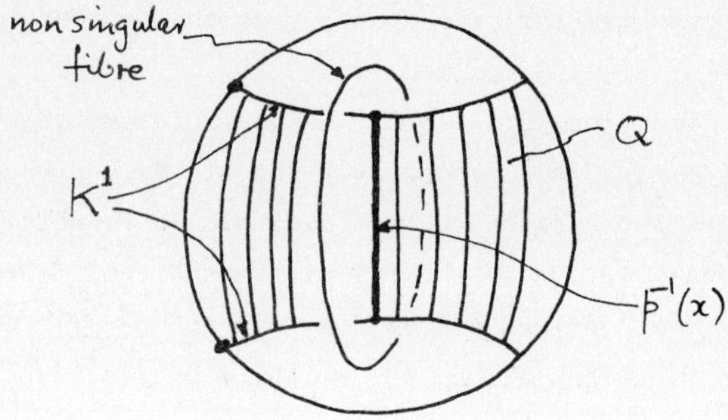

Figure 1

whose boundary ∂Q is K^1 union two arc fibres in ∂M. The non-singular fibres form a trivial bundle of circles foliating $M - Q$.

(2*) x *is a mirror point at boundary:*  x

The model is differentially half the above model (2).

(3) x *is a dihedral point with angle* π/α $(\alpha > 1)$:  x

The fibrations with this base are classified by a fraction β/α modulo 1.
Here, M is isomorphic to the quotient of $D^2 \times S^1$ by an action of the dihedral group $\Delta_{2\alpha}$ of order 2α, such that the generator of $\mathbb{Z}/\alpha \subset \Delta_{2\alpha}$ rotates D^2 by angle $2\pi/\alpha$ and S^1 by angle $-2\pi\beta/\alpha$, while each element of $\Delta_{2\alpha} - \mathbb{Z}/\alpha$ acts by reflection on D^2 and S^1. In particular, M is the quotient of a type (1) orbifold by an involution.

To describe M, let δ still denote the greatest common divisor of α and β. We introduce the conical orbifold D_δ obtained by quotienting D^2 by a $(2\pi/\delta)$-rotation. Then, if $\bar{\alpha} = \alpha/\delta$ and $\bar{\beta} = \beta/\delta$, M is the quotient of the orbifold $D_\delta \times S^1$ by $\Delta_{2\bar{\alpha}}$, the generator of $\mathbb{Z}/\bar{\alpha} \subset \Delta_{2\bar{\alpha}}$ rotating D_δ by (total angle) $(1/\bar{\alpha}) = 2\pi/\alpha$ and rotating S^1 by angle $-2\pi\bar{\beta}/\bar{\alpha} = -2\pi\beta/\alpha$ while the elements of $(\Delta_{2\bar{\alpha}}) - (\mathbb{Z}/\bar{\alpha})$ act as reflections on D_δ and S^1 ; here, the 'total angle' means the angle around the conical point of the orbifold D_δ , namely $2\pi/\delta$.

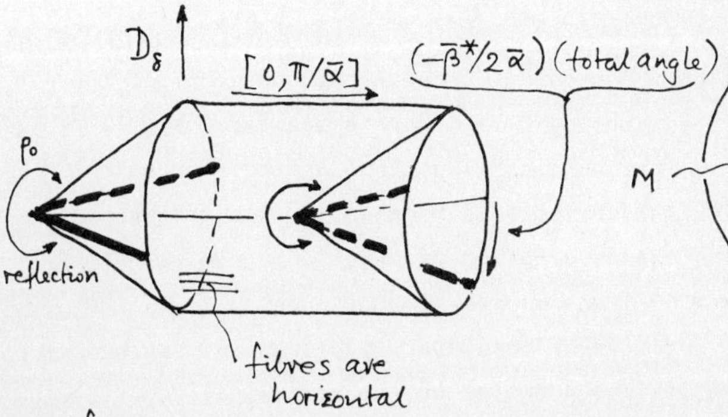

fig. 2

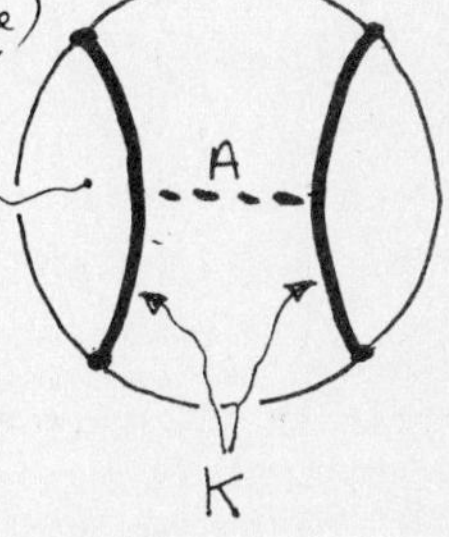

fig. 3

Take $D_\delta \times [0, \pi/\bar{\alpha}]$ as a fundamental domain in $D_\delta \times S^1$. The identifications yielding M come from reflections ρ_0 on $D_\delta \times 0$ and ρ_1 on $D_\delta \times (\pi/\bar{\alpha})$. Viewed in D_δ , the axis of ρ_1 is obtained from that of ρ_0 by rotation of

$$(\text{total angle}) \times (-\bar{\beta}^*/2\bar{\alpha}) = -\pi\beta^*/\alpha , \quad \text{where } \bar{\beta}^* = \beta^*/\beta ,$$

i.e.
$$\bar{\beta}^*\bar{\beta} \equiv 1 \bmod \bar{\alpha} .$$

It follows that $|M|$ is a 3-ball where the singularity set of M consists of the image K of the axes of ρ_0 and ρ_1 plus, when $\delta > 1$, the arc $A = |p^{-1}(x)|$.

This A forms a strut between the arcs such that $A \cup K$ has the form of an unknotted letter H , as in Figure 3. Viewing $(|M| , K)$ as a rational tangle of infinite slope in the terminology of Bonahon, L.C. & Siebenmann, L.C. (to appear) and Hatcher, A. & Thurston, W.P. (1978), i.e. parametrizing ∂M by the standard pillow orbifold $\mathbb{R}^2/\underline{\underline{R}}$ so that the boundary of the disc in $||M|$ separating the two components of K has infinite slope (see §1), the fibration of ∂M consists of arcs and circles of slope $-\bar{\beta}^*/\bar{\alpha} = -\beta^*/\alpha$ modulo 1 in $\partial M \cong \mathbb{R}^2/\underline{\underline{R}}$. Similarly, in the boundary of each of a shrinking family of regular ε-neighbourhoods N_ε of the strut A ; thus we can see the whole fibration at a glance. One can continuously follow an arc fibre in ∂N_ε as $\varepsilon \to 0$; it folds down onto A with geometric degree $\bar{\alpha}$ like a carpenter's measuring stick: $\wedge\!\!\!\wedge\!\!\!\wedge \to 1$. The projection $p : M \to B$, sends K onto the singular set Σ^1 of the base B ; if $\bar{\alpha}$ is odd, each component of K maps homeomorphically onto Σ^1 , whereas if $\bar{\alpha}$ is even each component of K is folded onto half of Σ^1 .

Geometric remark

The boundaries of the regular ε-neighbourhoods N_ε of the singular fibre, just described topologically, have a beautiful metric structure, that becomes more intrinsic and important in the geometric theory of S-fibrations discussed briefly in Bonahon, F. & Siebenmann, L.C. (to appear). Recalling the description of ∂N_ε as the quotient of the flat cylinder $[0,\pi/\bar{\alpha}] \times S^1_\varepsilon$, where $\pi/\bar{\alpha}$ is the length ℓ of the singular fibre and S^1_ε is a circle of perimeter $p_\varepsilon = 2\pi\ \varepsilon/\delta$, we see that ∂N_ε breaks up naturally into 4 copies of the (oriented) acute angled euclidean 2-simplex Δ^2 illustrated:

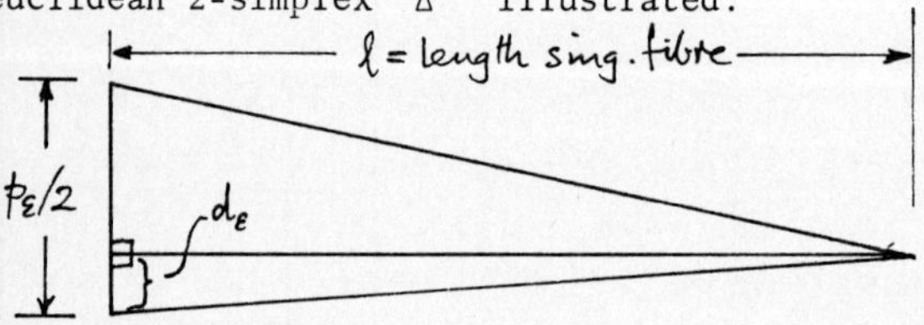

where $d_\varepsilon/(p_\varepsilon/2) \equiv -\beta^*/\alpha$ modulo 1 .

Thus ∂N_ε is a tetrahedron that is degree 1 isometric to the frontier $\delta\Delta^3$ of a 3-simplex Δ^3 in euclidean 3-space E^3 , having all four faces (degree +1) congruent to Δ^2 , and (hence) opposite edges equal in pairs. The shortest edge of $\delta\Delta^3 = \partial N_\varepsilon$ (for ε small) is of length $p_\varepsilon/2$; the fibres on ∂N_ε are perpendicular to it, and meridians are parallel to it.

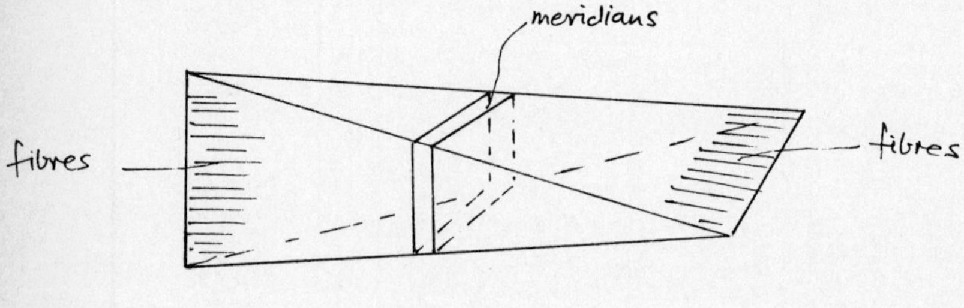

The universal cover of the orbifold ∂N_ε is the euclidean plane tesselated by copies of Δ^2. Note that the angle in E^3 between the two shortest edges of Δ^3 varies monotonically from 0 to π (with suitable orientation conventions) as $t = -\beta^*/\alpha$ modulo 1 varies from 0 to 1 ; the formula is $\cos^{-1}(1 - 2t)$, which is independent of ε .

Example

For the reader familiar with Bonahon, F. & Siebenmann, L.C. (to appear), in particular §9 therein (see also Montesinos, J.-M., 1973 & 1980; Thurston, W.P., 1976-79, §13-4), it should now be clear that a Montesinos knot, when viewed as an orbifold M with underlying topological space S^3 and singularity set the knot, on which the isotropy group is $\mathbb{Z}/2$ acting by rotation, admits a standard S-fibration. The base 2-orbifold B is topologically a 2-disc. Each ring is a singular fibre of type (1) with trivial monodromy and $\delta = 2$. Each rational tangle with non-integer slope β/α consistutes a fibred neighbourhood of a single fibre of type (3) with $\delta = 1$ and with monodromy β/α mod 1 . The singular fibres over mirror points constitute the band defined in Bonahon and Siebenmann, §9 (to appear). This accounts for all the singular fibres. The non-singular fibres are unknotted circles linking the band (like rings). This is illustrated by the example below:

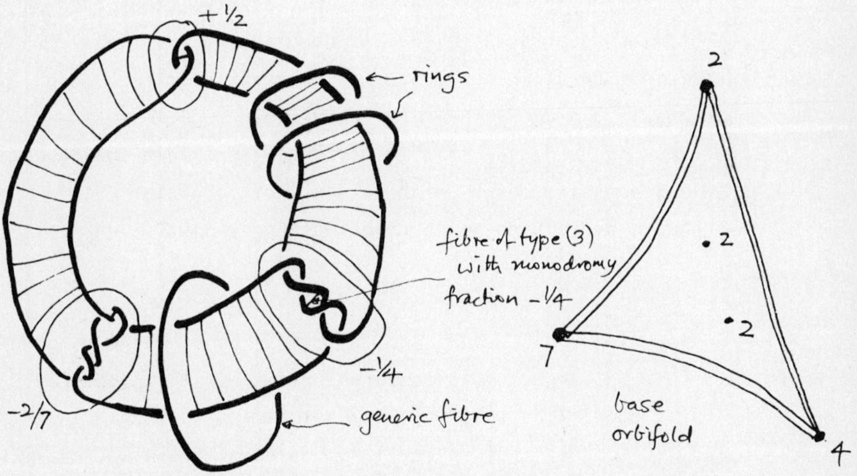

3 OBSTRUCTIONS AND INVARIANTS FOR ORIENTED FIBRATIONS

To give the oriented classification of S-fibrations $p : M^3 \to B^2$ over an arbitrary 2-dimensional orbifold $|B|$, we shall use obstructions lying in ordinary abelian cohomology groups of $|B|$, albeit with coefficients in abelian bundles of coefficients over $|B|$ (these coefficient bundles are not locally constant and do strongly reflect the orbifold structure of B ; they are really coefficient sheaves). This involves machinery not strictly necessary for the classification (cf. Seifert, H., 1932) ; but it seems conceptually helpful.

Consider an S-fibration $p : M \to B$. For each n , local (degree +1) automorphisms of the fibration define a bundle of coefficients $\pi_n\underline{\underline{LA}}(p)$ over the base $|B|$. The fibre of $\pi_n\underline{\underline{LA}}(p)$ above the point x in $|B|$ consists of the n-th homotopy group of the degree +1) orbifold automorphisms $p^{-1}(U) \to p^{-1}(U)$ respecting *each* fibre, where U is a regular neighbourhood of x in B . This is independent of the choice of U , but beware that $p^{-1}(x)$ cannot in general replace $p^{-1}(U)$ unless x is a nonsingular point of B .

When p is an S-fibration (still oriented) over a 2-orbifold, a case by case analysis calculates $\pi_n\underline{\underline{LA}}(p)$. At a point x in B such that $|p^{-1}(x)|$ is an interval, π_0 is $\mathbb{Z}/2$ and π_1 is 0 . At a point x such that $|p^{-1}(x)|$ is a circle, π_0 is 0 and π_1 is $\mathbb{Z}$.

It is now possible to define the cohomology groups $H^k(B;\pi_n\underline{\underline{LA}}(p))$, extending to orbifolds classical equivariant cohomology (cf. Bredon, G., 1967). A simplest definition uses a smooth triangulation of B such that the isotropy group is constant on the interior of each simplex; subsequent discussion assumes one has been chosen. In this context, each coefficient bundle we encounter is constant over each open simplex and so amounts to a *covariant functor from inclusions among the simplices of* B *to abelian groups*. This is obviously just what one needs to define a complex of cochains on B with values in the bundle.

The standard procedures of obstruction theory now let one show:

Fact

Two S-fibrations $p : M \to B$ and $p' : M' \to B$ over the same compact 2-orbifold B are (degree +1) isomorphic (fixing B) precisely if the following three conditions (a), (b), (c) hold:

(a) The fibrations are locally isomorphic, i.e. for each x in B the restrictions of p and p' over a small neighbourhood U of x are isomorphic.

(b) An obstruction $\omega_1(p,p')$ lying in $H^1(B;\pi_0\underline{\underline{LA}}(p))$ vanishes. The vanishing of $\omega_1(p,p')$ is equivalent to the existence of an isomorphism

from p to p' over a regular neighbourhood of the 1-skeleton of B.
(c) A second obstruction $\omega_2(p,p')$ lying in $H^2(B;\pi_1\underline{\underline{LA}}(p))$ vanishes; it is defined when $\omega_1(p,p') = 0$. □

Complement

Every $\omega \in H^1(B;\pi_0\underline{\underline{LA}}(p))$ is realized by a fibration $p_\omega : N \to B$ that is locally isomorphic to p and has $\omega_1(p,p_\omega) = \omega$, and the isomorphism classes of all such fibrations p' with $\omega_1(p,p') = \omega$ are in 1 - 1 correspondence with $H^2(B;\pi_1\underline{\underline{LA}}(p_\omega))$ by the map $p' \to \omega_2(p_\omega,p')$. □

We next discuss the obstruction groups, assuming B connected. $H^1(B;\pi_0\underline{\underline{LA}}(p))$ is $(\mathbb{Z}/2)^k$ where k is the number of circles (called mirror cycles) in the closure of the set of mirror points in B.

To calculate the group $H^2(B;\pi_1\underline{\underline{LA}}(p))$ first note that the coefficient bundle $\pi_1\underline{\underline{LA}}(p)$ coincides with the bundle that assigns to each x in B with regular neighbourhood U the group $H_1(|p^{-1}(x)|;\mathbb{Z})$, which is $\mathbb{Z}$ or 0, according as $|p^{-1}(x)|$ is a circle or an interval. It follows that $H^2(B;\pi_1 LA(p))$ is $H^2(|B|,\partial_m|B|;\mathbb{Z}^\omega)$ where $\partial_m|B|$ is the closure of the set of mirror points and the coefficient bundle $\mathbb{Z}^\omega$ is twisted by the orientation homomorphism $\omega : \pi_1|B| \to \mathbb{Z}/2$ of the (Seifert) fibration over $\text{int}|B|$. The orientation of M then gives an (preferred) isomorphism of this cohomology group with $\mathbb{Z}$ or 0 according as $\partial B = \phi$ (i.e. $\partial|B| = \partial_m|B|$) or not.

Thus, the above analysis provides an enumeration of all (oriented) S-fibrations p' over the compact connected 2-orbifold B with a given local isomorphism type. However, $\omega_1(p,p')$ and $\omega_2(p_\omega,p')$ are by no means invariants of p' alone as they depend on the choice of p and p_ω. It is certainly preferable to have at one's disposal canonically defined invariants of S^1-fibrations.

One basic invariant of a connected S^1-fibration M is the *Euler number* $e_o \in Q$ (defined below) which is the obstruction to finding a multifold section of $p : M \to B$. By an *n-fold section* of p, where $n \geq 1$, we mean a locally bicollared 2-suborbifold $F \subset M$ such that $p| : F \to B$ is an orbifold covering of geometric degree n (locally bicollared means that the pair (M,F) is locally isomorphic to (2-orbifold) $\times$ $(\mathbb{R},0)$ near F).

This Euler number was recognized by several connaisseurs of Seifert fibrations in the late 1970's: Neumann and Raymond (1978);

Macbeath (1980) and others; we first heard of it from Thurston in 1976. Related ideas of Wall (1961) and Sullivan (1981) concerning rational obstructions suggest its arrival was overdue. The most enlightening definitions we have encountered are in terms of cohomology classifying the corresponding fundamental group extension rationally, see Jankins, M. & Neumann, W. (1983), and in terms of general orbifold obstruction theory Haefliger, A. (1982) and Haefliger, A. & Quach, N.D. (1982). The definitions we use below are very ad hoc.

The reader may (or may not) find it helpful to define e_o using the same type of obstruction theory as for ω_1 and ω_2, as follows. There is a single obstruction to finding a multifold section

$$e_o(p) \in H^2(B;\pi_1\underline{\underline{LS}}(p)) ,$$

where the coefficient bundle is defined as follows: For $x \in B$, let $\underline{\underline{LS}}_n(p)_x$ denote the space of locally defined n-fold sections of p above a regular neighbourhood U of x. There are natural maps $\underline{\underline{LS}}_n(p)_x \to \underline{\underline{LS}}_{mn}(p)_x$, well-defined up to homotopy, that produce an mn-fold section from an n-fold section by replacing each sheet of the n-fold section by m parallel copies. Then the coefficient bundle is defined by considering for each such x the fundamental group of the space $\underline{\underline{LS}}(p)_x = \varinjlim \underline{\underline{LS}}_n(p)_x$, which is well-defined up to homotopy equivalence. Indeed, these groups $\pi_1\underline{\underline{LS}}(p)_x$ are abelian (see below) and constant on the interior of each simplex of the triangulation of B; they therefore induce a covariant functor from the category of simplices of B (with inclusions as morphisms) to that of abelian groups, which enables us to define the cohomology group $H^2(B;\pi_1\underline{\underline{LS}}(p))$. The class $e_o(p)$ is the only obstruction to the existence of a multifold section since, as we will see below, $\pi_0\underline{\underline{LS}}(p)_x = 0$.

When p has monodromy fraction β/α mod 1 at x ($\alpha = 1$ when x is regular, or a mirror point), $\underline{\underline{LS}}_n(p)_x \neq \phi$ precisely when $nm(x)\beta/\alpha$ is an integer, where $m(x)$ is $1/2$ at a dihedral or mirror point and is 1 otherwise. When $\underline{\underline{LS}}_n(p)_x \neq \phi$, one readily sees from the models of §2 that $\pi_0\underline{\underline{LS}}_n(p)_x = 0$ and that the evaluation map

$$\pi_1\underline{\underline{LS}}_n(p)_x \to \frac{1}{n}H_1(|p^{-1}(x)|;\mathbb{Z}) \subset H_1(|p^{-1}(x)|;\mathbb{Q})$$

is an isomorphism, compatible with the maps $\underline{\underline{LS}}_n(p)_x \to \underline{\underline{LS}}_{mn}(p)_x$. This yields a canonical isomorphism $\pi_1 LS(p)_x \cong H_1(|p^{-1}(x)|;\mathbb{Q}) \cong 0$ or $\mathbb{Q}$.

As for $H^2(B;\pi_1\underline{\underline{LS}}(p))$, the orientation of M then provides a preferred identification of $H^2(B;\pi_1\underline{\underline{LS}}(p))$ with Q when M is oriented and with 0 otherwise. The cohomology class $e_o(p)$ can therefore be considered as a well-defined rational number, called the *Euler number* of the fibration.

This definition applies even when M is not orientable and shows that $p : M \to B$ then admits a multifold section, as the corresponding obstruction group then turns out to be zero (see §§5-6).

Here is a humbler way to define $e_o \in Q$ for $p : M \to B$ (assuming M closed and oriented). The fibration p admits a multifold section F_o above the complement of a regular neighbourhood N(x) of some regular point $x \in B$; this can be easily constructed, starting from a neighbourhood of the singular points Σ of B . (In a manifold S^1-bundle, an n-fold section can be regarded as a 1-fold section of another S^1-bundle, namely the bundle of regular n-gons in the fibres of the first, for a suitable metric.) Express $\pm\,\partial F_o$ as $\alpha a + \beta b$ in $H_1(\partial p^{-1}N(x);\mathbb{Z})$ where a and b are meridian and fibre in $\partial p^{-1}N(x)$, so chosen that the intersection number $a\cdot b$ is +1 in the oriented boundary $\partial p^{-1}N(x)$. Then the Euler number $e_o = e_o(p)$ is the 'slope' $-\beta/\alpha$. Direct transversality considerations show that $-\beta/\alpha$ is an invariant of $p : M \to B$. See §4 for a still more practical way to calculate e_o .

Clearly e_o lies in $\frac{1}{n}\mathbb{Z} \subset Q$ where n is the least common multiple of $\{\alpha(x)/m(x)$; x singular in $B\}$; also reversing the orientation of M replaces e_o by $-e_o$.

The behaviour of e_o with respect to orbifold covering maps should be carefully noted : if $M \to M'$ is an oriented orbifold covering map of S^1-fibred oriented connected 3-orbifolds, and sends fibres to fibres mapping typical fibres with geometric degree $d \in \mathbb{N}$ and base spaces with geometric degree $\delta \in \mathbb{N}$, then

$$e_o(M') = (d/\delta)e_o(M) .$$

We call this the *covering formula*. Its proof is almost immediate from our definitions of e_o ; there are numerous alternatives, see Neumann, W.D. & Raymond, F. (1978); Gazolaz, M.D.C. (1977) and the remarks on classification of geometric S^1-fibrations in Bonahon, F. & Siebenmann, L.C. (1983).

4 THE PRACTICAL ORIENTED CLASSIFICATION

Here now is the practical combinatorial classification, in the spirit of Seifert, H. (1932). Given an oriented S-fibration $p : M^3 \to B$ and a certain number of arbitrary choices to be described later, we introduce a so-called data function Φ. Then, provided we consider it modulo an equivalence relation that "neutralizes" the choices made, this Φ will classify the fibration. The data function Φ consists of:

(i) a slope $\beta/\alpha \in Q$ attached at each conical point of B with angle $2\pi/\alpha$ or each dihedral point with angle π/α ;

(ii) a parity $\varepsilon \in \mathbb{Z}/2$ associated to each *mirror cycle* (i.e. circle in the closure of the mirror points of B) ;

(iii) lastly, when B is closed, a half-integer $e \in \frac{1}{2}\mathbb{Z}$ whose parity $2e \bmod 2$ is the sum over all mirror cycles of their parities ε .

The choices involved in the definition of Φ are the following: let $N(X^o)$ be a regular neighbourhood of the set X^o of conical or dihedral points of B . By §2, each component of $p^{-1}\partial N(X^o)$ is isomorphic to the torus $\mathbb{R}^2/\mathbb{Z}^2$ or to the pillow orbifold $\mathbb{R}^2/\underline{\underline{R}}$ (see §1). Then one chooses an oriented parametrization by $\mathbb{R}^2/\mathbb{Z}^2$ or $\mathbb{R}^2/\underline{\underline{R}}$ for each component of $p^{-1}\partial N(X^o)$ so that the fibres of p are vertical (slope ∞) in this parametrization.

With these choices fixed, Φ assigns slopes β/α , parities ε , and a semi-integer e as follows.

To each point $x \in X^o$, the function Φ assigns the slope in the parametrized boundary torus or Conway sphere $p^{-1}\partial N(x)$ of a 'meridian' multifold section of p above the component $N(x)$ of $N(X^o)$.

Consider a mirror cycle C^1 in B . Over the part of C^1 outside of $N(X^o)$ there lies a "band" of fibres that are topologically intervals. To this "band" one can add standard topological squares over $N(X^o)$ whose edges are of slope ∞ or 0 , so as to form an interval bundle E^2 over $C^1 \approx S^1$. Then Φ assigns to C^1 the parity $\varepsilon = 1$ or 0 according as this bundle is twisted (a Möbius band) or untwisted (an annulus).

As for $e \in \frac{1}{2}\mathbb{Z}$, it is the obstruction to extend a certain 2-fold partial section. Indeed, up to isotopy respecting each fibre, there is a preferred 2-fold section s of p above $N(X^o) - X^o$ that has slope 0 on each boundary component of $p^{-1}N(X^o)$. Then e is by definition the obstruction to extend s over all of $B - X^o$. Thus e

is e_o for the new S^1-fibration gotten by replacing each singular fibre by a less singular one that gives slope 0 for the chosen parametrization of $p^{-1}\partial N(X^o)$.

Warning

The isomorphism $\times 2 : \frac{1}{2}\mathbb{Z} \to \mathbb{Z}$ is needed to relate this e to the $e \in \mathbb{Z}$ of Bonahon, F. & Siebenmann, L.C. (§12, to appear).

Classification theorem 4.1.

When the 2-orbifold B is connected, the (oriented) classification fixing B of S-fibrations over B in terms of data functions is delightfully simple. It amounts to the equivalence of data functions Φ generated by the operation:

(*) Change the slope β/α at a conical or dihedral singular point $x \in X^o$ to $(\beta/\alpha) + 1$; in case x lies on a mirror cycle simultaneously change this cycle's parity ε to $\varepsilon + 1$; in case B is closed simultaneously change the semi-integer e then defined to $e + m(x)$ in $\frac{1}{2}\mathbb{Z}$ where by definition

$$m(x) = \begin{cases} \frac{1}{2} & \text{if } x \text{ is a dihedral point} \\ 1 & \text{if } x \text{ is a conical point.} \end{cases}$$

Note that Φ can be uniquely *normalized* within its equivalence class so that each slope lies in (say) the interval $(-\frac{1}{2},\frac{1}{2}]$.

The proof of this classification can be perfectly direct and elementary (Seifert's approach), or one can rely on the calculation of our obstruction cocycles as indicated below.

Suppose that two data functions Φ and Φ' give the same S-fibrations locally, i.e. that at each conical or dihedral point of B , we have $\beta/\alpha - \beta'/\alpha$ in $\mathbb{Z}$. Then one sees that $\omega_1(p,p')$ assigns to each mirror cycle $C^1 \subset |B|$ the parity in $\mathbb{Z}/2$ of

$$(\varepsilon - \varepsilon') + \Sigma\{\beta/\alpha - \beta'/\alpha \ ; \ \text{points in } C^1\}$$

which, if Φ , Φ' are normalized, is simply $\varepsilon - \varepsilon'$.

The Euler number $e_o \in Q$ of the S-fibration $M(\Phi) \to B$ is calculated to be

$$e_o = e - \Sigma\{m(x)\beta(x)/\alpha(x) \ ; \ x \in X^o\}$$

where, as before, m(x) is $\frac{1}{2}$ for x a dihedral point and $m(x) = 1$ otherwise.

Suppose now that $\omega_1(p,p') = 0$, i.e. that Φ and Φ' give S-fibrations isomorphic over the 1-skeleton of B . Then one calculates that $\omega_2(p,p') \in \mathbb{Z}$ is $e_o' - e_o$. An application of the obstruction theory of §3, clearly completes the proof of 4.1.

Moreover, we conclude that the Euler number e_o together with the type over the 1-skeleton characterize the (oriented) S-fibration M over B . Further, assuming B closed and connected, e_o *is any rational number such that the following expression be zero in* $\mathbb{Q}/\mathbb{Z}$ when calculated for (any specific) data function Φ describing the behaviour over the 1-skeleton:

$$e_o + \Sigma\{m(x)\beta(x)/\alpha(x) ;\ x \in X^o\} + \Sigma\{\tfrac{1}{2}\varepsilon(C^1) ;\ C^1 \text{ a mirror cycle}\} .$$

Remarks

a) This e_o corresponds to $\frac{1}{2}$ of the e_o of Bonahon, F. & Siebenmann, L.C., §12, to appear; the explanation is that the e_o of Bonahon and Siebenmann, §12, corresponds to the present e_o of a natural 2-fold branched cover that is Seifert fibred (cf. Montesinos, J.-M., 1980 and the covering formula for e_o in §3). Similarly for e , as already noted.

b) In case B is closed and connected, with one mirror cycle (as occurs for the S-fibrations of closed Montesinos pairs) *the cycle parity becomes redundant data:* Indeed ε is the parity of the integer 2e .

Since S-fibrations on a given 3-orbifold M tend to be unique up to isotopy (see Bonahon, L.C. & Siebenmann, L.C., to appear), a base-free classification is vital for the classification of M as orbifold. Fortunately, a satisfactory one follows immediately from the above.

Theorem 4.2

Two oriented compact connected S-fibred 3-orbifolds M , M' over different bases B , B' are fibred isomorphic if and only if one can find an isomorphism $f : B \to B'$ making the data functions Φ , Φ' correspond up to equivalence (or correspond exactly if they are normalized data functions) . When they correspond, f is covered by a fibred isomorphism $F : M \to M'$.

Change of ambient orientation for M corresponds to change of sign of e and of all slopes assigned by the data function Φ (not normalized); in particular, the invariants e_o and the slopes

$\beta(x)/\alpha(x) \in Q/\mathbb{Z}$ change sign. Thus the non-oriented classification of orientable S-fibred 3-orbifolds is settled by this theorem as well.

5 *THE NON-ORIENTABLE LOCAL TYPES*

From this point we abandon the convention that all 3-orbifolds and their isomorphisms be oriented.

Given a non-oriented S-fibred 3-orbifold M, we will occasionally consider its *orientation covering* $\hat{M} \to M$ to organize our study of M. Indeed, the S-fibration of M lifts to an S-fibration of the oriented orbifold $\hat{M}$. The covering $\hat{M} \to M$ induces a covering $\hat{B} \to B$ (2- or 1-fold) of fibration bases, and the covering involution ρ is fibred and induces on $\hat{B}$ an involution or the identity. Choosing x in B, we denote by f the fibre $p^{-1}(x) \subset M$ over x and we denote by $\hat{f}$ the fibre of $\hat{M}$ over f.

The fibred isomorphism class of a fibred regular neighbourhood of f in M is called the *type of* f. All the possible types of fibres in S-fibrations are thus the fibred isomorphism types of fibrations M *such that the base* B *is (abstractly) a regular neighbourhood of* $x \in B$. We assume this about B for our classification of fibre types.

The *orientable* local types have been arranged into classes (0), (1), (2), (2*), (3) in §2. Note however that the monodromies β/α in $Q/\mathbb{Z}$ used there for (1) and (3) depend by their sign on a choice of orientation. We ask the reader to tolerate our abuse of language in referring hereafter to the type classes (1) and (3) as types.

When M is non-orientable, $\hat{M}$ is connected and oriented and hence classified as above. By definition, $\hat{M}$ can therefore be identified with one of the model fibrations in §2 so that ρ is a fibred involution. For $x \in \text{int}(B)$, these non-orientable models are listed in Table 1 below; the type diagram occurring in this table will be explained later. There are similar models (5*), (6*), (12*), (15*) and (16*) for the case when $x \in \partial B$, that are differentiably half the corresponding models (5), (6), (12), (15) or (16).

An alternative to our use of $\hat{M}$ to sort out these models would be to use models that are local in M not B (cf. the introduction) together with suitable 'holonomies'.

Another alternative method, in essence Seifert's, uses the

type of f	type of B	type diagram	type of $\hat{f}$	type of $\hat{B}$	monodr. of $\hat{f}$	action of p on $\hat{B}$	action of p on $\hat{f}$	local decomp. models
(4)	2α 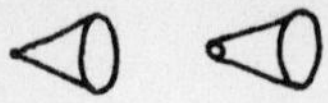		(1)	α 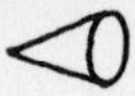	0	rot.	refl.	pin
(5)			(0)			refl.	Id	stick
(6)			(0)			refl.	rot.	stick
(7)	α 		(1)	α 	0	refl.	Id	stick
(8)	α 		(1)	α 	0	refl.	rot	stick
(9)	2α 		(1)	2α 	1/2	refl.	Id	stick
(10)	2α 		(2) or (3)	α 	0	refl.	Id	fold
(11)	2α 		(2) or (3)	α 	0	refl.	refl.	pin, fold

generic fibre topologically a circle above this line

---

generic fibre topologically an interval below this line

type of f	type of B	type diagram	type of $\hat{f}$	type of $\hat{B}$	monodr. of $\hat{f}$	action of p on $\hat{B}$	action of p on $\hat{f}$	local decomp. models
(12)			(0)			Id	refl.	stick
(13)	α 		(1)	α 	0	Id	refl.	stick
(14)	2α 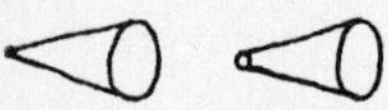		(1)	2α 	1/2	Id	refl.	pin
(15)			(2)			Id	Id	fold
(16)			(2)			Id	refl.	stick
(17)	α 		(3)	α 	0	Id	Id	stick
(18)	α 		(3)	α 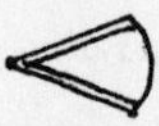	0	Id	refl.	fold
(19)	2α 		(3)	2α 	1/2	Id	Id	fold

Table 1

induced S-fibration of the 2-orbifold that is the frontier* δT of a fibred tubular neighbourhood $T = (S^1 \times D^2)/G$ of f in M, together with a 'Querschnitt' $Q = p^{-1}(x)$, where $p : T \to f$ is the projection and $x \in f$ is a nonsingular point of f. This tube frontier δT, its S-fibration, and Q together classify the fibre type.

Here are a few explanations: Each model M is of necessity the quotient of one of the standard orientable euclidean models for $\hat{M}$ in §2 by an orientation-reversing fibred isometric involution ρ the monodromy must be 0 or $\frac{1}{2}$. For type (9), $|\hat{M}|$ is a Seifert fibred solid torus with monodromy $\frac{1}{2}$ and ρ acts by reflection through a fibred Möbius band. In case (14), one obtains $|M|$ from $D^2 \times I$, fibred by the intervals $* \times I$, by identifying each $x \times 1$ with $(-x) \times 1$ on the end $D^2 \times 1$; the model $|M|$ in (4) is similarly obtained by performing the same identification on both ends $D^2 \times (\pm 1)$ of $D^2 \times I$. For type (19), the reader could profitably consider for $|\hat{M}|$ the tetrahedral model of Figure 5 in Bonahon, F. & Siebenmann, L.C., §17, (to appear), where ρ acts by reflection through a plane.

The last column of the table indicates the simplest model (or pair of models) from the introduction that suffice to locally describe the topological decomposition presented by the S-fibration near the fibre f.

The *type diagrams* appearing in Table 1 go a long way toward classifying these fibre types. By definition, the *type diagram* Δ associated to an S-fibration $M^3 \to B^2$ (over any base B) is the function that

(i) distinguishes the conical points in B of type (4) or (14); in practice by circling them.

(ii) assigns to each component of the set of mirror singularities of B the (one!) type of fibre over it; in practice by marking it $=\!=\!=$, $\underset{\sim\sim\sim}{0}$, $\underset{\sim\sim\sim}{\pi}$, $\underset{+\!+\!+\!+}{0}$, or $\underset{+\!+\!+\!+}{R}$ to indicate type (2), (5), (6), (15) or (16) respectively. Here the symbols 0, π, 0, R indicate the action of ρ on the corresponding fibre $\hat{f} \subset \hat{M}$ (0 meaning rotation of angle 0 or identity, π meaning rotation of angle π, and R meaning reflection).

---

*If $x \in \text{int}B$, one can use ∂M in place of δT.

Example:

This type diagram Δ is intended to be part of a so-called data function that will fully identify the fibration over B in Y .

Beware that the type diagram Δ does not pretend to fully specify the *system of fibre types* for the fibration $p : M \to B$, i.e. the function assigning to each point $x \in B$ the type of the fibre $p^{-1}(x) \subset M$. However, to do that one clearly needs only enhance Δ to

(iii) specify the generic fibre, as a circle or an interval, say by the symbol (S^1) or (I) .

(iv) specify the monodromy up to sign $\pm \beta/\alpha$ in $\mathbb{Q}/\mathbb{Z}$ at each point in B where the fibre is of orientable type (1) or (3) .

What functions, assigning to each $x \in B$ a fibre type, are realized by an S-fibration $p : M \to B$? Certainly the function must be locally coherent in the sense that the restriction of the function to a regular neighbourhood of each point of B be so realized. If $\partial|B| \neq \phi$, there is no further condition. But if $\partial|B|$ is empty there is exactly one *extra condition*, namely that there be an even number of singularities of type (4) or (14) . These facts will become clear in the next two sections.

Thus the term *type diagram* for a given base B is readily defined *abstractly* (without reference to S-fibrations), the one surprising restriction being that an even number of conical points be circled in case $\partial|B|$ is empty. Note also that intersection of distinct sorts of mirror occurs only at dihedral points of B of angle $\pi/2\alpha$, and that conical points circled in the type diagram have angle $2\pi/2\alpha$.

6 THE NON-ORIENTABLE CLASSIFICATION WHEN THE GENERIC FIBRE IS A CIRCLE

We classify in this section the S-fibrations $p : M \to B$ where *the 3-orbifold* M *is connected compact and non-orientable,* while the generic fibres, i.e. those over B-{singularities} , are *circles.*

Let X^1 be the closure in $|B|$ of mirror points over which the fibre in M is of type (2) . It is called the *Montesinos mirror complex.* A circle C^1 in X^1 is called a *Montesinos mirror cycle*; it is *orientation preserving* if a neighbourhood $p^{-1}N(C^1)$ is orientable, and *orientation reversing* if not; note that $p^{-1}\partial N(C^1)$ is a torus or a Klein bottle.

We start with the classification of such M up to fibred isomorphism *fixing the base* B. Much as in the oriented case, making some hidden choices (a), (b), (c), we will associate to the fibration a data function Φ which, modulo an equivalence relation neutralizing the choices, will classify the fibration. More precisely, we specify:

(i) Φ *assigns to* B *the type diagram for* M.

(ii) Φ *singles out a 1-submanifold* K^1 *of* $|B| - \text{int}N(X_4)$ where $N(X_4)$ is a regular neighbourhood in B of the finite set $X_4 \subset \text{int}B$ of singular points of type (4) in §5. This K^1 is subject to the conditions: K^1 avoids the conical and dihedral points of B; it meets each component of $\partial N(X_4)$ in exactly one point, and each component of $\partial|B|$ in ≤ 1 point; the suborbifold

$$M_K = p^{-1}(\text{int}B - N(X_4) - K)$$

is *orientable*, and this property fails if any component of K is suppressed. (Topologically $|M_K|$ is a manifold and this orientability of M_K amounts to orientability of $|M_K|$ as manifold.)

It is easily seen that the class of K^1 is well-defined in $H_1(|B| - \text{int}N(X_4), \text{boundary}; \mathbb{Z}/2)$. Indeed, K^1 represents the Poincaré dual of the sum of orientation classes

$$w_1(|B| - \text{int}N(X_4)) + w' \in H^1(|B| - \text{int}N(X_4); \mathbb{Z}/2)$$

where $w' : \pi_1(\text{int}|B| - N(X_4)) \to \mathbb{Z}/2$ gives the obstruction to coherently orienting the (circle) fibres of p over a loop in $\text{int}|B| - N(X_4)$. This sum clearly pulls back to the orientation class of the topological manifold $p^{-1}(\text{int}|B| - N(X_4))$.

The reason why we can make K meet each component γ of $\partial N(X_4)$ in exactly one point is that $p^{-1}(\gamma)$ is a Klein bottle, and consequently admits no orientable neighbourhood in M.

The first two hidden choices are:

(a) Choose an orientation for M_K.

(b) Choose a degree 1 parametrization of $p^{-1}\partial N(x) \subset M_K$ by $\mathbb{R}^2/\mathbb{Z}^2$ or $\mathbb{R}^2/R$ for each conical or dihedral point x with fibre type (1) or (3) as in the oriented case in §4.

(iii) Φ *assigns to each conical or dihedral point* x *in* B *with orientable fibre type* (1) *or* (3) *the slope* $\beta/\alpha \in Q$, in $p^{-1}\partial N(x)$ parametrized by (b) , of a (meridian) multifold section defined over $N(x)$.

(iv) Φ *assigns to each Montesinos cycle* $C^1 \subset |B|$ (orientable or not) *a parity* ε *in* $\mathbb{Z}/2$ defined just as in the orientable case.

The definition of Φ has one more part precisely in case

(*) $\partial|B|$ consists entirely of Montesinos cycles disjoint from K (i.e. orientation preserving), and no conical points with fibre type (4) occur (circled) in the type diagram (i) .

In this case a third hidden choice intervenes:

(c) Choose for each Montesinos cycle C^1 an oriented parametrization of $T(C^1) = p^{-}\partial N(C^1)$ by the standard torus $\mathbb{R}^2/\mathbb{Z}^2$ making the fibres of slope ∞ .

Observe that there is a standard 2-fold section t of p over

$$N(\partial|B|) - \{\text{dihedral points}\}$$

up to isotopy, namely one that is of slope 0 for the parametrizations (b) over $\partial N(x)$, x dihedral, cf. §4 . It is perhaps helpful to note that when $\varepsilon(C^1) = 0$, t is two (fibrewise) isotopic copies of a 1-fold section above $\partial N(C_1)$, and we then have a standard parametrization of $T(C^1)$ making t of slope 0 .

(v) *In case* (*) , Φ *assigns to each (orientation preserving) Montesinos cycle* C^1 *a weight* $\varepsilon' = \varepsilon'(C^1) \in \mathbb{Z}$ *of the same parity as* $\varepsilon(C^1) \in \mathbb{Z}/2$ *and* Φ *also assigns to* B *itself a parity* $\eta \in \mathbb{Z}/2$.

Namely $\varepsilon'(C^1)$ is twice the slope of t on $T(C^1)$ for the above parametrization, while η is the obstruction to obtaining a (1-fold) section of p over $|\beta| - \Sigma B$ of slope 0 on each parametrized torus making up $\partial p^{-1}N(\Sigma B)$ (corresponding to a Montesinos cycle or a conical point). Indeed, *because* M *is non-orientable*, the cohomology group where this obstruction lies is readily seen to be isomorphic to $\mathbb{Z}/2$.

The data function Φ is designed to be specified diagrammatically.

Example

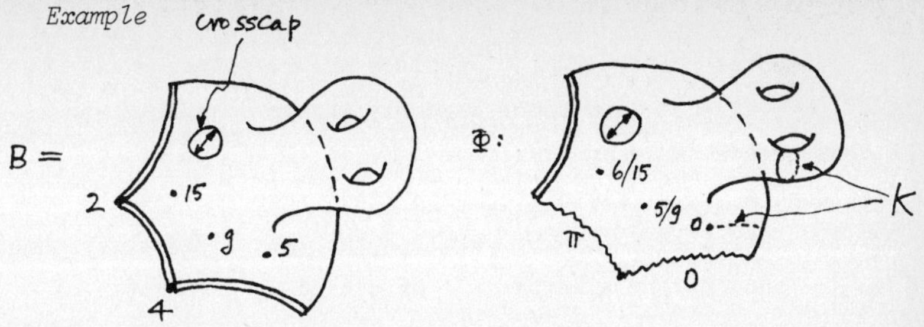

An *abstract data function* Φ is a function that assigns: a type diagram to B ; a 1-submanifold K^1 as described in (ii) above; slopes $\beta/\alpha \in Q$ to the conical and dihedral singular points where the type diagram requires a locally orientable fibration; parities $\varepsilon \in \mathbb{Z}/2$ to the Montesinos cycles; and also in case (*) integers ε' of parity ε to the Montesinos cycles, and η in $\mathbb{Z}/2$ to B . Since the fibrations are to be non-orientable, it is required that either there exist non-orientable mirror types, or K is non-zero in

$$H_1(B - \mathrm{int}N(X_4)\ ,\ \text{boundary};\ \mathbb{Z}/2)$$

(or both).

Given an abstract data function Φ it is not difficult to piece together a non-orientable S-fibration M over B giving rise to for suitable choices (a) , (b) , (c) .

It is thus clear that S-fibred orbifolds M over B up to fibred isomorphism correspond to abstract data functions Φ up to a *suitable equivalence*. This equivalence turns out to be the obvious one generated by the following moves $(\bar{a})$, $(\bar{b})$, $(\bar{c})$ and $(\bar{K})$ designed to neutralize the choices (a) , (b) , (c) and K^1 in the definition of Φ .

$(\bar{a})$ Change the signs of all slopes $\beta/\alpha \in Q$ and of all integers $\varepsilon' \in \mathbb{Z}$.

$(\bar{b}_1)$ Change the slope β/α at a conical point to $\beta/\alpha + n$ with $n \in \mathbb{Z}$. When $\eta \in \mathbb{Z}/2$ is defined, simultaneously change it to $\eta + n$.

$(\bar{b}_2)$ Change the slope β/α at a dihedral point to $\beta/\alpha + n$ with $n \in \mathbb{Z}$. When this dihedral point is contained in a Montesinos cycle, simultaneously change its parity $\varepsilon \in \mathbb{Z}/2$ to $\varepsilon + n$ and, when defined, $\varepsilon' \in \mathbb{Z}$ to

$\varepsilon' + n$.

($\bar{c}$) For a Montesinos cycle C^1 , replace the weight $\varepsilon' \in \mathbb{Z}$ (when defined) by $\varepsilon' + 2$, and simultaneously change $\eta \in \mathbb{Z}/2$ to $\eta + 1$. (As a consequence, only the class of ε' in $\mathbb{Z}/4$ counts.)

($\bar{K}$) Replace K by a K' of the same kind, representing the same class in $H_1(|B| - N(X_4), \partial ; \mathbb{Z}/2)$. Realize the mod 2 relative homology between K and K' by a union L^2 of closed-up components of $(|B| - N(X_4)) - K \cup K'$ and, at each conical or dihedral point of B contained in L , reverse the signs of the slope $\beta/\alpha \in Q$; simultaneously, at each Montesinos cycle contained in L , change the weight $\varepsilon' \in \mathbb{Z}$ (if any) to $-\varepsilon'$.

Normalization

By ($\bar{K}$) we can and usually will normalize K so that K either consists of arcs only, or consists of a single non-separating circle, or again is empty.

We can now formally state

Classification theorem 6.1 (fixed base)

For a given compact connected base 2-orbifold B , the S-fibred non-orientable 3-orbifolds M over B with typical fibre a circle are classified up to fibred isomorphism fixing B by their associated data functions Φ modulo the equivalence $\approx$ generated by ($\bar{a}$) , ($\bar{b}$) , ($\bar{c}$) , ($\bar{K}$) above.

Proof of Theorem 6.1

We know already that, up to equivalence, Φ is an invariant of M up to fibred isomorphism fixing B, since the choices in the construction of Φ have been neutralized. It remains to check that if $p : M \to B$ and $p' : M' \to B$ yield $\Phi \approx \Phi'$, then there is a fibred isomorphism $M \to M'$ inducing the identity on B . This can be checked by a strictly elementary argument in the spirit of Seifert, H. (1932), but prefer to make use of orbifold obstruction theory.

First, consider a sequence of steps of types ($\bar{a}$) , ($\bar{b}$) , ($\bar{c}$) , ($\bar{K}$) converting the data function Φ to Φ' . By making corresponding changes in the hidden geometric choices K^1 , (a) , (b) , (c) defining Φ for $p : M \to B$, one sees that, for suitable such choices, the new Φ is identical to Φ' . Thus, we may assume $\Phi = \Phi'$.

Imitating the oriented case (see §4) we can use the equality

$\Phi = \Phi'$ to build a fibred isomorphism from p to p' over a neighbourhood N_1 of the 1-skeleton of B, say $f : p^{-1}N_1 \to p'^{-1}N_1$, which is degree $+1$ with respect to the orientations of choice (a) wherever these are defined. The obstruction to find f is a certain $\omega_1^+(p,p') \in H^1(B;\pi_0\underline{\underline{LA}}^+(p))$ where the coefficient bundle $\pi_0\underline{\underline{LA}}^+(p)$ is defined using only local automorphisms respecting the orientations of the generic fibres (i.e. restricting to orientation-preserving automorphisms at fibres of orientable type). As in the orientable case, this obstruction is readily seen to vanish when $\Phi = \Phi'$.

Next, we consider the obstruction $\omega_2(p,p')$, *the one obstruction to extending over all of* B *a given local isomorphism of* p *and* p' *above a neighbourhood of the* 0-*skeleton that is known to extend to one above a neighbourhood of the* 1-*skeleton.* It lies in $H^2(B;\pi_1 LA(p))$, which, as M is connected and non-orientable, is $\mathbb{Z}/2$ when $\partial|B|$ consists entirely of Montesinos cycles (orientable or not!), and is 0 otherwise. By changing the given local isomorphism between p and p' near the 0-skeleton of B one can modify $\omega_2(p,p')$ precisely when either some component of $\partial|B|$ is an orientation-reversing Montesinos cycle C^1 or else some conical point x in B has fibre type (4). The required change over the 0-skeleton is the nontrivial 'antipodal' automorphism near C^1 or x, which induces the antipodal map on each generic fibre, a reflection on fibres of type (2) or (4), and a reflection or the identity on fibres of type (3) (according to the parity of the denominator of the corresponding slope, put in reduced form). That this alters ω_2 follows from the fact that the induced 'antipodal' fibred automorphism of the Klein bottle $p^{-1}\partial N(C^1)$ or $p^{-1}\partial N(x)$ is isotopic to a Dehn twist along one of its fibres. It follows that $\omega_2(p,p')$ can be made 0 whenever η and η' in $\mathbb{Z}/2$ are not defined. On the other hand, when η, η' are defined, there is no indeterminacy, and $\omega_2(p,p') = \eta - \eta'$ provided Φ and Φ' otherwise coincide. This completes our proof of Theorem 6.1.

Remark

Had we naively treated $\omega_2(p,p')$ as a primary obstruction (i.e. with no interminacy), we would have been led to define fallaciously $\eta \in \mathbb{Z}/2$ as part of Φ whenever $\partial|B|$ consists entirely of Montesinos cycles, orientation preserving or not. In fact ω_2 is met as a *secondary* obstruction and the calculation of its indeterminacy is vital.

We turn to the *base-free* classification, i.e. the classification of S-fibred connected compact non-orientable 3-orbifolds up to fibred isomorphism inducing any isomorphism of base spaces whatever. In principle it is provided by the following immediate corollary of 6.1.

Classification theorem 6.2 (with varying base)

Consider two S-fibrations $p : M \to B$ and $p' : M' \to B'$ of compact connected non-orientable 3-orbifolds having generic fibre S^1 and associated data functions Φ and Φ'. There exists a fibred isomorphism $M \to M'$ inducing an isomorphism $g : B \to B'$ precisely if the transported data function $g(\Phi)$ is equivalent to Φ'. □

However, this result is not immediately very useful in deciding whether there exists some fibred isomorphism $M \to M'$, in case there are up to isotopy a great many isomorphisms $g : B \to B'$ to be tested.

In general we proceed as follows. We note that, up to isotopy, there are only finitely many isomorphisms

$$g_0 : N(\partial B \cup \Sigma) \to N(\partial B' \cup \Sigma') \subset B' .$$

One easily catalogues them and determines which of them arise from some isomorphism $g : B \to B'$. In general a great many g (up to isotopy) give the same g_0. From Φ and Φ' we shall define *weak data functions* Ψ, Ψ', and *weak equivalence* $\sim$ thereof so that the transported weak data function $g(\Psi)$ depends only on the isotopy class of g_0. Thus the following result that we shall prove with care is adequate to determine the base-free classification.

Improved classification theorem 6.3 (for data above)

Given an isomorphism $g : B \to B'$, (which can be thought of as the identity), one can find a fibred isomorphism $F : M \to M'$ inducing an isomorphism $f : B \to B'$, equal to $g : B \to B'$ on $\Sigma \cup \partial B$, precisely if the transported weak data function $g(\Psi) = g(\Psi)$ is weakly equivalent to Ψ'.

Corollary 6.4

The base-free fibred classification of non-orientable S-fibrations $p : M \to B$ with generic fibre S^1 is by the orbits of the weak equivalence classes of their weak data functions under the finite action of $\mathrm{Aut}(B)$. □

Indeed, as indicated above, the action of Aut(B) on the set of weak data functions will factor through the finite group of isotopy classes of restrictions of automorphisms of B to $N(\partial B \cup \Sigma)$.

Given a data function Φ on B , one ingredient of the associated weak data function Φ is the *Seifert class of the pair* $(|B|, K^1)$, where K^1 is the 1-submanifold $K^1 \subset |B| - \mathrm{int}N(X_4)$ which was part (ii) of Φ . This Seifert class is defined only when K is closed, in which case it amounts to prescribing one of the five symbols

$$\underline{\underline{O}}\ ,\quad \underline{\underline{No}}\ ,\quad \underline{\underline{NnI}}\ ,\quad \underline{\underline{NnII}}\ ,\quad \underline{\underline{NnIII}}\ .$$

The meaning and assignment of these symbols, which were introduced by Seifert, H., (§8, 1932), is clearest when K *is normalized to be a circle or empty*. Indeed they then determine (normalized) K^1 up to homeomorphism of $|B|$ and can be assigned as follows. The letters O , N indicate whether $K \subset B$ is empty or not (i.e. whether $p^{-1}(\mathrm{int}B)$ is orientable or not for a corresponding fibration $p : M \to B$) . The letters o , n indicate whether the topological manifold $|B|$ is orientable or not. If N holds, K IS A CIRCLE, and one defines conditions:

I : $|B| - K$ is orientable.

II : $|B| - K$ is non-orientable and K is 1-sided.

III : $|B| - K$ is non-orientable and K is 2-sided.

Juxtaposition of letters indicates conjunction of conditions.

To make this definition homological, consider the topological 2-manifold $\bar{B}$ obtained by capping off each boundary component of $|B|$ with a 2-disc. Let κ be the class of K^1 in $H_1(\bar{B};\mathbb{Z}/2)$ and let $\omega \in H_1(\bar{B};\mathbb{Z}/2)$ be the Poincaré dual of the orientation class $\omega_1(\bar{B}) \in H^1(\bar{B};\mathbb{Z}/2) = \mathrm{Hom}(\pi_1 B, \mathbb{Z}/2)$.

Then the homological conditions are:

$$O : \kappa = 0\ ;\quad o : \omega = 0\ ;\quad N : \kappa \neq 0\ ;\quad n : \omega \neq 0\ ;$$

$$I : \kappa = \omega\ ;\quad II : \kappa \neq \omega\ ,\ \kappa\cdot\kappa \neq 0\ ;\quad III : \kappa \neq \omega\ ,\ \kappa\cdot\kappa = 0$$

where $\kappa\cdot\kappa \in \mathbb{Z}/2$ is homological intersection number.

In particular, *only* O *can be realized when* $\bar{B} \cong S^2$, i.e. $|B|$ is planar. Similarly, the *only possible Seifert classes when* $\bar{B} \cong \mathbb{R}\mathbb{P}^2$ *are* O *and* NnI , *while* NnIII *is excluded when* $\bar{B}$ *is a Klein bottle*. In all other cases, all of the five Seifert classes can occur.

Observe also that, when Φ arises from $p : M \to B$, the Poincaré dual of $[K]$ in $H^1(|B| ; \mathbb{Z}/2) = \mathrm{Hom}(\pi_1 |B| ; \mathbb{Z}/2)$ classifies the 2-fold covering of $\mathrm{int}|B|$ given by the 2-fold orbifold covering $\hat{B} \to B$ induced by the orientation covering $\hat{M} \to M$ discussed in §1 .

Given a data function Φ on B, we now define the associated *weak data function* Ψ by replacing the 1-submanifold $K^1 \subset |B| - \mathrm{int}\, N(X_4)$ which was part (ii) of Φ by one or two things as follows. (We continue to normalize K^1 to be one essential circle, or arcs, or again empty.)

$(ii)_1$ (in case $\partial K^1 = \emptyset$) *the Seifert class of the pair* $(|B| , K^1)$.

$(ii)_2$ (in case $\partial K^1 = \emptyset$ and $|B| - K^1$ is orientable but $|B|$ is non-orientable). *The orientation up to sign of* $\partial|B|$ *induced by the orientation up to sign of the orientable* 2-*manifold* $|B| - K$.
This is clearly significant only when $\partial|B|$ has ≥ 2 components.

$(ii)^*_1$ (in case $\partial K^1 \neq \emptyset$) *The points* $K^1 \cap \partial|B|$.
These points will in practice be marked by stars * on the type diagram.

$(ii)^*_2$ (in case $\partial K^1 \neq \emptyset$, $|B|$ is non-orientable, and $X_4 = \emptyset$). *The residue class of* $[K^1]$ *in* $H_1(|B| , \partial|B| ; \mathbb{Z}/2)$ *modulo the image of the orientation preserving* 1-*cycles on* $|B|$.
Since the boundary of K^1 is specified in $(ii)^*_1$, there are just two residue classes in question, namely that of $[K^1]$ and that of $[K^1] + [C^1]$ where C^1 is an orientation reversing cycle on $|B|$.

Let $\partial^*|B|$ denote the union of those components of $\partial|B|$ meeting K (those starred in $(ii)^*_1$) .

Assertion 6.5

Let us associate to K any orientation of $\partial^*|B|$ extending to a neighbourhood of $K \cup \partial^*|B|$. Then, this defines an equivalence between the residue class of $[K^1]$ specified in $(ii)^*_2$ and an orientation of $\partial^*|B|$ modulo orientation reversal at an even number of components.

Proof of 6.5

The obstruction to extending such an orientation of $\partial^*|B|$ over all of $|B|$ is a cohomology class $\tilde{\omega} \in H^1(|B| , \partial^*|B| ; \mathbb{Z}/2)$ that extends $\omega \in H^1(|B| ; \mathbb{Z}/2)$, the orientation class of $|B|$. Consider another 1-submanifold K' of $|B|$ with the same boundary $\partial K' = \partial K \subset \partial^*|B|$ and no closed components. Then $K \cup K'$ can be regarded as a 1-cycle on $|B|$ and we have

$$\tilde{\omega}([K']) + \tilde{\omega}([K]) = \omega([K \cup K']) \in \mathbb{Z}/2 .$$

For each arc $A \subset |B|$ with $\partial A \subset \partial^*|B|$, e.g. a component of K or K' , the obstruction to extending the given orientation of $\partial^*|B|$ to a neighbourhood of $(\partial^*|B|) \cup A$ is $\tilde{\omega}([A]) \in \mathbb{Z}/2$. Thus $\tilde{\omega}([K]) = 0$. Hence $\omega([K \cup K'])$ equals $\tilde{\omega}([K'])$, which in turn is the number modulo 2 of components of $\partial^*|B|$ where the orientation extending over $\partial^*|B| \cup K$ differs from one extending over $\partial^*|B| \cup K'$. The assertion is an immediate consequence. □

An *abstract weak data function* Ψ on B is no more nor less than one that arises as above from an abstract data function Φ . However, the term abstract is morally justified by the following serviceable combinatorial characterization. An *abstract weak data function* Ψ on a compact connected 2-orbifold B consists precisely of:

(i) *A type diagram on* B (see §5) .

(ii)* *A finite number* n_* *of points on* $\partial|B|$ *(marked by stars).* These are not dihedral points; there is no more than one per component of $\partial|B|$; and the number n_4 of conical points of type (4) specified by the type diagram is congruent to n_* modulo 2 .

$(ii)^*_2$ (in case $n_4 = 0$, but $n_* > 0$ and $|B|$ is non-orientable) *An orientation of the union* $\partial^*|B|$ *of the* n_* *starred components of* $\partial|B|$, *modulo simultaneous orientation reversal on any even number of these components.*

$(ii)_1$ (in case $n_* = 0$) *A Seifert class among:*

$$0 , \underline{\underline{\mathrm{No}}} , \underline{\underline{\mathrm{NnI}}} , \underline{\underline{\mathrm{NnII}}} , \underline{\underline{\mathrm{NnIII}}} .$$

This class must be realizable on $|B|$ (see conditions above). Further, Seifert class $\underline{\underline{0}}$ is not admissible unless the type diagram (i) specifies points of non-orientable mirror type (5) $\underset{0}{\sim\!\sim\!\sim}$ or (6) $\underset{\pi}{\sim\!\sim\!\sim}$ of §5 . (This restriction occurs because we do not want data functions corresponding to orientable S-fibrations.)

$(ii)_2$ (in case $n_* = 0$, and the Seifert type specified is $\underline{\underline{\mathrm{NnI}}}$) *An orientation up to sign* $\partial|B|$.

(iii) *A slope* β/α *in* $\mathbb{Q}$ *for each conical or dihedral point of* B *near which the type diagram dictates an orientable (local)* S-*fibration.*

(iv) *A parity* $\varepsilon(C^1)$ *in* $\mathbb{Z}/2$ *for each Montesinos cycle* $C^1 \subset \partial|B|$ *indicated by the type diagram.*

(v) (in case $N_4 = n_* = 0$ and $\partial|B|$ consists entirely of Montesinos cycles, necessarily orientation preserving) *A weight* $\varepsilon'(C^1) \in \mathbb{Z}$ *of parity* $\varepsilon(C^1)$ *for each Montesinos cycle* C^1 *, and a weight* $\eta \in \mathbb{Z}/2$ *for* B *itself.*

Weak equivalence $\sim$ of weak data functions Ψ on B is the (tightest) one compatible with the established equivalence of data functions. It is generated by moves $(\bar{a})$, $(\bar{b}_1)$, $(\bar{b}_2)$, $(\bar{c})$, formulated exactly as for data functions, plus, *in case* B *is not assigned symbol* 0 *in part* (ii_1), the following three moves replacing $(\bar{K})$:

$(\bar{K}_1)$ Slide a star in $\partial|B|$ over a dihedral point x ; if this x is assigned slope $\beta/\alpha \in Q$ replace this slope by $-\beta/\alpha$.

$(\bar{K}_2)$ For any one conical point that is assigned a slope $\beta/\alpha \in Q$, reverse the sign of this slope.

$(\bar{K}_3)$ On any one component C of $\partial|B|$ simultaneously reverse the signs of all slopes, and of the weight $\varepsilon' \in \mathbb{Z}$ (if ε' is defined for C) ; further, in the case when $n_* = 0$ and the Seifert symbol is $\underline{\underline{\underline{\text{NnI}}}}$, reverse the orientation of C to get a new orientation of $\partial|B|$ (up to sign) for part $(ii)_2$ of Ψ .

Proof of the improved classification theorem 6.3

We are given data functions Φ and Φ' for $M \to B$ and $M \to B'$ with associated weak data function Ψ and Ψ' , together with an isomorphism $g : B \to B'$. We must show that $g\Psi \sim \Psi'$ if and only if there exists a fibred isomorphism $f : M \to M'$ inducing $f : B \to B'$ equal to g on $\Sigma \cup \partial B$. Without loss of generality, we assume $B = B'$ and g = identity.

Suppose F exists. Then $f\Phi \approx \Phi'$ by Theorem 6.1, which by 'forgetting structure' implies $f\Psi \sim \Psi'$. We have $f\Psi = \Psi$ because f fixes $\Sigma \cup \partial B$. Indeed, this is just a matter of definition of Ψ as regards everything but datum $(ii)_2^*$; but Assertion 6.5 shows that even this datum is unchanged by f . So $\Psi' \sim \Psi$ as required.

Conversely, suppose $\Psi \sim \Psi'$. We seek $f : B \to B$ fixing $\Sigma \cup \partial B$ such that $f\Phi \approx \Phi'$; then the classification theorem 6.1 provides an isomorphism F as required. Thus it suffices to find this f .

The weak equivalence $\Psi \sim \Psi'$, viewed as a sequence of weak equivalence moves altering Ψ to get Ψ' , lifts to a sequence of equivalence moves that alters Φ to Φ'' where Φ'' yields exactly Ψ' by forgetting. But Φ' too yields Ψ' by forgetting. Hence Φ'' and Φ' differ only in the submanifolds K'' and K' specified in $|B| - \mathrm{int}N(X_4)$. Further, $\partial K' = \partial K''$.

We now distinguish cases.

First case: $\partial K' = \partial K'' \neq \emptyset$, but $X_4 = \emptyset$. Then, we know that $K' \cup K''$ is an orientation-preserving cycle on $|B|$; indeed, in case $|B|$ is not orientable, this follows from the coincidence of the data $(ii)_2^*$ associated to K' and K'' . If Σ_c denotes the set of conical points of B , choose an (orientation-preserving) simple closed curve C in $|B| - N(\Sigma_c)$ such that $[C] = [K' \cup K'']$ in $H_1(|B| - N(\Sigma_c) , \partial^*|B|; \mathbb{Z}/2)$; connected summing C with copies of components of $\partial^*|B|$ if necessary, C can be chosen to intersect K'' transversely in exactly one point. Then C is bicollared and, as a first approximation to f , consider a Dehn twist T around C . By construction, $[TK''] = [K']$ in $H_1(|B| - N(\Sigma_c) , \partial^*|B|; \mathbb{Z}/2)$.

We can assume TK'' is transversal to K' away from $\partial^*|B|$. Then the homology between K' and TK'' rel $\partial^*|B|$ offered by closures of certain components of $|B| - (K' \cup TK'')$ may contain whole components of $\partial^*|B|$. To remedy this, consider the composition f of T with Dehn twists around slightly pushed in copies of the components of $\partial^*|B|$ contained in this homology, and take this as a final choice of $f : B \to B$. Now we have a new homology L^2 between fK'' and K' meeting $\Sigma \cup \partial B$ only at $\partial K' = \partial fK''$ such that L^2 contains no conical points, dihedral points or Montesinos cycles. By move $(\bar{K})$, it follows that $\Phi' \approx f\Phi''$. Since $\Phi'' \approx \Phi$, we have $f\Phi'' \approx f\Phi$ and therefore $\Phi' \approx f\Phi$ as required.

Second case: $X_4 \neq \emptyset$ (and $\partial K' = \partial K'' \neq \emptyset$) . If the cycle $K' \cup K''$ is orientation preserving in $|B|$, we proceed exactly as in the previous case. Otherwise, let $f_1 : B \to B$ glide one point x of X_4 around an orientation-reversing loop $C \subset |B|$ with $C \cap \Sigma = x$; this f_1 fixes x and the complement of a tubular neighbourhood of C (but reverses orientation near x) . Then the cycle $K' \cup f_1K''$ is orientation-preserving and we are back to

Third case: $\partial K' = \partial K'' = \emptyset$. The proof in this case is based on the following elementary fact: Let F' and F'' be two compact connected surfaces of the same topological type, and consider a homeomorphism $f : \partial F'' \to \partial F'$. Then, if F' and F'' are non-orientable, f always extends to a homeomorphism $F'' \to F'$; if F' and F'' are orientable, f extends if and only if it respects the orientations up to sign of $\partial F'$ and $\partial F''$ defined by orientations of F' and F''. We will apply this fact to the surfaces F' and F'' obtained by splitting $|B|$ along the circles K' and K'' respectively. These two surfaces are homeomorphic because K' and K'' have the same Seifert symbol, namely that assigned by Ψ'.

We assume $K' \neq \emptyset \neq K''$ (i.e. the Seifert symbol is not $\underline{\underline{0}}$) as the theorem is vacuous for Seifert symbol $\underline{\underline{0}}$.

When $|B| - K'$ is non-orientable, i.e. when the Seifert symbol is $\underline{\underline{\mathrm{NnII}}}$ or $\underline{\underline{\mathrm{NnIII}}}$, choose a homeomorphism $h : N(K'') \to N(K')$, which exists since K' and K'' have the same Seifert symbol. Then h together with the identity of $\partial|B|$ induce a homeomorphism $f : \partial F'' \to \partial F'$ which, by our preliminary remark, extends to a homeomorphism $f : F'' \to F'$. By connectivity of F', we may assume that f fixes Σ, and therefore induces an automorphism $f : B \to B$ fixing $\Sigma \cup \partial|B|$ and with $fK'' = K'$. Then $f\Phi'' = \Phi'$. Since $\Phi \approx \Phi''$, we have $f\Phi \approx f\Phi'' = \Phi'$ as required.

When $|B|$ is orientable, i.e. when the Seifert symbol is $\underline{\underline{\mathrm{No}}}$, choose an orientation-preserving homeomorphism $h : N(K'') \to N(K')$, and let $f : \partial F'' \to \partial F'$ be induced by h and the identity of $\partial|B|$. Then, f is orientation-preserving and extends to a homeomorphism $f : F'' \to F'$. We then conclude as above.

The last subcase is when $|B|$ is non-orientable but $|B| - K'$ (and $|B| - K''$) is orientable, i.e. when the Seifert symbol is $\underline{\underline{\mathrm{NnI}}}$. Then, by equality of the data $(\mathrm{ii})_2$, we know that F' and F'' admit orientations inducing the same orientation on $\partial|B|$; fix such orientations. Now, the boundary orientations on the 1 or 2 components of $\partial F'$ corresponding to K' fit together to give a preferred orientation of K'; similarly for K''. Start with any oriented homeomorphism $h : K'' \to K'$; arbitrarily extend it to $h : N(K'') \to N(K')$ (there are two different choices up to isotopy if K'' is bicollared); and let this extension together with the identity on $\partial|B|$ define $f : \partial F'' \to \partial F'$. By construction, this f respects orientations induced on $\partial F''$ and $\partial F'$, and consequently extends to $f : F'' \to F'$. We conclude as in previous subcases.

This completes the proof of Theorem 6.3. □

Concluding remarks

(a) The reader should pause to convince himself that Theorem 6.3 offers a *finite* algorithm to determine the finite set of (base-free) fibred isomorphism classes of non-orientable S-fibrations having a base isomorphic to a given 2-orbifold B .

A vital point is that slopes may as well be normalized to lie in $[-\frac{1}{2}, \frac{1}{2}]$; then for fixed B and fixed stars there are only finitely many weak data functions.

(b) The improved classification Theorem 6.2 will be used in a final chapter §9 to give an overview of all isomorphism classes of non-oriented S-fibrations.

7 CLASSIFICATION OF S-FIBRATIONS WHOSE GENERIC FIBRE IS TOPOLOGICALLY AN INTERVAL

This is the classification of S-fibrations whose generic fibre is the orbifold $[[0,1]]$. It is quite similar to the one where the generic fibre is a circle, although considerably simpler.

Remark

Using a silvering process, this classification is readily seen to be equivalent to the fibred classification of 3-orbifolds equipped with an orbifold bundle map $p' : M' \to B'$ with generic fibre the manifold $[0,1]$ rather than the orbifold $[[0,1]]$. (Hint: Consider $\partial_0 M' \subset \partial M'$ the union of the boundaries of all fibres; this is where silvering will change the orbifold structure. Let DM' be the double obtained by gluing two copies of M' along $\partial_0 M'$ and consider the quotient 'silvered' orbifold $M = DM'/(\mathbb{Z}/2)$ which is the quotient of DM' by the involution exchanging the two halves; of course $|M| = |M'|$ but $\Sigma M = \Sigma M' \cup |\partial_0 M'|)$.

Over a fixed connected compact base 2-orbifold B , such orbifold fibrations $p : M \to B$ are classified using data functions Φ on B consisting of

(i) *A type diagram on* B *for typical fibre* $[[0,1]]$ (see §5) .

(ii) *A class* $\omega \in H^1(|B| - X_{14} ; \mathbb{Z}/2)$ *where* X_{14} *is the finite set of conical points of type* (14) *indicated in the type diagram; this* ω *is required to be nontrivial near each point* $x \in X_{14}$.

This class ω is $\omega' + \omega''$ where ω' is the (restriction of the) orientation class of $|B|$, and ω'' is the first Stiefel-Whitney class of the topological I-bundle

$$p^{-1}(\mathrm{int}|B| - X_{14}) \xrightarrow{p} \mathrm{int}|B| - X_{14} .$$

The peculiarity rôle of fibres of type (14) is much the same as for fibres of type (4) . If γ is a small loop around such a point, the restriction of p above γ is topologically a non-trivial I-bundle, namely a Möbius band.

Fibred isomorphism classes *fixing* B of such fibrations then stand in one-to-one correspondence with such data functions; (no equivalence on the data functions is needed). The proof is an exercise; it is like that of 6.1 but much simpler.

The base-free classification involves degrading data functions Φ on B to weak data functions Ψ on B as follows. Ψ is derived from Φ by replacing ω which is part (ii) of Φ by

(ii)$_1$ (in case $X_{14} = \emptyset$ and $\omega = 0$ on $\partial|B|$). The Seifert symbol among

$\underline{\underline{O}}$, $\underline{\underline{No}}$, $\underline{\underline{NnI}}$, $\underline{\underline{NnII}}$, $\underline{\underline{NnIII}}$

for the pair $(|B|, K)$ where K is empty if $\omega = 0$ and otherwise K is circle inside $|B|$ Poincaré dual to ω .

(ii)$^*_1$ Prescription of the set of components of $\partial|B|$ on which ω is non-zero (usually by placing stars on these components in the type diagram).

(ii)$^*_2$ (in case $\omega \neq 0$ on $\partial|B|$, and $|B|$ is non-orientable, and $X_{14} = \emptyset$). The residue class of ω modulo the elements of $H^1(|B|, \partial|B|; \mathbb{Z}/2)$ whose cup-product with the orientation class of $|B|$ is trivial. This is simply the Poincaré dual of the corresponding datum when the generic fibre is a circle. As in Assertion 6.5, this is equivalent to an orientation of the union of components of $\partial|B|$ where ω is non-trivial, modulo orientation reversal at an even number of them.

The required equivalence $\sim$ of weak data functions is $=$, i.e. 'equality'.

Then the verbatim statement of the Improved Classification

Theorem 6.3 holds true in the present context; its proof is an exercise, similar to but easier than the given proof of 6.3. *Thus the base-free fibred classification of fibrations* $p : M \to B$ *with generic fibre topologically* I *is by the finite orbits of their weak data functions under the finite action of* Aut B .

8 *S-FIBRATIONS WITH A GLOBAL TRIVIALITY CHART*

This section shows that in spite of all their complexities, non-orientable S-fibrations $M \to B$ are for simple reasons all finite orbifold coverings of $B \times [[0 , 1]]$. They also have finite regular coverings of the form $B' \times S^1$.

A *triviality chart* for an orbifold bundle $p : M \to B$ over an open set $B_o \subset B$ is to consist of finite regular orbifold coverings g and G in a commutative diagram

$$\begin{array}{ccc} \bar{B}_o \times F & \xrightarrow{\text{proj}} & \bar{B}_o \\ G \big\downarrow & & \big\downarrow g \\ p^{-1}B_o & \xrightarrow[p]{} & B_o \end{array}$$

such that the covering translation group Γ for G respects the product structure of $\bar{B}_o \times F$.

The chart is *global* if $B_o = B$.

Theorem 8.1 (proven below)

For an S-fibration $p : M \to B$ the following three conditions are equivalent:

(a) p admits a multifold section.

(b) p admits a global fibration chart.

(c) There exists an orbifold bundle map $q : M \to [[0 , 1]]$ to the interval with ends silvered such that $(p , q) : M \to B \times [[0 , 1]]$ mapping $x \to (p(x) , q(x))$ is an orbifold covering.

Remark

The covering in (c) cannot in general be regular, essentially because the only geometric degrees occurring for regular coverings of $[[0 , 1]]$ by itself are 1 and 2 .

Recall from §3 and §6 that for M of dimension 3 a multifold section of p exists unless B is closed, M is orientable, and the Euler number $e_o \in Q$ of p is non-zero.

Beware that multifold sections and fibrations $q : M \to [[0,1]]$ as described in (c) are in general not unique, even up to fibred automorphism of M fixing B. For example, when M is a locally trivial circle bundle over B, an n-fold section gives via a 'holonomy' construction a homomorphism $\pi_1|B| \to \Delta_{2n}$ to the dihedral group of symmetries of the regular n-gon; and any such homomorphism is realized by an n-fold section *provided* composition with the orientation homomorphism $\Delta_{2n} \to \mathbb{Z}/2$ gives $\omega_1 : \pi_1|B| \to \mathbb{Z}/2$, the first Stiefel-Whitney class of the circle bundle.

Algebraic corollary of 8.1

Given the base 2-orbifold B, the abstract groups arising as orbifold fundamental group $\pi_1 M$ (see Thurston, W.P., 1976-79, Chapter 13) where $p : M \to B$ is any S-fibration with a multifold section, are (up to group isomorphism) precisely those subgroups of finite index in $(\pi_1 B) \times ((\mathbb{Z}/2) * (\mathbb{Z}/2))$ that project onto $\pi_1 B$. □

The proof of this uses the notion of an orthogonal orbifold S^1-bundle, i.e. one with structural group $O(2)$, in (essentially) the sense of Ehresmann or Steenrod. By definition, this is given by an atlas A of triviality charts such that, with the notation above 8.1 :

(i) For each fibration chart in A, the model fibre F is S^1 and the covering group Γ acts orthogonally on it.

(ii) If $q_i : S^1 \to p^{-1}(x)$, $i = 1,2$, are two coverings induced by two charts in A, there exists an orthogonal map $\theta : S^1 \to S^1$ such that $q_2 = q_1\theta$.

Note that each fibre is then naturally a Euclidean 1-orbifold of length $2\pi/n$ for some n.

Lemma 8.2

Every orbifold S^1-fibration $p : M \to B$ admits an $O(2)$ structure.

For an ad hoc proof in dimension 3 observe that the data function Φ for p can be realized by an orthogonal S-fibration; then use our classification to conclude that this fibration is isomorphic to $p : M \to B$. □

A better proof, valid in all dimensions, introduces an orbifold Riemannian metric along the fibres, then regulates fibre length with a smooth function so that all generic circle [resp. interval] fibres have length 2π [resp. π] . This argument also shows that the (simplicial) space of orthogonal structures (with this length condition) is contractible. □

Proof of Theorem 8.1 (valid in all dimensions)

It is an easy exercise to prove that (b) $\Longrightarrow$ (c) $\Longrightarrow$ (a) , so we begin with (a) $\Longrightarrow$ (c) .

We deal first with the case where the generic fibre is topologically an interval. The orthogonal structure provided by Lemma 8.2 makes each interval fibre metric with length π or $\pi/2$; and there is a canonical orbifold fibration $q : M \to [[0 , \pi/2]]$ which on generic fibres gives the distance from the middle of the fibre. The map $(p , q) : M \to B \times [[0 ,\pi/2]]$ is clearly an orbifold submersion, hence a covering.

In the case of generic fibre S^1 , let F^2 be an n-fold section $(n \geq 1)$ of $p : M \to B$ and split M at F to get an orbifold I-bundle $\bar{p} : \bar{M} \to B$. Interpreting it momentarily as an S-fibration by a doubling trick we get from the above an orbifold bundle map $\bar{q} : \bar{M} \to [[0 ,\pi/2]]$. This induces an orbifold bundle map $q : M \to [[0 ,\pi/2]]$ such that $(p, q) : M \to B \times [[0 ,\pi/2]]$ is an orbifold covering.

To conclude, we check that (c) implies (b) . The orbifold bundle map $q : M \to [[0 ,1]]$ regarded as a flat connection on the orbifold bundle $p : M \to B$ induces a holonomy homomorphism $\theta : \pi B \to O(2)$, where $O(2)$ is the isometry group of a generic fibre for the metric pulled back by q . The image of θ is finite since it respects the finite non-empty subset in $q^{-1}(1)$. Let $\tilde{B}$ be the finite regular covering B whose fundamental group is kernel (θ) . The induced regular covering of M is the global fibration chart that (b) requires.

□

9 A FLORAL ATLAS OF THE UNORIENTED S-FIBRATIONS

The classification of non-oriented S-fibrations given in §6 is, as it stands, not much more than a finite decision procedure to decide whether two S-fibrations are isomorphic. In this final section, we go on to a systematic survey of all isomorphism classes of compact unoriented S-fibrations; it is a reasonably satisfactory extension of Seifert's 1932 classification (Seifert, H., 1932). We emphasize that this section deals with the base-free classification, that is, there is no assumption that isomorphisms induce a specified mapping of base orbifolds. Also, the unoriented S-fibrations classified may or may not be orientable.

As our methodology is reminiscent of botany we have allowed ourselves to borrow a few botanical terms. Although all proofs will follow trivially from the work in earlier sections, the resulting classification will involve some surprising quirks.

By a *flower* we shall mean an S-fibration F whose base B is of one of two sorts:

(1) B is a conical orbifold  , i.e. an abstract regular neighbourhood of a conical point.

(2) |B| is an annulus and B is an abstract regular neighbourhood of one of the two boundary circles of |B| , e.g. B =  .

If B is of type (1) , we call the flower *simple*; if it is of type (2) , we say the flower is *composite*.

A flower F^3 may come equipped with one or more orientations. A *bloom orientation* is an orientation of the complement $F - \Sigma(F)$ of the 3-orbifold singularity set $\Sigma(F)$. ($|F|$ is then a topological manifold and the bloom orientation amounts to giving it an orientation as topological manifold.) A *base orientation* is an orientation of $B - \Sigma B$; it induces an orientation of $\partial|B|$. A *fibre orientation* is a coherent orientation of all generic fibres viewed as 1-manifolds.

Two of the three sorts of orientation on a given flower uniquely determine an orientation of the third sort, as is familiar in bundle theory.

Beware that although all flowers are base orientable, some are not bloom orientable: namely, those simple flowers of types (4) and (14) in §5, and those composite flowers for which any weak data function Ψ of §6 involves a star on each component of $\partial|B|$.

Although we are ultimately interested principally in the unoriented classification of S-fibrations, it is vital to know the orientation preserving fibred classification of flowers that are equipped with various combinations of orientations: namely, bloom orientation, base orientation and both these orientations.

These classifications clearly include answers to all questions concerning *amphicheirality*. Here are the relevant definitions. A (bloom-) orientable flower F is *amphicheiral* if it admits a fibred automorphism $f : F \to F$ reversing bloom orientation; in case f can preserve [respectively reverse] base orientation, F is said to be (+)-*amphicheiral* [respectively (-)-*amphicheiral*].

The classification of flowers, possibly with bloom and/or base orientation is as follows.

For simple flowers (over a conical base), the bloom-oriented [respectively unoriented] classification was presented in §2 [respectively §5]. As for base orientation, one notes that every reflection of the base can be covered by a fibred order 2 automorphism of the flower, and in the bloom orientable case even a bloom orientation preserving one.

The classification of composite flowers is made possible by §4, §6 and §7, using data functions and equivalences thereof. Here are the results.

For the bloom-oriented case the result is very simple: if we normalize a data function Φ of §4 for a flower so that slopes lie in $(-\frac{1}{2}, \frac{1}{2}]$, then Φ becomes a complete invariant when taken up to isomorphism of the base (preserving base orientation if one is present).

In the absence of a bloom-orientation, we use §§6-7 as follows. Given composite flowers F and F' with weak data functions Ψ and Ψ', we seek a fibred isomorphism $F \to F'$ (perhaps preserving given orientations of the bases B and B'). Here is an efficient sequence of tests to decide whether this is possible.

(a) Are the type diagrams isomorphic?

(b) Is bloom orientability of F and F' the same?

(c_+) In the bloom-orientable case, is there an isomorphism $f : B \to B'$ making assigned slopes correspond up to a simultaneous reversal of their signs?

(c_-) In the non-orientable case, can * be slid around $\partial|B|$ using move ($\bar{\kappa}_1$) of §6 so that an isomorphism $f : B \to B'$ exists making stars correspond and slopes modulo 1 correspond possibly after a simultaneous reversal of signs using move ($\bar{a}$) of §6?

The above are tests that must be passed, i.e. they represent necessary conditions.

At this point, we can use move ($\bar{b}_2$) of §6 to change Ψ so that all slopes in Q (if any) correspond. In case no parity ε is defined by Ψ and Ψ' , this suffices by 6.1 or 6.2 to build a fibred isomorphism $F \to F'$ over f .

The case where a parity ε is defined by Ψ and Ψ' is the case (see §6) where the fibre types were first studied by Montesinos (Montesinos, J.-M., 1973); and so we call such flowers *Montesinos flowers*. The test to decide whether at this point there exists a fibred isomorphism $F \to F'$ inducing f involves a subtlety; indeed, coincidence of the parities ε assigned by Ψ and Ψ' , is now sufficient but *not* in general necessary. Indeed supposing the assigned parities are distinct, the flowers F and F' are nevertheless isomorphic, precisely if, in a certain sense, ε *is negligible for* F (respecting base orientation if one is present). This negligibility is an intrinsic property of F that we proceed to define and analyse.

Definition

Let F and $F^{\#}$ be composite flowers over a common base B that have weak data functions Ψ , $\Psi^{\#}$ coinciding except for the parity ε , which is different. Let $f : B \to B$ be an automorphism. Then ε *is negligible for* F *over* f if there exists an isomorphism $F \to F^{\#}$ inducing f .

Negligibility of ε for F *respecting base orientation* means negligibility over some $f : B \to B$ respecting orientation of $|B|$. Similarly one defines negligibility of ε for F *reversing base orientation* or *fixing* B . Likewise negligibility without further qualification.

Notice that $F^{\#}$ can always be derived from F as follows. Cut F open over any bicollared arc $L^1 \subset B$, that avoids dihedral points and cuts the annulus $|B|$ open to give a square. Then glue the cut-open F back together in the essentially unique way that reverses the orientation of the one interval fibre in the cut

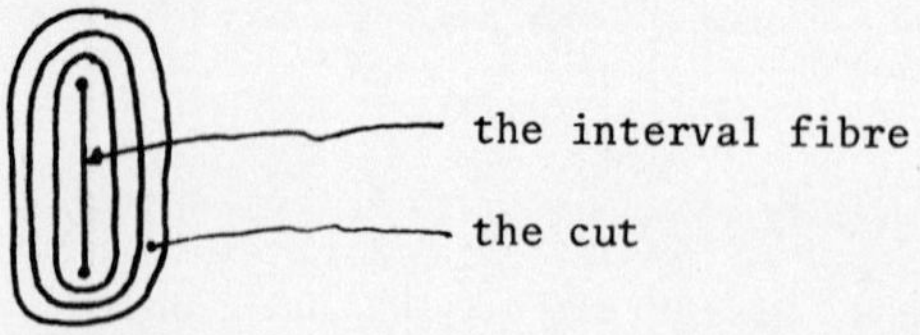

while preserving an orientation of the cut. This incidentally shows that the definition of negligibility is independent of the choice of Ψ for F (§6 also shows this).

Here are some examples to focus the reader's ideas. In each, the Montesinos flower F , which may or may not be bloom-orientable, is described by the given data function Φ . It is understood that the parity ε is 0 in each example, and that the vertical arcs in the last two examples are the K^1 prescribed by Φ (showing F is non-orientable).

Example. Φ:

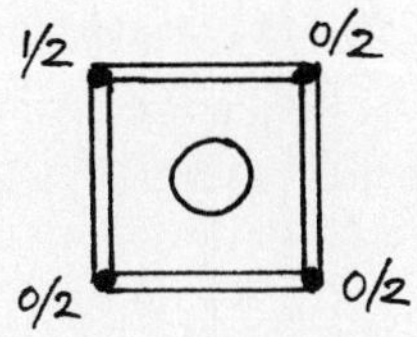

($|B|$ an annulus). Here ε is negligible fixing B .

Example. Φ:

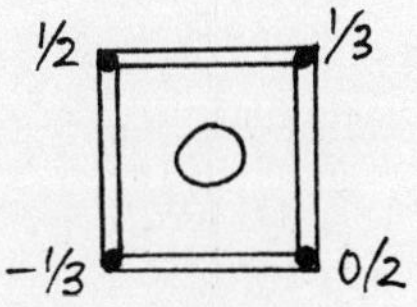

Here ε is negligible reversing base orientation but not preserving it.

Example. Φ:

Here ε is not negligible.

Example. Φ:

Here ε is negligible preserving and reversing base orientation, but not fixing the base.

Example. Φ:

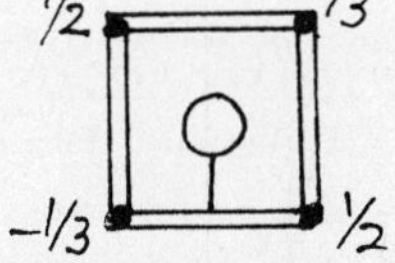

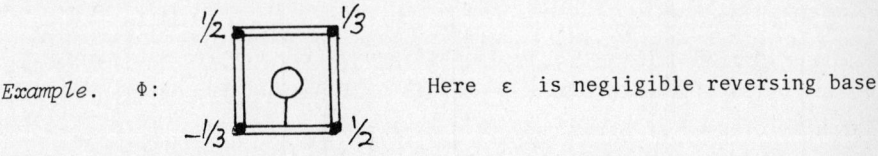

Here ε is negligible reversing base orientation but not preserving it.

Proposition 9.1

Let F be a (bloom-) orientable Montesinos flower with weak data function Ψ. Then:

(a) ε is negligible respecting orientation of $|B| \Longleftrightarrow \varepsilon$ is negligible fixing $B \Longleftrightarrow$ all slopes are 0 or $\frac{1}{2}$ modulo 1 and an odd number are equal $\frac{1}{2}$ modulo 1.

(b) ε is negligible reversing the orientation of B precisely if some reflection f of B sends each slope modulo 1 to its negative, and an odd number of slopes $\frac{1}{2}$ modulo 1 occur (exactly one being fixed by f). □

Note that the (bloom-) orientable Montesinos flowers F for which ε is negligible are *nearly* amphicheiral in the sense that the cycle of slopes satisfies an amphicheirality condition (geometrically said, F is amphicheiral after removal of a finite number of interval fibres of the simplest type). On the other hand no such F is itself amphicheiral. Thus negligibility of ε is a sort of 'false amphicheirality' in this case.

Proposition 9.2

Let F be a non-orientable Montesinos flower with weak data function Ψ; suppose Ψ marks the point $*$ of a mirror in the mirror cycle $C^1 = \partial|B| - \partial B$. The following is a necessary and sufficient condition for negligibility of ε over $f : B \to B$:

There exists an embedded topological arc $J \subset C^1$ joining $*$ to a point $*'$ that lies in the same mirror as $f(*)$, so that:

(i) J contains an odd number of dihedral points assigned slope $\frac{1}{2}$ modulo 1, and

(ii) sliding $*$ along J to $*'$ and changing the signs of slopes at the dihedral points in J according to move $(\bar{\kappa}_2)$ of §6, produces a new data function Ψ' such that the slopes modulo 1 of $f\Psi$ and Ψ' are identical. □

Having settled the classification of flowers (possibly with bloom and/or base orientations), we can go on to survey the unoriented isomorphism classes of all (compact) S-fibrations.

We propose to systematically identify up to isomorphism an arbitrary (compact) S-fibration M with base orbifold B by seizing on its restriction over a regular neighbourhood of $\partial B \cup \Sigma B$ together with a minimum of auxiliary information. The components of this restricted

S-fibration are by definition flowers! Taking due note of certain orientability properties of M (there are nine possibilities indicated in the first column of Tables A and B), we form the flowers into a *bouquet* by choosing appropriate orientations (deriving from §4, §6 or §7), and perhaps fixing an extra piece of global information (notably e_0 of §3 in case Seifert's $\underline{\underline{0}}$ symbol appears in the first column). See the second column of Tables A and B for the orientation choices appropriate in each case. The orientations are understood to be ones naturally encountered in the analysis given in §4, §6 or §7.

As the formation of the bouquet involves some unforced choices, these must be compensated by allowing an appropriate rearrangement of the bouquet; more precisely the bouquet is to be viewed through the equivalence specified in the third and last column of Tables A and B.

We assert that the homeomorphism class of $|B|$ *the orientability class* (among the nine in column 1) *and the equivalence class of the bouquet together form a complete invariant identifying the isomorphism class of the* S-*fibration* M (modulo the possible ambiguity indicated for the bottom half of Table A, which will be resolved presently). The proof is an exercise using previous sections.

Thus we have a key for the identification of S-fibrations of the form one finds for plants in a botanical atlas. Note incidentally that our keys work backwards as well. Indeed, given any compact surface $|B|$, orientation class among the nine, and bouquet of the corresponding sort, these three subject only to obvious compatibility conditions, it is a trivial matter to construct a compact S-fibration of this class, that gives rise to this bouquet by the procedure described above. As obvious conditions we need only mention the compatibility of Seifert symbols with $|B|$ as discussed in §6, and the condition that there be as many flowers as boundary components of $|B|$.

We conclude by resolving the possible ambiguity occurring for the part of the S-fibration classification in the lower half of Table A. Recall that it comes from the datum $\eta \in \mathbb{Z}/2$ in weak data functions.

Given an S-fibration M with weak data function Ψ, over a base B, for which η is defined, we say that η is *negligible* when there exists an automorphism $f : B \to B$ such that $f\Psi$ is equivalent to the weak data function $\Psi^{\#}$ obtained by replacing η by $\eta + 1$ in Ψ, keeping all other data unchanged. We will also say that η is

TABLE A

S-FIBRATIONS WITH $\partial K = \emptyset$ (i.e. $n_* + n_4 = 0$)

Seifert symbols (giving orientability class)	*Bouquet constitution* (all flowers are bloom orientable)	*Equivalence of bouquets* (beyond isom. respecting orientations)
$K = \emptyset$, $\underline{\underline{O}}$ … $\lbrace$ $\underline{\underline{o}}$ …… $\lvert B \rvert$ orientable	Doubly oriented flowers, and $e_o \in Q$. This e_o is 0 unless M is orientable as orbifold and $\partial M = \emptyset$; it is subject to a congruence from §4.	Simultaneous reversal of all base orientations. Simultaneous reversal of all slopes.
$K = \emptyset$, $\underline{\underline{O}}$ … $\lbrace$ $\underline{\underline{n}}$ …… $\lvert B \rvert$ non-orientable	Bloom-oriented flowers and $e_o \in Q$ as in the above case.	Simultaneous reversal of all orientations, and the sign of e_o.

Below this level the classification is up to one possible ambiguity if all simple flowers are Seifert flowers and all composite flowers are orientable Montesinos flowers, i.e. if η is defined in §6; this ambiguity is resolved in §9 below.

$K \neq \emptyset$, $\underline{\underline{N}}$ …… $\lbrace$ $\underline{\underline{o}}$ …… $\lvert B \rvert$ orientable	Base oriented flowers	Simultaneous reversal of all orientations.
$K \neq \emptyset$, $\underline{\underline{N}}$ …… $\lbrace$ $\underline{\underline{n}}$ …… $\lvert B \rvert$ non-orient. $\lbrace$ $\underline{\underline{I}}$ …… $\lvert B \rvert$ - K orientable	Fibre oriented flowers (a fibre orientation amounts to giving a bloom orientation and a base orientation up to simultaneous reversal of these two).	Simultaneous reversal of all fibre orientations.
$K \neq \emptyset$, $\underline{\underline{N}}$ …… $\lbrace$ $\underline{\underline{n}}$ …… $\lvert B \rvert$ non-orient. $\lbrace$ $\lvert B \rvert$ - K …… non-orient. $\lbrace$ $\underline{\underline{II}}$ …… K bicollared	Flowers with no orientation	None
$K \neq \emptyset$, $\underline{\underline{N}}$ …… $\lbrace$ $\underline{\underline{n}}$ …… $\lvert B \rvert$ non-orient. $\lbrace$ $\lvert B \rvert$ - K …… non-orient. $\lbrace$ $\underline{\underline{III}}$ …… K not bicollared		

TABLE B

S-FIBRATIONS WITH $\partial K \neq \emptyset$ (i.e. $n_* + n_4 \neq 0$)

Orientability class	*Bouquet constitution* (the number $n_* + n_4$ of non bloom-orientable flowers is always even)	*Equivalence of bouquets* (beyond isom. respecting orientations)
$\lvert B \rvert$ orientable	Base-oriented flowers	Reversal of all orientations
$\lvert B \rvert$ non-orientable ..... $n_4 \neq 0$	Flowers with no orientation	None
$\lvert B \rvert$ non-orientable ..... $n_4 = 0$	Base oriented flowers	Base orientation reversal at any one unstarred flower Base orientation reversal at any pair of starred flowers

Tables similar to A and B exist for the S-fibrations with generic fibre $[[0, 1]]$; however they nearly degenerate since the following radical simplifications occur: The Euler number e_o is always 0 ; bloom and fibre orientations are irrelevant (however base orientations *are* relevant) ; there is no ambiguity below the middle level of Table A. Each flower is determined by its type diagram of §5 plus a parity which decides whether its total space is orientable as a topological manifold.

negligible for f *if we want to mention* f . Thus, *an equivalence class of bouquets in the bottom half to Table A corresponds to exactly one isomorphism class of* S-*fibration if* η *is negligible or undefined, and corresponds to two isomorphism classes otherwise.*

We want an efficient algorithm, in terms of flowers, to decide when η is negligible. Recalling how η was defined in §6, the relevant property of flowers is fertility:

Definition

A bloom-orientable Seifert or Montesinos flower F is *fertile* if it admits a fibred automorphism $f : F \to F$ whose restriction to the boundary torus ∂F acts non-trivially on $H_1(\partial F ; \mathbb{Z}/2)$. We will also say that F is *fertile for* f if we want to mention the automorphism f .

Note that there are only two automorphisms of $H_1(\partial F ; \mathbb{Z}/2)$ respecting the homology class of the fibre. Thus, one can rephrase the above condition by saying that, for any one 1-fold section s_F of the fibration ∂F , the mod 2 intersection number of s_F and $f(s_F)$ in ∂F is 1 .

Given an orientation of F , normalization of slope(s) to lie in $(-\frac{1}{2}, \frac{1}{2}]$ provides a standard section of the fibration on ∂F . This implies

Lemma 9.3

If the flower F is fertile for f , the automorphism f necessarily reverses bloom-orientation. □

Note two corollaries of 9.3:

(a) A fertile flower must be amphicheiral.

(b) If a flower F is fertile for some f , it is fertile for any other bloom-orientation reversing automorphism. □

Recall that, for an S-fibration $M \to B$, the datum η (if any) in a data function was defined with the help of a hidden choice of a 1-fold section s of the fibration restricted to the boundaries of all flowers of M .

The way fertility leads to negligibility of η is controlled by the following lemma applied to the closed up complement of all flowers (this lemma is implicit in the proof of 6.3).

Lemma 9.4

Let M_0 be a non-orientable compact connected circle fibred 3-manifold (all fibres generic) with base B_0, such that ∂M_0 consists of 2-tori. Consider a fibred automorphism g of ∂M_0 such that:

(a) If B_0 is orientable (i.e. the Seifert symbol of M_0 is $\underline{\underline{\text{No}}}$), then g induces on all components of ∂B_0 a map of the same degree.

(b) If there is fibre orientation for M_0 (i.e. the Seifert symbol is $\underline{\underline{\text{NnI}}}$), then on all components of ∂M_0, the map g maps fibres with the same degree.

Then g extends to a fibred automorphism of the complement of a fibre in int M_0. Further, g extends to an automorphism of M_0 itself if and only if for some section s of ∂M_0 the $\mathbb{Z}/2$ intersection number $g(s)\cdot s$ is zero in ∂M_0. □

Proposition 9.5

Let F be a bloom-orientable flower with base B, associated to the weak data function Ψ.

If F is a Seifert flower (with one singular circle fibre, with type (1) of §2), it is fertile precisely when the slope assigned by Ψ is $\frac{1}{2}$ modulo 1.

If F is a Montesinos flower, it is fertile precisely when it is amphicheiral and

$$s(\Psi) + \varepsilon = 1 \text{ modulo } 2 ,$$

where $s(\Psi)$ is the sum of all slopes in Ψ. When F is amphicheiral and all slopes in Ψ are normalized to lie in $(-\frac{1}{2}, \frac{1}{2}]$, this $s(\Psi)$ is also half the number of slopes $\frac{1}{2}$. □

We are now equipped to determine when the datum η of a weak data function Ψ over B is negligible. (Recall that η is defined precisely when each flower of a corresponding S-fibration is a Seifert flower or an orientable Montesinos flower.)

Theorem 9.6

Consider an S-fibration M with weak data function Ψ, for which the datum η is defined.

When the Seifert symbol of Ψ is $\underline{\underline{\text{No}}}$, i.e. $|B|$ is orientable, η is negligible if and only if either

(i) some flower F is (+)-amphicheiral and fertile, or

(ii) the flowers of the bouquet (equipped with base orientations) can be presented as

$$F_1,\ldots,F_p \ , \quad G_1,\bar{G}_1,\ldots,G_q,\bar{G}_q \ , \qquad p \geq 1 \ , \quad q \geq 0 \ ,$$

where G_j and $\bar{G}_j$ are isomorphic reversing base orientations, and there exist automorphisms $f_i : F_i \to F_i$ reversing base orientation for all $1 \leq i \leq p$, so that the number of indices i for which F_i is fertile for f_i is odd.

When the Seifert symbol of Ψ is $\underline{\underline{\underline{NnI}}}$, the condition for negligibility of η is exactly the same, replacing "base orientation" by "fibre orientation" everywhere.

When the Seifert symbol of Ψ is $\underline{\underline{\underline{NnII}}}$ or $\underline{\underline{\underline{NnIII}}}$, η is negligible if and only if some flower is fertile.

Remark

The classification of S-fibrations fixing the base given in Theorem 6.1 raises the question of *negligibility* of η *fixing the base* B (i.e. negligibility for the identity of B). The criterion for this is very simple: (data of 9.6) η *is negligible fixing* B *if and only if all slopes assigned are* 0 *or* $\frac{1}{2}$ *modulo* 1 *and an odd number of flowers of the bouquet are fertile.*

Proof of 9.6

Let $p : M \to B$ and $p^{\#} : M^{\#} \to B^{\#}$ be two S-fibrations over the same base B, whose associated weak data functions Ψ and $\Psi^{\#}$ differ (only) as regards η. Then it is possible and convenient to identify the complement of a generic fibre in int M to the complement of a generic fibre in int $M^{\#}$ (cf. definition of η). Further we can assume that the bouquets associated to M and $M^{\#}$ are identical.

Our task is to decide precisely when M itself is fibred isomorphic to $M^{\#}$.

First focus attention on the case when the Seifert symbol is $\underline{\underline{N}}$.

Supposing condition (i), there exists a flower F in the bouquet associated to M, that is fertile for some automorphism $f : F \to F$ that respects base orientation. Then consider the automorphism f_+ of the disjoint sum of all the flowers of the bouquet associated to M which is f on F and the identity elsewhere. In view of the definition of η, Lemma 9.4 then shows that f_+ extends to an isomorphism $M \to M^{\#}$ (and not to an automorphism $M \to M$). Thus (i) implies negligibility.

Similarly, (ii) is a sufficient condition by consideration of the automorphism of $p^{-1}N(\Sigma B)$ defined by

$$F_i \xrightarrow{f_i} F_i \ , \quad G_j \xrightarrow{g_j} \bar{G}_j \ , \quad \bar{G}_j \xrightarrow{g_j^{-1}} G_j$$

as prescribed in (ii), with the g_j reversing base orientations. Note that the fibred isomorphism $M \to M^{\#}$ so constructed reverses the orientation of $|B|$.

To show that (i) or (ii) necessarily holds if η is negligible, consider a fibred isomorphism $h : M \to M^{\#}$. After a fibred isotopic deformation, h respects the disjoint sum of all flowers of M since this bouquet is a fibred regular neighbourhood of $p^{-1}N(\partial B \cup \Sigma B)$, which clearly is respected. (All this makes sense because of our near identification of M with $M^{\#}$.) Under the action of h this bouquet breaks up into a finite number of cycles of the form:

$$E_1 \xrightarrow{h_1} E_2 \xrightarrow{h_2} \cdots \xrightarrow{h_{n-1}} E_n \quad (\text{with } h_n : E_n \to E_1 \text{ closing the cycle})$$

where the E_i are flowers and the h_i are restrictions of h.

Considering a section of ∂E_1 and translating it around the cycle, one sees using Lemma 9.4 that, for an odd number of such cycles, it will happen that $h_n h_{n-1} \cdots h_1$ acts non-trivially on $H_1(\partial E_1 ; \mathbb{Z}/2)$, i.e. that E_1 is fertile for $h_n h_{n-1} \cdots h_1$.

If M admits a (+)-amphicheiral fertile flower, then negligibility of η is already proved. So we can assume the contrary. Then f necessarily reverses; the orientation of $|B|$. Rearrange each cycle into new cycles of length 1 or 2 thus:

$$E_1 \underset{h_1^{-1}}{\overset{h_1}{\rightleftarrows}} E_2 \ , \quad E_3 \underset{h_2^{-1}}{\overset{h_3}{\rightleftarrows}} E_4 \ , \ldots , \ E_{n-1} \underset{h_{n-1}^{-1}}{\overset{h_{n-1}}{\rightleftarrows}} E_n \qquad \text{when } n \text{ is even,}$$

$$E_1 \xrightarrow{h_n h_{n-1} \cdots h_1} E_1 \ , \quad E_2 \underset{h_2^{-1}}{\overset{h_2}{\rightleftarrows}} E_3 \ , \ldots , \ E_{n-1} \underset{h_{n-1}^{-1}}{\overset{h_{n-1}}{\rightleftarrows}} E_n \qquad \text{when } n \text{ is odd.}$$

These reveal that condition (ii) holds.

This completes the proof of Proposition 9.5 when the Seifert symbol is N̲o̲ . The proof is verbatim the same when the Seifert symbol is N̲n̲I̲ , after we replace base orientations by fibre orientations.

As for the case of the Seifert symbols N̲n̲I̲I̲ and N̲n̲I̲I̲I̲ , the proof is again similar, but simpler. □

9.7 Algebraic interpretation

One of the main theorems of the theory initiated by Conner and Raymond asserts (when restricted to dimension 3) that if we choose a base orbifold B with infinite orbifold fundamental group π , then the S-fibrations M with base B are classified up to fibred isomorphism (not fixing B) by their fundamental groups viewed as extensions of π (extension equivalence not fixing π) ; Conner, P.E. & Raymond, F. (19) and Lee K.B. & Raymond, F. (1983). Further (see also Zieschang, H. & Zimmerman, B. (1982)), all extensions of π by $\mathbb{Z}$ and by $(\mathbb{Z}/2) * (\mathbb{Z}/2)$ so occur. What is more, when B is hyperbolic, the fibre subgroup $(\mathbb{Z}/2) * (\mathbb{Z}/2)$ or $\mathbb{Z}$ of $\pi_1 M$ is characteristic; thus, in this case the above extensions of π are equivalent if and only if they are isomorphic as abstract groups. This provides a significant interpretation of a large part of our atlas of S-fibrations as a classification up to isomorphism of a family of abstract groups. (In this vein, see also the algebraic remark in §8 , and the relation of tame decompositions to S-fibrations explained in the introduction.)

9.8 Concluding remarks

Theorem 6.3 would let one prove Theorem 9.6 in purely combinatorial fashion. The geometric proof we have given is intended to suggest that the base free classification of this chapter could well be attached directly after the local analysis of §5 . As written, this article shows the growth rings of our understanding of S-fibrations; the notions of flower and bouquet arose when we realized that negligibility of η (and even ε) posed subtle residual problems at the end of §6 .

The reader interested in exploring algebraic methods of classification will we expect find that the data functions Φ of §6 are intimately related to cocycles classifying (fixing base) the corresponding fundamental group extension, cf. Jankins, M. & Neumann, W. (1983). The final classification in Zieschang, H. & Zimmerman, B. (1982) using the language of combinatorial group theory is perhaps closest to our Theorem 6.2 (or 6.3?) ,

REFERENCES

Boileau, M. & Siebenmann, L.C. (1980). A planar classification of pretzel knots and Montesinos knots, preprint, Orsay.

Bonahon, F. & Siebenmann, L.C. (To appear). A do-it-yourself crystal classification.

Bonahon, F. & Siebenmann, L.C. Geometric splittings of classical knots, and the algebraic knots of Conway. To appear in LMS Lecture Note Series, in preparation.

Bonahon, F. & Siebenmann, L.C. (To appear). Seifert 3-orbifolds and characteristic splittings of 3-orbifolds.

Bonahon, F. & Siebenmann, L.C. (1983). Seifert 3-orbifolds and their role as natural crystalline parts of arbitrary compact irreducible 3-orbifolds, preprint IHES and Orsay.

Bredon, G. (1967). Equivariant cohomology theories. Springer Lecture Notes in Mathematics 34.

Bredon, G.E. (1972). Introduction to compact transformation groups. Academic Press.

Conner, P.E. & Raymond, F. Actions of compact Lie groups on aspherical manifolds, in Topology of Manifolds (Athens Ca. Conference). Ed. J. Cantrell & C.H. Edwards, Markham, Chicago, Ill., 227-264.

Conner, P.E. & Raymond, F. (1977). Deforming homotopy equivalences to homeomorphisms in aspherical manifolds. Bull. Amer. Math. Soc. 83, 36-85.

Dunbar, W.D. (1981). Fibered orbifolds and crystallographic groups. Thesis, Princeton.

Dunbar, W.D. (1983). Geometric orbifolds, preprint, Rice University.

Edwards, R.D., Millett, K.C. & Sullivan, D. (1977). Foliations with all leaves compact. Topology 16, 13-32.

Epstein, D.B.A. (1972). Periodic flows on 3-manifolds. Ann. of Math. 95, 58-82.

Epstein, D.B.A. (1976). Foliations with all leaves compact. Ann Inst. Fourier 26, 265-282.

Epstein, D.B.A. (1981). Pointwise periodic homeomorphisms. Proc. London Math. Soc. 42, 415-460.

Gazolaz, M.d.C. (1977). Thesis, University of Chicago.

Griffith, P.A. & Morgan, J. (1981). Rational homotopy theory and differential forms. Birkhauser.

Haefliger, A. (1982). Groupoides d'holonomie et classifiants, prépublication, Genève.

Haefliger, A. & Quach, N.D. (1982). Une présentation du groupe fondamental d'une orbifold, prépublication, Genève.

Hatcher, A. & Thurston, W.P. (1978). Incompressible surfaces in 2-bridge knot complements, preprint.

Holmann, H. (1964). Seifertsche Faserräume. Math. Ann. 157, 138-166.

Illman, S. (1978). Smooth equivariant triangulations of G-manifolds for G a finite group. Math. Ann. 233, 199-220.

Jankins, M. & Neumann, W. (1983). Lectures on Seifert manifolds. Brandeis Univ. Lecture Notes 2 (mimeo).

Kobayashi, S. & Nomizu, K. (1963). Foundations of differential geometry. Interscience, New York.

Lashof, R. (To appear). Equivariant isotopies and submersions. Pacific J. Math.

Lee, K.B. & Raymond, F. (1983). Geometric realization of group extensions by the Seifert construction, preprint.

MacBeath, A.M. (1980). On Seifert fibre groups, I, polycopied, Univ. of Pittsburg.

Montesinos, J.-M. (1973). Variedades de Seifert que son recubridores ciclicos ramificados de dos hojas. Bol. Soc. Mat. Mexicana 18, 1-32.

Montesinos, J.-M. (1980). Revêtements ramifiés de noeuds, espaces de Seifert et scindements de Heegaard, prepublication Orsay.

Montgomery, D. (1937). Pointwise periodic homeomorphisms. Amer. J. Math. 59, 118-120.

Neumann, W.D. & Raymond, F. (1978). Seifert manifolds, plumbing, μ-invariant and orientation reversing maps. Proc. Conf. Alg. & Geom., Top., Santa Barbara 1977. Springer Lecture Notes in Mathematics 644.

Orlik, P. (1972). Seifert manifolds. Springer Lecture Notes in Mathematics 291.

Orlik, P., Vogt, E. & Zieschang, H. (1967). Zur Topologie gefaserter dreidimensionaler Mannigfaltigkeiten. Topology 6, 49-64.

Reef, G. (1952). Sur certaines propriétés topologiques des variétés feuillettées. Act. Sci. et. Ind. No. 1183, Hermann, Paris.

Satake, I. (1956). On a generalization of the notion of manifold. Proc. Nat. Acad. Sci. USA 42, 359-363.

Satake, I. (1957). The Gauss-Bonnet theorem for V-manifolds. J. Math. Soc. Japan 9, 464-492.

Scott, G.P. (1983). The geometries of 3-manifolds. Bull. London Math. Soc. 15, 401-487.

Seifert, H. (1932). Topologie dreidimensionaler gefaserter Räume. Acta Math., 147-238.

Seifert, H. & Threlfall, W. (1930 & 1932). Topologische Untersuchung der Diskontinuitätsbereiche endlicher Bewegungsgruppen der dreidimensionalen sphärischen Räumes. I, Math. Ann. 104, 1-70; II, Math. Ann. 107, 88-110.

Seifert, H. & Threlfall, W. (1934). Lehrbuch der Topologie. Teubner. Chelsea 1945.

Seifert, H. & Threlfall, W. (1980). A textbook of topology. Academic Press.

Siebenmann, L.C. (1975). Exercices sur les noeuds rationnels, polycopié, Orsay.

Siebenmann, L.C. (1978). Une introduction aux espaces fibrés de Seifert, polycopié, Orsay.

Thurston, W.P. (1976-79). The geometry and topology of 3-manifolds. Lecture Notes, Princeton University. Part 1 being §§1-7 is to appear as an Ann. of Math. Study; Part 2 is §§8-13.

Thurston, W.P. (1980). Hyperbolic structures on 3-manifolds; article in four or more parts to appear in Ann. of Math. Part I: Deformation of an annular manifolds; Part II: Surface groups and 3-manifolds which fiber over the circle (preprints, Princeton University).

Thurston, W.P. (1982). Three-dimensional manifolds, Kleinian groups and hyperbolic geometry. Bull. Amer. Math. Soc. 6, 357-381.

Thurston, W.P. (1982). Hyperbolic geometry and 3-manifolds (notes written by P. Scott). Proceedings of the Conference on Topology in Low Dimension, Bangor 1979. Cambridge University Press.

Waldhausen, F. (1967). Eine Klasses von 3-dimensionalen Mannigfaltigkeiten. Invent. Math. 3, 308-333 und 87-117.

Wall, C.T.C. (1961). Rational Euler characteristics. Proc. Camb. Phil. Soc. 57, 182-184.

Zieschang, H., Vogt, E. & Coldewey, H. (1970). Flächen und ebene discontinuierliche Gruppen. Springer Lecture Notes in Mathematics 122. Updated English edition: Springer Lecture Notes in Mathematics (1980) 835.

Zieschang, H. & Zimmerman, B. (1982). Über erweiterungen von $\mathbb{Z}$ und $\mathbb{Z}_2 * \mathbb{Z}_2$ durch nicht euklidische kristallographische Gruppen. Math. Ann. 259, 29-51.

EXCHANGEABLE BRAIDS

H.R. Morton

Abstract. Two-component links in S^3 are studied in which each component is unknotted and lies like a closed braid having the other component as axis. These are termed 'exchangeably braided' links. The set of such links with a given linking number is shown to form a proper subset of an explicit finite list, constructed from a list of braids described by Stallings. Some sufficient conditions for a link on the finite list to be exchangeably braided are given.

1 *INTRODUCTION*

Representations of the unknot as a closed braid have been studied variously by Stallings, Goldsmith and others, and can be used in describing a number of constructions for fibred knots, (Goldsmith, D, 1975; Morton, H.R., 1978).

To represent a knot $K \subset S^3$ as a closed braid a disjoint unknotted curve L must be chosen as axis. The complement $S^3 - L$ is then a product, $\dot{D}^2 \times S^1$. If we can choose a projection $P_L : S^3 - L \to S^1$ which gives a regular n-fold covering of S^1 when restricted to K then K is called a *closed braid relative to axis* L. It is common in this case to view K as the 'closure', $\hat{\beta}$, of an n-string braid β, an element of the braid group B_n, see Birman (1974). The essential geometric information about the braid is contained in the link $K \cup L$, from which the element $\beta \in B_n$ can be recovered up to conjugacy in B_n, (Morton, H.R., 1978).

In this paper, I shall study links $K \cup L$ for which K is a closed braid relative to L, and in addition L is a closed braid relative to K. I propose to call such a link *exchangeably braided*. A braid $\beta \in B_n$ for which $\hat{\beta} \cup$ axis $(= K \cup L)$ is exchangeably braided, and hence isotopic to axis $\cup \hat{\gamma}$ for some other braid $\gamma \in B_n$, which will be termed *exchangeable*.

Two features of an exchangeable braid $\beta \in B_n$ are then apparent.

1. The closure $\hat{\beta}$ is unknotted, for it is required to form the axis for the closed braid $\hat{\gamma}$.
2. There is a disc D spanning $\hat{\beta}$ which meets the axis, L, in exactly n points.

This follows by choosing D as any one of the level discs for a projection $S^3 - \hat{\beta} \to S^1$ under which L is seen as a closed braid relative to $\hat{\beta}$. (The existence of such a projection can be interpreted as straightening out $\hat{\beta}$ so that L appears to wind 'monotonically' about it.)

Here $n = \mathrm{lk}(\hat{\beta}, L)$ is the minimum possible number of transverse intersections of L with a disc spanning $\hat{\beta}$, since the linking number is equal to the algebraic number of these intersections, taken with their sign. Property 2 can then be thought of as requiring the 'geometric' linking number of $\hat{\beta}$ and L to equal their algebraic linking number.

An exchangeable braid will satisfy rather more than Property 2, for the level discs of the projection above provide a whole 1-parameter family of spanning discs, each meeting L in n points.

Stallings (1978), describes for each n a finite set of braids on n strings with unknotted closure, which I shall call *Stallings braids*. These and their conjugates form a proper subset of the braids with unknotted closure, [K]. In this paper I shall show that exchangeable braids, up to conjugacy, form a proper subset of the set of Stallings braids.

This follows from two main results.

Theorem 1

Let $\beta \in B_n$. Then β is conjugate to a Stallings braid if and only if it satisfies Properties 1 and 2 above; i.e. $\hat{\beta}$ is unknotted, and is spanned by a disc meeting the axis in exactly n points.

Corollary 1.1

Every exchangeable braid is conjugate to a Stallings braid.

Corollary 1.2

There are only a finite number of exchangeably braided links with linking number n , since each link can then be realised as the closure of a Stallings n-braid, with its axis, and there are only finitely many such braids.

Theorem 2

There exist Stallings braids which are not exchangeable.

An exchangeable braid β determines a link $\hat{\beta} \cup L = K \cup \hat{\gamma}$, and, up to conjugacy, another braid γ whose closure is L . It is not always true that β and γ are conjugate, so the link $K \cup L$ is not necessarily *interchangeable* in the sense there is an isotopy interchanging the components.

Goldsmith's construction of fibred knots starts with an unknotted closed braid $\hat{\beta}$ with axis L , and observes that in the k-fold cyclic cover of S^3 branched over $\hat{\beta}$, (which is again S^3) , the preimage of L will be a fibred knot or link, the fibres being simply the preimages of the family of discs which span L and meet $\hat{\beta}$ in n points.

If β is exchangeable this preimage is simply the closure of the kth power of the corresponding braid γ . Now the braid γ is itself exchangeable, for the roles of β and γ can be exchanged.

Remark

If β is exchangeable then β^k is fibred for each k .

Goldsmith extends her construction to the case where L is a 'generalised axis' for an unknotted curve K . The curve L is called a 'generalised axis' for K if (a) the complement of L is fibred, and (b) there is a fibre projection $p_L : S^3 - L \to S^1$ under which K covers S^1 .

It would be interesting to study exchangeability in the wider context of links $K \cup L$ where each component forms a generalised axis for the other. The closure of an homogeneous braid, with its axis, gives an example of such a link.

2 CHARACTERISATION OF STALLINGS BRAIDS

Stallings defines the elementary braid $\sigma_{i,j} \in B_n$, for $1 \le i < j \le n$ by interchanging the ith and j^{th} strings with a single positive crossing, in front of any intermediate strings, and leaving all other strings alone, see Figure 1.

Figure 1

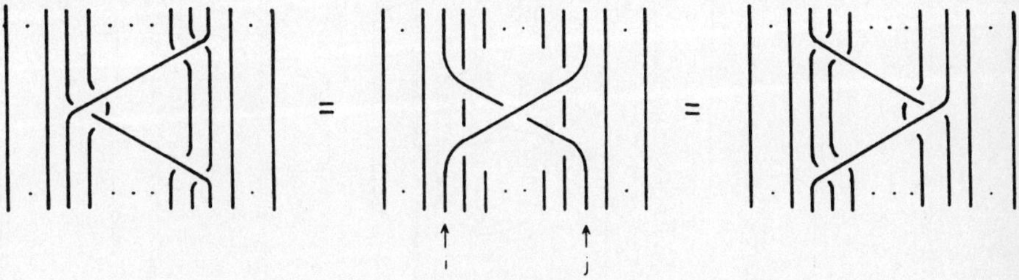

The elementary braid $\sigma_{i,j}$

A *Stallings braid* is defined to be a product of $n-1$ elementary braids, or their inverses, whose closure has only one component.

Lemma 1

Every Stallings braid has unknotted closure, with a spanning disc meeting the axis in n points.

Proof

Construct a spanning surface for the closure of a product of elementary braids as prescribed by Stallings. To do this, start with the closure of the trivial braid spanned by n disjoint discs labelled $1,\dots,n$ corresponding to the string labelling, and each meeting the axis in one point. The complement of an open tubular neighbourhood of the axis is a closed solid torus $D^2 \times S^1$ which can be arranged to meet these n discs in n annuli, each of the form arc $\times S^1$. Imagine the discs stacked vertically, and refer to the D^2 factor in $D^2 \times S^1$ as horizontal. Where the elementary braid $\sigma_{i,j}$ or its inverse occurs in the given product connect the ith and j^{th} discs by a ribbon in $D^2 \times S^1$, avoiding the other $n-2$ discs and any previous ribbons. Choose the core of the ribbon to lie at one horizontal level, and arrange the cores in order along the S^1 factor according to the order of the elementary braids in the product. The ribbon itself is given a left or right-hand half-twist, so that no edge meets the same horizontal level more than once, see Figure 2. Up to isotopy the ribbons can be made to lie

Figure 2

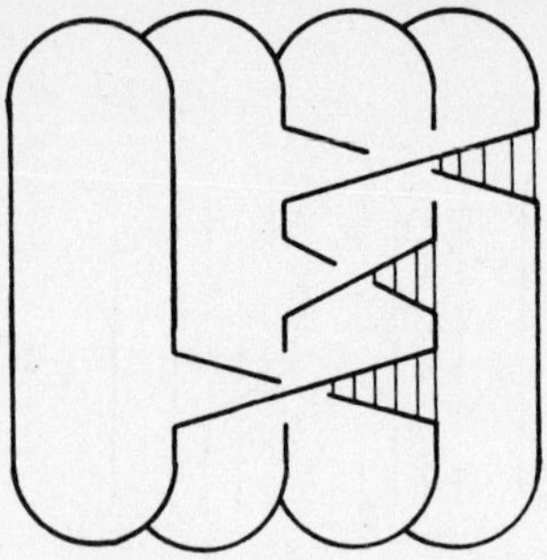

A Stallings disc for $w = \sigma_{2,4}\sigma_{2,3}\sigma_{1,3}$

arbitrarily close to their core levels, so that different ribbons lie in completely different levels. Each ribbon is determined up to isotopy by i, j and the sign of the twist. Together with the n discs they form an oriented surface spanning the closure of the given braid and meeting the axis in n points.

For a Stallings braid this surface will be connected and from its Euler characteristic it must be a disc, meeting the axis as required. I shall refer to such a spanning surface as a *Stallings disc.*

Lemma 2

Any braid $\beta \in B_n$ whose closure is spanned by a disc meeting the axis in n points is conjugate to a Stallings braid.

Proof

It will be enough to exhibit a Stallings braid whose closure is isotopic to $\hat{\beta}$ in the complement of the axis L. Such a braid is available if we can produce a disc spanning $\hat{\beta}$ made up from n discs each meeting L in one point, joined by $n-1$ ribbons each with a horizontal core and a single half-twist. The n discs taken in order around L then prescribe the labelling of strings in the braid, and the ribbons determine the elementary braids in the product, with their order determined by the core levels.

Take a disc spanning $\hat{\beta}$ which meets L in $c_1, \ldots, c_n$. Under some projection $p_L : S^3 - L \to S^1$ the curve $\hat{\beta}$ is a regular n-fold cover, while small circles around each c_i can be chosen which project homeomorphically. We can assume that curve parallel to $\hat{\beta}$ in some collar also cover S^1 regularly, so that two pieces of a level curve for p_L never meet at a point of $\hat{\beta}$, or of $L \cap D$.

Isotop D, leaving a neighbourhood of $D \cap L$ and $\hat{\beta}$ unchanged, so that the number of non-degenerate critical points for p_L on $D - L$ is as small as possible, and study the behaviour of p_L on $D - L$.

The level curve through any saddle point locally forms a cross. All four arms of this cross may continue to meet $\hat{\beta}$ or $D \cap L$, forming an *essential saddle*, or two of these arms may meet, forming an *inessential saddle*.

Where the number of critical points is minimal there can be no inessential saddles. For the two arms which meet provide a closed curve C in D. Now the intersections of L with D all have the same sense, so we can calculate the degree of p_L when restricted to any closed curve in $D - L$ by counting the number of points of $D \cap L$ enclosed by the curve. The map p_L is constant, and hence of degree 0, on the curve C. The subdisc of D bounded by C then contains no points of L. D can now be isotoped in the complement of L across a ball bounded by this subdisc and the disc bounded by C in the critical level so as to cancel the saddle with a local extremum, possibly after an innermost disc argument, as in Morton, H.R. (1979), or Bonahon, F. (preprint).

It follows that there will be no local extrema either among the minimal number of critical points, otherwise as the level of p_L is changed the expanding system of level circles surrounding the critical point will eventually encounter an inessential saddle - they cannot reach either $\hat{\beta}$ or $D \cap L$ intact, since two pieces of a level curve never meet at such points.

Every non-critical component of a level curve must then be an arc. Furthermore, these arcs must each join a point of $D \cap L$ to a point of $\hat{\beta}$. For if such an arc were to join two points of $\hat{\beta}$ then the direction around $\hat{\beta}$ of increase of level at the two points would be in opposite senses, while level increases monotonically on travelling in one sense around $\hat{\beta}$. Similarly two points p, q of $D \cap L$ cannot be joined, for we can choose small circles around p and q on which p_L changes monotonically. Since L crosses D in the same sense at p

as at q, the direction of increase of p_L around each of these circles will be in the same sense relative to their orientation in D. The two circles cannot then be joined by a non-critical level curve of p_L.

The surface $D - L$ has Euler characteristic $1 - n$. Since the function p_L Is suitably behaved on the boundary and round the missing points it follows that p_L has exactly $n - 1$ saddle points. These saddles are all essential. One pair of opposite arms must join points of $\hat{\beta}$ while the other pair joins two (distinct) points of $D \cap L$, otherwise some neighbouring level curve would join points of the wrong king.

Look now at the graph Γ embedded in D whose vertices are the points of $D \cap L$ and whose edges are the $n - 1$ pairs of opposite arms of the saddles joining such points. The graph Γ cannot separate D, for there will be points in every component of $D - \Gamma$ which are joined to $\hat{\beta}$ by a non-critical arc, necessarily lying itself in $D - \Gamma$. Hence by virtue of its Euler characteristic Γ must be a tree.

Construct a neighbourhood of Γ in D consisting of a disc round each vertex joined by narrow ribbons around each edge. These ribbons can be chosen so that p_L is monotone on their edges.

The boundary of this neighbourhood is then exhibited as the closure of a Stallings braid with axis L, where the ribbon cores arise from the critical curves joining points of $D \cap L$, and the sign of the half-twist depends on the direction of the positive normal to D at the critical point. Since $\hat{\beta}$ is isotopic through $D - L$ to this curve Lemma 2 is established.

The characteristic of Stallings braids given in Theorem 1 is then complete.

3 NON-EXCHANGEABLE BRAIDS

The simplest example of a non-exchangeable Stallings braid is the braid $\omega = \sigma_{2,4}\sigma_{2,3}\sigma_{1,3} = \sigma_3\sigma_2\sigma_3^{-1}\sigma_2\sigma_1^{-1}\sigma_2\sigma_1 \in B_4$, whose closure is shown in Figure 2. Here σ_i, $i = 1,\ldots,n-1$ denote the standard generators of the braid group B_n, where σ_i is the elementary braid $\sigma_{i,i+1}$ which interchanges the strings i and $i+1$.

The elementary braid $\sigma_{i,j}$ can be written in terms of the standard generators as $\sigma_i^{-1}\sigma_{i+1}^{-1} \cdots \sigma_{j-2}^{-1}\sigma_{j-1}\sigma_{j-2} \cdots \sigma_i$, or equally $\sigma_{j-1}\sigma_{j-2} \cdots \sigma_i \sigma_{i+1}^{-1} \cdots \sigma_{j-1}^{-1}$ as illustrated in Figure 1. The non-exchangeability of ω follows from a calculation of the two variable Alexander polynomial $\Delta(x,t)$ for the link $\hat{\omega} \cup$ axis L, where t, x denote generators of $H(S^3 - (\hat{\omega} \cup L))$ represented by meridians of $\hat{\omega}$, L respectively.

This calculation can be made most readily by a slight extension of Birman's observations on the Alexander polynomial of a closed braid, (Birman, J.S., 1974), which is proved in the final section of this paper.

Theorem 3

Suppose that $\beta \in B_n$, with reduced Burau matrix $B(t) \in GL(n-1, \mathbb{Z}[t, t^{-1}])$. Then the Alexander polynomial, $\Delta(x,t)$, of the braided link $\hat{\beta} \cup$ axis is given by $\Delta(x,t) = \det(B(t) - xI)$, the characteristic polynomial of $B(t)$, where x, t are represented by meridians of the axis and $\hat{\beta}$ respectively.

Remark

The reduced Murau matrix $B(t)$ is the image of β under the reduced Burau representation $\rho : B_n \to GL(n-1, \mathbb{Z}[t, t^{-1}])$, (see Birman, J.S., 1974). In this representation

$$\rho(\sigma_i) = \begin{bmatrix} 1 & & & & & \\ & \ddots & & & & \\ & & 1 & 0 & 0 & \\ & & t & -t & 1 & \\ & & 0 & 0 & 1 & \\ & & & & & \ddots \\ & & & & & & 1 \end{bmatrix} \leftarrow i^{th} \text{ row}$$

truncated appropriately where $i = 1$ or $i = n-1$.

The polynomial $\Delta(x,t)$ is defined up to multiplication by $\pm x^r t^s$. Its extreme powers of x for the braided link in Theorem 3 are x^{n-1} with coefficient 1 and x^0 with coefficient $(-1)^{n-1} \det B(t)$. Now $\det \rho(\sigma_i) = -t$, so $\det B(t) = (-t)^{|\beta|}$, where $|\beta|$ = algebraic number of crossings in β, counting $+1$ for each σ_j and -1 for each σ_j^{-1}. Even after multiplication by a unit $\pm x^r t^s$, the coefficients of the extreme powers of x will each remain a power of t.

If a braid β is to be exchangeable then $\hat{\beta} \cup$ axis, $= K \cup L$ say, must also be braided relative to K, and so its Alexander polynomial will have a similar form with the roles of x and t reversed. Thus the coefficients of the extreme powers of t must be powers of x, and further, the 'degree' in t, that is the difference in degree of the extreme powers of t, must also be $n - 1 = \mathrm{lk}(K, L) - 1$.

Corollary 3.1

The Stallings braid ω above is not exchangeable.

Proof

Its reduced Burau matrix $\omega(t)$ is

$$\omega(t) = \begin{pmatrix} 1 - t & -t^{-1} & t^{-1} \\ 2t - 3t^2 + 2t^3 - t^4 & -2 + t & -t^{-1} + 3 - 3t + t^2 \\ t^2 - 2t^3 + t^4 - t^5 & -t + t^2 & t - 2t^2 + t^3 \end{pmatrix}$$

having characteristic polynomial

$$\det(xI - \omega(t)) = x^3 + p_1(t)x^2 + p_2(t)x + t^3,$$

where $p_1(t) = -\mathrm{tr}\,\omega(t) = 1 - t + 2t^2 - t^3$. This polynomial has degree 3 in t, but the coefficients of the extreme powers of t are not powers of x. For example, the coefficient of t^3 is $1 + ax - x^2$, where a is its coefficient in $p_2(t)$. Consequently the axis of $\hat{\omega}$ is not braided relative to $\hat{\omega}$.

In trying to decide whether a Stallings braid β is exchangeable it can be helpful to look at a schematic diagram, in which the n strings are drawn vertically, and $\sigma_{i,j}^{\pm 1}$ is represented by a horizontal line with a sign attached, joining string i to string j and lying in front of the intervening strings. The braid ω is represented in this way in Figure 3.

This representation corresponds closely to the graph Γ constructed at the end of §2 from a disc D meeting an unknotted curve L in n points, when $K = \partial D$ is braided relative to L, and $n = \mathrm{lk}(L, K)$. In such a representation the vertical strings correspond to small circles, one around each vertex of Γ in D, i.e. each intersection of L with D, and the horizontals correspond to the

Figure 3

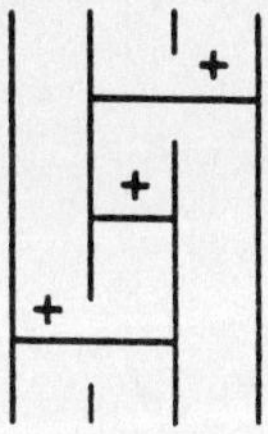

edges of Γ outside these circles.

A Stallings disc D for β can be constructed by clothing the horizontal lines with half-twisted ribbons to join discs spanning the vertical strings. The axis L will meet D once in each vertical disc, and is then separated into n arcs, $L_1, \ldots, L_n$. From the embedding of $D \cup L$ in S^3 we can decide whether L is braided relative to $\hat{\beta}$, and so whether β is exchangeable, as follows.

Construction

Split S^3 open along D to give $D \times I$, i.e. choose an explicit map $q : D \times I \to S^3$ which identifies only $D \times \{0\}$ and $D \times \{1\}$ with D, and $\partial D \times I$ with ∂D. Then $D \times I$ contains n arcs, $A_i = q^{-1}(L_i)$, formed from the pieces of the axis L.

Theorem 4

L is braided relative to ∂D if and only if there is an isotopy of these arcs in $D \times I$, rel boundary, to n arcs running monotonically from $D \times \{0\}$ to $D \times \{1\}$.

Proof

Such an isotopy, followed by projection to I, will determine via q a projection from $S^3 - \partial D$ to S^1, whose restriction to L is a regular n-fold covering map.

Coversely, if L is braided relative to ∂D, and $p : S^3 - \partial D \to S^1$ is a suitable projection then there is an isotopy of S^3 carrying $D_0 \cup L$ to $D \cup L$, where $D_0 = p^{-1}(1)$. This isotopy is analogous to one between two minimal genus spanning surfaces for a fibred knot, and can be found by examining the inverse images of D_0,

D and L in the universal cover of $S^3 - \partial D$.

Remark

We can deduce an algebraic criterion for exchangeability from the fact, (Hempel, J., 1976 (Theorem 10.2)), that the inclusion $(D \times \{0\}$, n points) $\subset$ $(D \times I$, n arcs) induces an isomorphism of fundamental groups if and only if the arcs lie monotonically up to isotopy rel boundary. On returning to D in S^3 , this result combined with Theorem 4 gives the following test:

Corollary 4.1

Let L meet a disc D in $n = \text{lk}(L, \partial D)$ points. Then L is braided relative to ∂D if and only if the inclusion $D^+ - L \subset S^3 - (D \cup L)$ induces a fundamental group isomorphism, where D^+ denotes a translate of D through a small distance in the direction of the positive normal to D .

Corollary 4.2

With L, D as in 4.1, a necessary, but not sufficient, condition for L to be braided relative to ∂D is that any k of the arcs A_i can be isotoped to lie monotically in $D \times I$. Consequently $\pi_1(S^3 - (D \cup k$ arcs of $L))$ must be free on k generators if L is braided.

We can use 4.2 in the simplest case $k = 1$ to construct Stallings braids, based on a diagram such as Figure 4, which are not exchangeable for any choice of signs in the ribbons. This is in contrast to the braid ω in Figure 2, where a change of signs on the ribbons to give $v = \sigma_{2,4}^{-1}\sigma_{2,4}^{-1}\sigma_{1,3}^{-1}$ yields an exchangeable braid. (See Figure 8 and the end of §4 .)

Theorem 5

None of the braids $\sigma_{2,4}^{\pm1}\sigma_{3,5}^{\pm1}\sigma_{4,6}^{\pm1}\sigma_{1,5}^{\pm1}\sigma_{3,6}^{\pm1} \in B_6$ are exchangeable.

Proof

The axis is broken into six arcs by a Stallings disc D , and for one of these arcs L_1 we have $\pi_1(S^3 - (D \cup L_1)) \ncong \mathbb{Z}$. The arcs L_i are shown in Figure 5. In the calculation of π_1 we can replace $D \cup L_1$ by the curve in Figure 6, which forms a non-trivial knot.

Figure 4

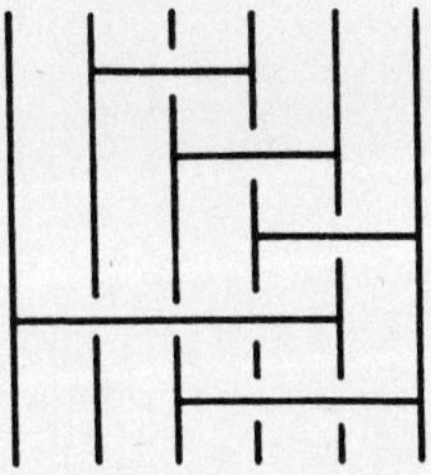

Figure 6

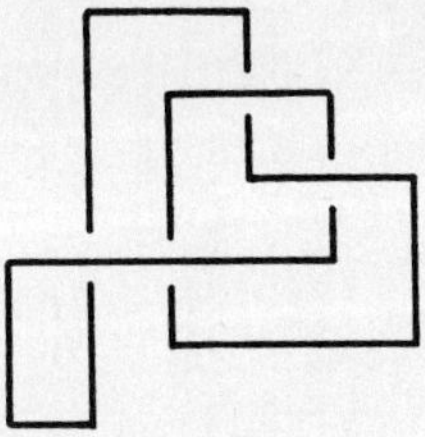

Figure 5

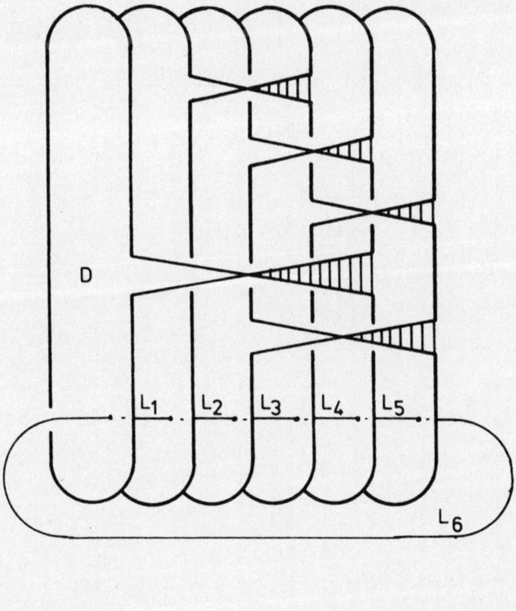

4 COMPOUND BRAIDS

By way of giving some sufficient geometric conditions for exchangeability, I shall describe two ways in which a Stallings braid may decompose into simpler Stallings braids, and prove that the original braid is exchangeable if and only if its constituents are.

(a) Murasugi Sums

The first decomposition is a counterpart of the generalised plumbing, or 'Murasugi sum' of two surfaces, originally described in (Murasugi, K., 1963). Here the construction is extended to apply to pairs (Stallings disc, axis), in a similar vein to the description given in (Gabai, D., 1982) or (Morton, H.R., 1983).

Given a Stallings braid $\beta \in B_n$, with Stallings disc D and axis L, I shall say that (D, L) is a *Murasugi sum* of (D_1, L_1) and (D_2, L_2) if (i) D_1 and D_2 are subdiscs of D each lying in one half-space of $\mathbb{R}^3$, and meeting only in a disc $D_0 = D_1 \cap D_2$ in the common plane $\mathbb{R}^2$, (ii) the axis L meets the separating plane $\mathbb{R}^2$ in just two points, $d \in D_0$ and $c \notin D_0$, (iii) the disc D_1 forms a Stallings disc with axis L_1, where L_1 consists of the part of L in one half-space, completed by an unknotted arc cd in the other half-space, and similarly for (D_2, L_2).

This sort of decomposition can be seen when none of the horizontal bands in the disc for β pass over the kth string, say, i.e. no generators $\sigma_{i,j}$ with $i < k < j$ occur in β. Then the vertical plane $\mathbb{R}^2$ containing the kth vertical disc will separate the Stallings disc D spanning $\hat{\beta}$ into two discs, D_1 consisting of the first k vertical discs with the ribbons joining them, and D_2 consisting of the last $n-k+1$ vertical discs and joining ribbons. These form Stallings discs, relative to the curve L as axis, for braids β_1 on the first k strings, and β_2 on the last $n-k+1$ strings, see for example Figure 7.

Theorem 6

The Stallings braid β is exchangeable if and only if β_1 and β_2 are exchangeable.

Figure 7

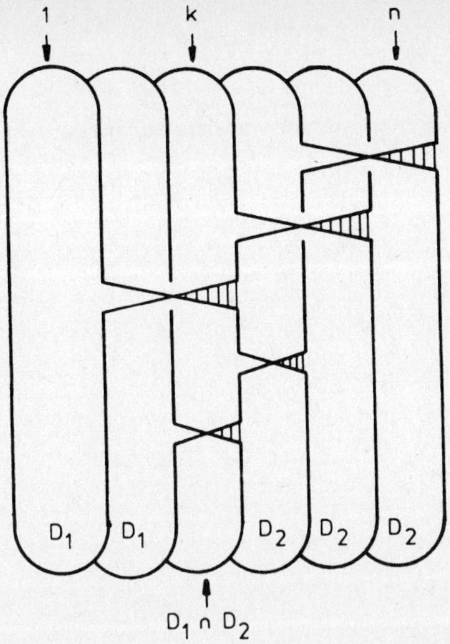

A Murasugi sum of two Stallings braids

Corollary 6.1

Every Stallings braid $\beta \in B_n$ in which one string, the kth say, is joined to each of the others by a ribbon is exchangeable.

Proof

If $k \neq 1, n$ then β is a Murasugi sum of two braids with the same property, on fewer strings. The result then follows by induction on n, using Theorem 6 and the fact that the two Stallings braids on two strings are exchangeable.

Otherwise let $k = 1$, ($k = n$ is similar). The conjugate braid $\beta^1 = \gamma^{-1}\beta\gamma$, where $\gamma = \sigma_{n-1} \cdots \sigma_2\sigma_1$, given by moving the last string of β over the others to become the first, is again a Stallings braid of the same form as β, with $k = 2$, and has thus already been shown to be exchangeable.

Proof of Theorem 6

Let D be the Stallings disc for $\hat{\beta}$. Write $c_1, \ldots, c_n$ for the points of D L in order along L, so that the separating plane $\mathbb{R}^2$ meets L in $c_k = d$, and in a point c between c_n and c_1.

Suppose first that β_1 and β_2 are exchangeable. We must show that there is a fibration for $S^3 - \partial D$ in which L lies monotonically. Now the construction in (Morton, H.R., 1983) of a fibration for $S^3 - \partial F$ where F is a Murasugi sum of two fibre surfaces F_1 and F_2 will apply here to give a fibration of $S_3 - \partial D$ arising explicitly from the given fibration of $S^3 - \partial D_1$ and $S^3 - \partial D_2$. It follows from this construction that each arc $c_i c_{i+1}$, for $1 \leq i < n$, and also the arcs $c\,c_1$ and $c_n c$ lie monotonically in this fibration, since their counterparts did in the constituent fibrations. Here L lies monotonically in the fibration, and so β is exchangeable.

Conversely, suppose that β is exchangeable. Choose an arc α in D_2 joining c_k to c_n. This arc α followed by the arc on L from c_n to c_1 gives an arc from c_k to c_1 which can be altered slightly by pushing gradually off D_2 so as to give an arc from c_k to c_1 which lies monotonically in the fibration of $S^3 - \partial D$. The closed curve C made up of this arc, with the arcs $c_1c_2, \ldots, c_{k-1}c_k$ of L is then braided relative to ∂D.

I claim that $\partial D \cup C$ is isotopic to $\hat{\beta}_1 \cup$ axis, showing that β_1 is exchangeable.

It is clear that $D \cup C$ is isotopic to $D_1 \cup C$, since C does not meet $D - D_1$. It remains to show that C is isotopic, rel D_1, to the axis for $\hat{\beta}_1 = \partial D_1$. Up to isotopy rel D_1, the curve C can be taken as   together with the part of L from c_n through c to c_k. It will then be enough to show that the arc $\alpha \cup c_n c$ is unknotted in the half-space which contained D_2, for this piece can then be moved to form the remaining part of the axis. Complete this arc to a curve C' by joining the arc $c\,c_1$ from L and an arc γ in D_1 joining c_1 to c_k. This curve C' then consists of one arc $c_n c_1$ on the axis of D with ends joined by an arc lying in D, and hence unknotted, by the exchangeability of β, since its complement will have free fundamental group (Corollary 4.2). Hence $\alpha \cup c_n c$ is unknotted in its half-space, since it forms a connected summand of the unknotted curve C'. This completes the proof of Theorem 6.

(b) Satellites

The other possible decomposition of a Stallings braid which I would like to describe is related to the construction of satellite knots and links. Generally, to construct a satellite of a link $L = L_1 \cup \ldots \cup L_r$ we start with another link, $C = C_1 \cup \ldots \cup C_k$, in which one unknotted component, C_k say, is selected. Choose one component, L_1 say, of the original link, and replace a solid torus neighbourhood V of L_1 by the complementary solid torus W to a neighbourhood of C_k. This replacement is by means of a faithful homeomorphism $h : W \to V$, i.e. one which carries a longitude of W to a longitude of V. The satellite link then consists of $h(C_1) \cup \ldots \cup h(C_{k-1}) \cup L_2 \cup \ldots \cup L_r$, and contains a 'splitting torus', $T = \partial V = h(\partial W)$.

I shall only be concerned here with the case $r = k = 2$ in which $C_1 \cup C_2$ and $L_1 \cup L_2$ each consist of a closed braid, together with its axis. It is not difficult to see that the resulting satellite $h(C_1) \cup L_2$ is a closed braid with axis L_2 on mn strings, when the constituent braids have m and n strings respectively. In fact a converse of this can be proved, either as a special case of results about fixed satellite links, or by bare-hands isotopy of a splitting torus for the satellite. The result can be stated as follows.

Theorem 7

A satellite link $K_1 \cup K_2 = h(C_1) \cup L_2$ arising from links $C_1 \cup C_2$ and $L_1 \cup L_2$ consists of a closed braid K_1 with axis K_2 if and only if C_1 is a closed braid with axis C_2 and L_1 is a closed braid with axis L_2.

As a consequence we have the following result about exchangeable braids.

Theorem 8

Let the satellite link $K_1 \cup K_2$ consist of a closed braid β with its axis. Then β is exchangeable if and only if β_1 and β_2 are exchangeable, where $C_1 \cup C_2 = \beta_1 \cup$ axis, $L_1 \cup L_2 = \beta_2 \cup$ axis, as in Theorem 7.

Proof

If β is exchangeable then $K_1 \cup K_2$ forms a closed braid K_2 with axis K_1. The splitting torus must be unknotted, so the link K can also be viewed as a satellite constructed from the links C and L with the roles of C_2 and L_1 interchanged and also the roles of C_1 and L_2. By Theorem 7 it follows that C and L also consist of closed braids C_2, L_2 with axes C_1 and L_1 respectively; thus β_1 and β_2 are exchangeable. Conversely, if β_1 and β_2 are exchangeable the homeomorphism $h : W \to V$ used in constructing the satellite, which carries the exterior of C_2 to the neighbourhood of L_1 can be extended to a homeomorphism $h : S^3 \to S^3$, with h^{-1} carrying the exterior of L_1 to a neighbourhood of C_2. The satellite of C constructed using h^{-1}, which consists of $h^{-1}(L_2) \cup C_1$, also forms a closed braid with axis C_1, since C_2 and L_2 are braided relative to C_1, L_2 respectively. This satellite, however, is equivalent under h to the link $L_2 \cup h(C_1)$, which forms the original link $K_2 \cup K_1$. Thus K_1 is braided relative to K_2, and β is exchangeable.

One of the simplest non-trivial examples of this construction is illustrated by the braid $v \in B_4$ shown in Figure 8, where the link $\hat{v} \cup$ axis results from the satellite construction with $\beta_1^{-1} = \beta_2 = \sigma_1 \in B_2$. Examples of this sort with $\beta_1 \neq \beta_2^{\pm 1}$ can be used to show that exchangeable braids do not always yield symmetric links, i.e. there need not be an isotopy interchanging the two components of the link $\hat{\beta} \cup$ axis for an exchangeable braid β.

Figure 8

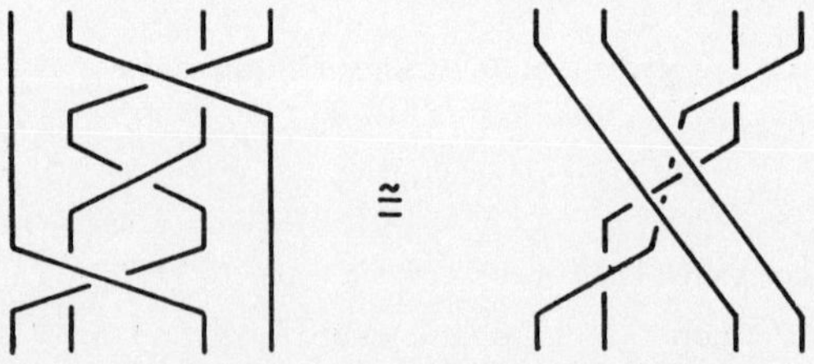

The Stallings braid $v = \sigma_{2,4}^{-1}\sigma_{2,3}\sigma_{1,3}^{-1}$

Conclusion

It would be nice to have an effective geometric test to decide which Stallings braids are exchangeable. Although I have developed some necessary and some sufficient conditions for exchangeability there is still a considerable gap between the two.

As a final section, I include a brief review of the properties of the Alexander polynomial for a link, and its calculation, leading to the relation with the Burau matrices as described in Theorem 3.

5 *ALEXANDER POLYNOMIALS OF CLOSED BRAIDS*

In Fox's theory of Alexander polynomials, (Fox, R.H., 1953 & 1954), a presentation of a group G, with n generators and r relations, yields an $r \times n$ matrix A with entries in the group ring $\mathbb{Z}[H]$, where H is the abelian group $G/[G,G]$, written multiplicatively. The ideals $E_k(A)$ in $\mathbb{Z}[H]$ generated by the $(n-k) \times (n-k)$ minors of A are invariants of the group G. In the case when G is the group of a link with μ components the group H is free abelian of rank μ.

Fox shows that the ideal

$$\begin{aligned} E_1(G) &= D \quad \mu = 1 \\ &= D.I \quad \mu > 1, \end{aligned}$$

where I is the augmentation ideal, and D is a principal ideal, with generator Δ. The element $\Delta \in \mathbb{Z}[H]$, determined up to unit multiple, can be viewed as an integer polynomial in μ variables, and is defined to be the Alexander polynomial of the link. Fox shows, in (Fox, R.H., 1954 (6.4)), that if the column of A corresponding to a generator $g \in G$ is deleted to give a matrix B, then $E_0(B)$ is a principal ideal, with generator

$$\Delta.(\langle g \rangle - 1), \qquad \text{for } \mu > 1,$$

or $$\Delta.\frac{\langle g \rangle - 1}{t - 1}, \qquad \text{for } \mu = 1,$$

where $\langle g \rangle \in H$ is the image of g, and t is a generator of H in the case $\mu = 1$.

One immediate consequence is that if $r = n - 1$, we can find

Δ readily as det $B . \dfrac{1}{<g> - 1}$, when $\mu > 1$ and B is given as above by deleting a column of A.

Proof of Theorem 3

A closed braid $\hat{\beta}$ on n strings with axis L forms a link with group G generated by $t_1, \ldots, t_n$ and x and relations $\beta(t_i) = x^{-1} t_i x$ for each i, where β is an automorphism of the free group F_n (see Birman, J.S., 1974).

The group F_n appears here as $\pi_1(D^2$ - n pts.), where D^2 is a spanning disc for L, meeting $\hat{\beta}$ in n points.

For a knot $\hat{\beta}$ the abelianisation $H = G/[G, G]$ is free abelian on two generators $t = <t_i>$, $i = 1, \ldots, n$, corresponding to meridians of $\hat{\beta}$, and $x = <x>$, corresponding to meridians of L.

Birman shows that the matrix $\left(\dfrac{\partial\beta(t_i)}{\partial t_j}\right)_H = \tilde{B}(t)$ of free derivatives evaluated in $\mathbb{Z}[H]$ is just the full $(n \times n)$ Burau matrix of the braid $\beta \in B_n$.

Applying Fox's free calculus to the presentation of G given above yields the $n \times (n+1)$ matrix $A = \left(\begin{array}{c|c} \tilde{B}(t) - xI_n & \begin{array}{c} t-1 \\ \vdots \\ t-1 \end{array} \end{array}\right)$,

where the last column corresponds to the generator $x \in G$. The Alexander polynomial $\Delta(x, t)$ of $\hat{\beta} \cup L$ is then given by deleting this last column, so that $\Delta(x,t) = \det(\tilde{B}(t) - xI_n)/x - 1$. Now $\tilde{B}(t)$ is conjugate in $GL(n, \mathbb{Z}[t, t^{-1}])$ to $\left[\begin{array}{c|c} B(t) & \vdots \\ \hline 0 \ldots 0 & 1 \end{array}\right]$ where $B(t) \in GL(n-1, \mathbb{Z}[t, t^{-1}])$ is the *reduced Burau matrix* of β, (Birman, J.S., 1974). Hence $\det(B(t) - xI_{n-1}) = \Delta(x, t)$ is as claimed in Theorem 3.

REFERENCES

Birman, J.S. (1974). Braids, links and mapping class groups. Annals of Maths. Studies 82, Princeton University Press.

Bonahon, F. & Otal, J.-P. (Preprint) Scindements dc Heegaard des espaces lenticulaires. Orsay.

Fox, R.H. (1953). Free differential calculus I. Annals of Math. 57, 547-560.

Fox, R.H. (1954). Free differential calculus II. Annals of Math. 59, 196-210.

Gabai, D. (1982). The Murasugi sum is a natural geometric operation, preprint.

Goldmsith, D. (1975) Symmetric Fibred Links. In Knots, groups and 3-manifolds, ed. L.P. Neuwirth, 3-23. Annals of Maths. Studies 84, Princeton University Press.

Hempel, J. (1976). 3-manifolds. Annals of Maths. Studies 86, Princeton University Press.

Morton, H.R. (1978). Infinitely many fibred knots having the same Alexander polynomial. Topology 17, 101-4.

Morton, H.R. (1979). Closed braids which are not prime knots. Math. Proc. Camb. Phil. Soc. 86, 421-426.

Morton, H.R. (1983). Fibred knots with a given Alexander polynomial. L'Enseignement Mathématique, 31, 'Noeuds, Tresses et Singularités', ed. C. Weber, 205-222.

Murasugi, K. (1963). On a certain subgroup of the group of an alternating link. Amer. J. Math. 85 (1963), 544-550.

Stallings, J. (1978). Problems in low dimensional manifold theory. ed. R. Kirby. Proc. Symposia in Mure Math. Amer. Math. Soc. Stamford 1976, 32 part 2, 274-312.

NILPOTENT COVERINGS OF LINKS AND MILNOR'S INVARIANT

Kunio Murasugi

Abstract. The Milnor $\bar{\mu}$-invariants of a link L are shown to be the covering linkage invariants of appropriate covering spaces of S^3 branched along L .

1 *INTRODUCTION*

Let R be a commutative ring with identity 1 and $M_n(R)$ the ring of $n \times n$ matrices over R . The set of all upper triangular matrices whose diagonal entries are all 1 forms a subgroup N of the group of units in $M_m(R)$, and moreover, any finitely generated subgroup of N is nilpotent. (See Proposition 2.1.) Therefore, if R is a finite ring, then a homomorphic image of a finitely generated group G in N is a finite nilpotent group. In particular, if G is the group of a link L in S^3 then a nilpotent branched covering space $\tilde{M}$ of L will be associated with a homomorphism from G to a finite subgroup of N . Therefore, if $\tilde{M}$ is, for instance, a Q-homology sphere, we can define the covering linkage invariants of L (R. Hartley & K. Murasugi, 1977), i.e. the linking number between various components of the lift of L in $\tilde{M}$.

We will show in this paper that Milnor's $\bar{\mu}$-invariants $\bar{\mu}(j_1 j_2 \cdots j_k)$ are, in fact, the covering linkage invariants in appropriate nilpotent covering spaces of S^3 branched along L (Theorem 7.1). This interpretation of $\bar{\mu}$ will be used to clarify the connections with other invariants found in the literatures (D.L. Goldsmith, 1977; R. Holmes & N. Smythe, 1966; H. Lanfer, 1971).

It is also easy to see from this interpretation that $\bar{\mu}(j_1 j_2 \cdots j_k)$ is an isotopy invariant, and further, if $j_1, \ldots, j_k$ are all distinct, then $\bar{\mu}(j_1 \cdots j_k)$ is a homotopy invariant (J. Milnor, 1954). (See also §11).

2 *PRELIMINARIES*

We will begin with the following.

Proposition 2.1

Let N *be the subgroup of the group of units in* $M_n(R)$ *defined in* §1. *Then any finitely generated subgroup* G *of* N *is nilpotent of class at most* $n-1$.

Proof. Let $\{G_q\}$, $q = 1,2,\ldots$ be the lower central series of G, where $G = G$ and $G_{k+1} = [G, G_k]$, $k \geq 1$, is the commutator subgroup of G and G_k.

We will show that $G_n = 1$.

Let $\{x_1, x_2, \ldots, x_r\}$ be a set of generators of G. Write $x_i = \| a_{p,q}^{(i)} \|$, where $a_{p,q}^{(i)} \in R$, $a_{p,q}^{(i)} = 0$ for $p > q$, and $a_{p,p}^{(i)} = 1$, $1 \leq p, q \leq n$.

Then the proposition will follow from the lemma below.

Lemma.

Let $g = \| b_{ij} \|$ *be an element of* G, *and hence,* $b_{i,j} = 0$ *for* $i > j$ *and* $b_{i,i} = 1$.

If $g \in G_m$, *then*

$$b_{i,i+p} = 0, \quad 1 \leq i \leq n-1, \quad 1 \leq p \leq m-1, \quad 2 \leq i+p \leq n. \tag{2.1}$$

Proof. Proof will be done by induction on m. For $m = 1$, the lemma holds trivially. Assume inductively that the lemma holds for any element $g \in G_{m-1}$, $m \geq 2$. Now noting that $g^{-1} \in G_{m-1}$, a simple direct computation shows that $[g, x_p] = g x_p g^{-1} x_p^{-1}$ has property (2.1). This proves the lemma.

3 *NILPOTENT REPRESENTATIONS*

A matrix A in N is equivalently represented as an action on an n-tuple $(a_1, \ldots, a_n)$, $a_i \in R$, and we find that such a representation is much more convenient, particularly, to define Milnor's $\bar{\mu}$-invariant, and therefore, we will use such a representation in the remaining sections.

First we need few definitions.

Let R^k be the set of all k-tuples $\{(a_1, \ldots, a_k) \mid a_i \in R, 1 \leq i \leq k\}$. Let F be the free group with n free generators

$x_1, x_2, \ldots, x_n$.

Given an ordered sequence $\xi = \{j_1, j_2, \ldots, j_k\}$ of k positive integers between 1 and n , we define an action ϕ_ξ of F on R^k as follows.

$$\phi_\xi(x_i)(a_1, \ldots, a_k) = (b_1, \ldots, b_k), \tag{3.1}$$

where $b_m = a_m + \delta_{i,j_m} a_{m-1}$, $a_0 = 1$ and $\delta_{i,q}$ is the Kronecker's delta. We write sometimes $\phi_\xi(x_i)(a_j) = b_j$, $1 \le j \le k$.

Example 3.1

Let $n = 2$ and $k = 5$. Let $\xi = \{1,2,1,1,2\}$. Then

$$\phi_\xi(x_1)(a_1, a_2, a_3, a_4, a_5) = (a_1+1, a_2, a_3+a_2, a_4+a_3, a_5)$$

$$\phi_\xi(x_2)(a_1, a_2, a_3, a_4, a_5) = (a_1, a_2+a_1, a_3, a_4, a_5+a_4)$$

where $j_1 = j_3 = j_4 = 1$ and $j_2 = j_5 = 2$.

Remark 3.1

In the matrix notation,

$$\phi_\xi(x_i) = \begin{bmatrix} 1 & \delta_{i,j_1} & & & & \\ & 1 & \delta_{i,j_2} & & 0 & \\ & & 1 & \ddots & & \\ & & & \ddots & \ddots & \\ & 0 & & & 1 & \delta_{i,j_k} \\ & & & & & 1 \end{bmatrix} \in M_{k+1}(R) .$$

We should note that this matrix corresponds to an action on (k+1) - tuple

$$\phi_\xi(x_i)(a_0, a_1, \ldots, a_k) = (b_0, b_1, \ldots, b_k) ,$$

where a_0 and b_0 are always 1 .

Let Ω be the orbit of $(0, \ldots, 0)$ under the action ϕ_ξ on F , and $S(\Omega)$ the group of permutations on Ω . Then ϕ_ξ , in fact, defines a homomorphism

$$\phi_\xi \quad : \quad F \to S(\Omega) \ .$$

(We will use the same notation.) Since it is transitive, ϕ_ξ gives a transitive representation of F on $S(\Omega)$. Furthermore, ϕ_ξ is a nilpotent representation (Proposition 2.1).

Now in order to study ϕ_ξ , we need a explicit formula for the action ϕ_ξ on Ω .

For an element g in F , we denote by $\frac{\partial g}{\partial x_i}$ the Fox free derivative (R.H. Fox, 1953), and by 0 the trivializer, i.e. the ring homomorphism from the integer group ring $\mathbb{Z}F$ to $\mathbb{Z}$.

Proposition 3.1

Let $\xi = \{ j_1 j_2 , \ldots, j_k \}$, $1 \le j_1, j_2, \ldots, j_k \le n$. *For* $g \in F$, *write* $\phi_\xi(g)(a_q) = b_q$, $1 \le q \le k$. *Then*

$$b_q = a_q + a_{q-1} \left(\frac{\partial g}{\partial x_{j_q}}\right)^0 + a_{q-2} \left(\frac{\partial^2 g}{\partial x_{j_{q-1}} \partial x_{j_q}}\right)^0 +$$

$$\ldots + a_{q-p} \left(\frac{\partial^p g}{\partial x_{j_{q-p+1}} \ldots \partial x_{j_q}}\right)^0 + \ldots + \left(\frac{\partial^q g}{\partial x_{j_1} \ldots \partial x_{j_q}}\right)^0 .$$

Corollary 3.1

Using the same notation,

$$\phi_\xi(g)(1,0,\ldots,0) = \left[\left(\frac{\partial g}{\partial x_{j_1}}\right)^0 , \left(\frac{\partial^2 g}{\partial x_{j_1} \partial x_{j_2}}\right)^0 , \right.$$

$$\left. \ldots, \left(\frac{\partial^k g}{\partial x_{j_1} \ldots \partial x_{j_k}}\right)^0 \right] .$$

Proof of Proposition 3.1 . A proof will be done by induction on $|g|$, the length of g .

First, suppose $|g| = 1$. Hence, g is either x_i or x_i^{-1} . Then a direct computation shows that

$$b_q = \phi_\xi(x_i)(a_q) = a_q + \delta_{i,j_q} a_{q-1} = a_q + \left(\frac{\partial x_i}{\partial x_{j_q}}\right)^0 a_{q-1} , \qquad (3.3)$$

and

$$b_q = \phi_\xi(x_i^{-1})(a_q) = a_q + \sum_{\lambda=1}^{q} (-1)^\lambda \delta_{j_q,i}\, \delta_{j_{q-1},i} \cdots \delta_{j_{q-\lambda+1},i}\, a_{q-\lambda} .$$

This is equivalent to (3.2), since $\left(\dfrac{\partial x_i}{\partial x_{j_q}}\right)^0 = \delta_{i,j_q}$,

$$\left(\frac{\partial^p x_i}{\partial x_{j_{q-p+1}} \cdots \partial x_{j_q}}\right)^0 = 0 \quad \text{for } p \geq 2 , \quad \text{and}$$

$$\left(\frac{\partial^p x_i^{-1}}{\partial x_{j_{q-p+1}} \cdots \partial x_{j_q}}\right)^0 = (-1)^p\, \delta_{j_{q-p+1},i} \cdots \delta_{j_q,i} .$$ (See R.H. Fox, 1953.)

Now inductively, assume that (3.2) holds for any element in F whose length is less than ℓ. Let h be an element in F with $|h| = \ell$. Then $h = g\,x_i$ or $g\,x_i^{-1}$ for some g F with $|g| = \ell - 1$. Write $\phi_\xi(g)(a_q) = b_q$ and $\phi_\xi(x_i)(b_q) = c_q$. Then by induction assumption,

$$b_q = \sum_{p=0}^{q} a_{q-p} \left(\frac{\partial^p g}{\partial x_{j_{q-p+1}} \cdots \partial x_{j_q}}\right)^0 .$$

Note that we always assume $a \ = 1$. Also

$$c_q = b_q + b_{q-1}\left(\frac{\partial x_i}{\partial x_{j_q}}\right)^0 = b_q + b_{q-1}\,\delta_{j_q,i} .$$

Therefore,

$$c_q = \sum_{p=0}^{q} a_{q-p}\left(\frac{\partial^p g}{\partial x_{j_{q-p+1}} \cdots \partial x_{j_q}}\right)^0 + \sum_{p=0}^{q-1} a_{p-1-p}\left(\frac{\partial^p g}{\partial x_{j_{q-p}} \cdots \partial x_{q-1}}\right)^0 \delta_{j_q,i}$$

$$= a_q + \sum_{p=1}^{q} a_{q-p}\left(\frac{\partial^p g}{\partial x_{j_{q-p+1}} \cdots \partial x_{j_q}}\right)^0 + \left(\frac{\partial^{p-1} g}{\partial x_{j_{q-p+1}} \cdots \partial x_{j_{q-1}}}\right)^0 \delta_{j_q,i}$$

$$= a_q + \sum_{p=1}^{q} a_{q-p} \left(\frac{\partial^p (gx_i)}{\partial x_{j_{q-p+1}} \cdots \partial x_{j_q}} \right)^0 .$$

Next, we write $\phi_\xi (x_i^{-1})(b_q) = d_q$, $1 \le q \le k$. Then

$$d_q = b_q + \sum_{\lambda=1}^{q} (-1)^\lambda \delta_{j_q,i} \delta_{j_{q-1},i} \cdots \delta_{j_{q-\lambda+1},i} b_{q-\lambda}$$

$$= \sum_{p=0}^{q} a_{q-p} \left(\frac{\partial^p g}{\partial x_{j_{q-p+1}} \cdots \partial x_{j_q}} \right)^0$$

$$+ \sum_{\lambda=1}^{q} (-1)^\lambda \delta_{j_q,i} \delta_{j_{q-1},i} \cdots \delta_{j_{q-\lambda+1},i} \sum_{r=0}^{q-\lambda} a_{q-\lambda-r} \left(\frac{\partial^r g}{\partial x_{j_{q-\lambda-r+1}} \cdots \partial x_{j_{q-\lambda}}} \right)^0$$

$$= a_q + \sum_{p=1}^{q} a_{q-p} \left\{ \left(\frac{\partial^p g}{\partial x_{j_{q-p+1}} \cdots \partial x_{j_q}} \right)^0 - \delta_{j_q,i} \left(\frac{\partial^{p-1} g}{\partial x_{j_{q-p+1}} \cdots \partial x_{j_{q-1}}} \right)^0 \right.$$

$$\left. + \delta_{j_q,i} \delta_{j_{q-1},i} \left(\frac{\partial^{p-2} g}{\partial x_{j_{q-p+1}} \cdots \partial x_{j_{q-2}}} \right)^0 + (-1)^\lambda \delta_{j_q,i} \cdots \delta_{j_{q-\lambda+1},i} \right\}$$

$$= \sum_{p=0}^{q} a_{q-p} \left(\frac{\partial^p (gx_i^{-1})}{\partial x_{j_{q-p+1}} \cdots \partial x_{j_q}} \right)^0 .$$

This proves (3.2).

Proposition 3.2
$\phi_\xi(F)$ *is nilpotent of class at most* k .

Proof. If $g \in F_{k+1}$, then $\left(\frac{\partial^p g}{\partial x_{j_{q-p+1}} \cdots \partial x_{j_q}} \right)^0 = 0$ for $1 \le p \le k$ (R.H. Fox, 1953), and hence $\phi_\xi (F_{k+1}) = (\phi_\xi(F))_{k+1} = 1$.

(Of course, Proposition 3.2 follows from Proposition 2.1 .)

4 *MILNOR'S INVARIANTS*

We quickly review the definition of Milnor's $\bar{\mu}$-invariant $\bar{\mu}(j_1 j_2 \cdots j_k)$ for a sequence of integers, $1 \le j_1, \ldots, j_k \le n$. (See J Milnor, 1957; or K. Murasugi, 1966).

Let $L = K_1 \cup K_2 \cup \ldots \cup K_n$ be an oriented link of n components in the oriented 3-sphere S^3.

Let $G(L) = \langle x_{i,j} \mid r_{i,j} \rangle$ be a "modified" Wirtinger presentation of the link group $\pi_1(S^3 - L)$ in the following sense:

Generators $x_{i,1}$ correspond to prescribed meridian elements of K_i, $i = 1,2,\ldots,n$, and relators are of the form:

$$r_{ij} = u_{i,j}\, x_{i,1}\, u_{i,j}^{-1}\, x_{i,j+1}^{-1}, \quad j = 1,2,\ldots, \nu_i - 1$$

and

$$r_{i,\nu_i} = \eta_i\, x_{i,1}\, \eta_i^{-1}\, x_{i,1}^{-1},$$

where $u_{i,j}$ are words in $\{x_{p.q}\}$ and η_i represents the longitude of K_i, $i = 1,2,\ldots,n$. $\{x_{i,1}, \eta_i\}$, $1 \le i \le n$, forms a peripheral subgroup of $\pi_1(S^3 - L)$.

Consider two free groups F^* and F freely generated by $\{x_{i,j}\}$ and $\{x_i\}$, respectively, i.e.

$$F^* = \langle x_{ij} \mid \ \rangle, \quad 1 \le i \le n, \quad 1 \le j \le \nu_i$$

and

$$F = \langle x_1, x_2, \ldots, x_n \mid \ \rangle .$$

Let $\rho : F^* \to F^*$ and $\nu : F^* \to F$ be homomorphisms defined by

$$\rho(x_{i,1}) = x_{i,1}$$

$$\rho(x_{i,j+1}) = u_{i,j}\, x_{i,1}\, u_{i,j}^{-1}, \quad 1 \le j \le \nu_i - 1$$

and

$$\nu(x_{i,j}) = x_i, \quad 1 \le i \le n, \quad 1 \le j \le \nu_i .$$

Using ρ and ν, we define the third homomorphism

$$\theta_k = \nu\rho^{k-1} : F^* \to F ,$$

for $k \geq 1$.

Now, given a sequence $\xi = \{ j_1, j_2, \ldots, j_k \}$, $1 \leq j_1, \ldots, j_k \leq n$, we define an integer

$$\mu(j_1 j_2 \cdots j_{k-1} j_k) = \left(\frac{\partial^{k-1} \theta_{k-1}(\eta_{j_k})}{\partial x_{j_1} \partial x_{j_2} \cdots \partial x_{j_{k-1}}} \right)^0 .$$

Let $\bar{\Delta}(i_1 i_2 \ldots i_r) = \text{g.c.d.}\ \mu(\lambda_1, \ldots, \lambda_s)$, where $\lambda_1, \ldots, \lambda_s$ is to range over all sequences obtained by cancelling at least one of the indices $i_1, i_2, \ldots, i_r$ and permuting the remaining indices cyclically. Milnor then defined an integer

$$\bar{\mu}(j_1 j_2 \cdots j_k) \equiv \mu(j_1 j_2 \cdots j_k) \bmod \bar{\Delta}(j_1 j_2 \cdots j_k) ,$$

called now Milnor's invariant. It is an isotopy invariant of a link L , (J. Milnor, 1957).

Now, it is easy to see that

$$\bar{\mu}(i\,j)\ (i \neq j) \text{ is the linking number } \ell k(K_i, K_j) \text{ between two components } K_i , K_j . \tag{4.1}$$

Also, $\bar{\mu}(i\,i \ldots i) = 0$ for a sequence of a single number.

It is less trivial but was proved in (J. Milnor, 1957) that

$$\bar{\mu}(j_1 j_2 \cdots j_k) = \bar{\mu}(j_2 j_3 \cdots j_k j_1) . \tag{4.2}$$

Since $\bar{\mu}(ij)$ is the linking number, it would be natural to expect that $\bar{\mu}(j_1 j_2 \cdots j_k)$ could be interpreted as the linking number between two covering knots in a suitable branched covering space of L .

In the following sections, we will prove that this is certainly true.

5 *NILPOTENT COVERINGS OF LINKS*

Let F be the free group generated by $x_1, \ldots, x_n$ as was given in §4, and F_q the q-th member of the lower central series of F. Note that $F_1 = F$.

Proposition 5.1

For $j \geq 0$,

$$\theta_q(x_{i,j+1}) \equiv \theta_{q+1}(x_{i,j+1}) \bmod F_{q+1}.$$

Proof. A proof will be done by induction on q. When $q = 1$, by definition, $\theta_1(x_{i,j+1}) = x_i$, while $\theta_2(x_{i,j+1}) = \nu(u_{i,j}\, x_{i,1}\, u_{i,j}^{-1}) = \nu(u_{i,j})\, x_i\, \nu(u_{i,j})^{-1}$. Since $\nu(u_{i,j})\, x_i\, \nu(u_{i,j})^{-1} \equiv x_i \bmod F_2\ (= [F,F])$, it follows that $\theta_1(x_{i,j+1}) \equiv \theta_2(x_{i,j+1}) \bmod F_2$.

Now assume inductively that $\theta_q(x_{i,j+1}) \equiv \theta_{q+1}(x_{i,j+1}) \bmod F_{q+1}$. Then

$$\begin{aligned}\theta_{q+2}(x_{i,j+1}) &= \theta_{q+1}(u_{i,j}\, x_{i,1}\, u_{i,j}^{-1}) \\ &= \theta_{q+1}(u_{i,j})\, \theta_{q+1}(x_{i,1})\, \theta_{q+1}(u_{i,j})^{-1}\end{aligned}$$

and

$$\begin{aligned}\theta_{q+1}(x_{i,j+1}) &= \theta_q(u_{i,j}\, x_{i,1}\, u_{i,j}^{-1}) \\ &= \theta_q(u_{i,j})\, \theta_q(x_{i,1})\, \theta_q(u_{i,j})^{-1}.\end{aligned}$$

Since $\theta_{q+1}(u_{i,j}) \equiv \theta_q(u_{i,j}) \bmod F_{q+1}$, by induction assumption, it follows that

$$\theta_{q+1}(u_{i,j})\, x_i\, \theta_{q+1}(u_{i,j})^{-1} \equiv \theta_q(u_{i,j})\, x_i\, \theta_q(u_{i,j})^{-1} \text{ Mod } F_{q+2}$$

and hence

$$\theta_{q+2}(x_{i,j+1}) \equiv \theta_{q+1}(x_{i,j+1}) \bmod F_{q+2}.$$

Remark 5.1

Proposition 5.1 was originally proved by Milnor (1957, (6_k)) in the form: $\theta_q(x_{i,j+1}) \equiv \theta_{q+1}(x_{i,j+1}) \bmod F_q$. An almost identical proof provided above improves formula (6_k) to the present form.

Proposition 5.1 combined with Proposition 3.2 implies that

$$\mu(j_1 j_2 \cdots j_k) = \left(\frac{\partial^{k-1} \theta_r(\eta_{j_k})}{\partial x_{j_1} \partial x_{j_2} \cdots \partial x_{j_{k-1}}} \right)^0 \quad \text{for } r \geq k-1 . \quad (5.1)$$

Now suppose we are given a sequence $\hat{\xi} = \{ j_1 j_2 \cdots j_k j_{k+1} j_{k+2} \}$ of $k+2$ integers between 1 and n. Let $\xi = \{ j_1 j_2 \cdots j_k \}$ be the subsequence of the first k integers in $\hat{\xi}$ and let $m = \Delta(\hat{\xi})$ be the integer defined in §4 for $\hat{\xi}$.

Take $\mathbb{Z}_m \left(= \mathbb{Z}/_{(m)} \right)$ as a commutative ring R considered in §1 and define a homomorphism

$$\hat{\phi}_\xi = \phi_\xi \theta_{k+1} : F^* \to S(\mathbb{Z}_m^k) .$$

Proposition 5.2

For any relator $r_{i,j}$ *in the modified Wirtinger presentation given in* §4, $\hat{\phi}_\xi(r_{i,j}) = 1$. *Therefore,* $\hat{\phi}_\xi$ *defines a homomorphism from the link group* $G(L)$ *to* $S(\mathbb{Z}_m^k)$.

Proof. First consider a relator $r_{i,j}$ with $j < \nu_i$. This is of the form:

$$r_{i,j} = u_{i,j} x_{i,1} u_{i,j}^{-1} x_{i,j+1}^{-1} .$$

Then by Proposition 5.1, we have

$$\begin{aligned} \hat{\phi}_\xi(r_{i,j}) &= \phi_\xi(\theta_{k+1}(r_{i,j})) \\ &= \phi_\xi(\theta_{k+1}(u_{i,j} x_{i,1} u_{i,j}^{-1}) \theta_{k+1}(x_{i,j+1})^{-1}) \\ &= \phi_\xi(\theta_{k+2}(x_{i,j+1}) \theta_{k+1}(x_{i,j+1})^{-1}) \in \phi_\xi(F_{k+2}) . \end{aligned}$$

Since $\phi_\xi(F_{k+2}) = 1$ by Proposition 3.2, it follows that $\hat{\phi}_\xi(r_{i,j}) = 1$ for $j < \nu_i$.

Now consider the meridian-longitude relator

$$r_{i,\nu_i} = [\eta_i , x_{i,1}] , \quad i = 1,2,\dots,n .$$

Write

$$\phi_\xi (\theta_{k+1}[\eta_i , x_{i,1}])(a_1 ,\dots, a_k) = (b_1 ,\dots, b_k) .$$

We have to show that $b_q = a_q$.

Let $g = \theta_{k+1}[\eta_i , x_{i,1}]$. Since we have, from (3.2),

$$b_q - a_q = a_{q-1}\left(\frac{\partial g}{\partial x_{j_q}}\right)^0 + a_{q-2}\left(\frac{\partial^2 g}{\partial x_{j_{q-1}}\partial x_{j_q}}\right)^0 + \dots + \left(\frac{\partial}{\partial x_{j_1}\dots\partial x_{j_q}}\right)^0 ,$$

it is enough to show that

$$\left(\frac{\partial^{q-p} g}{\partial x_{j_{p+1}} \dots \partial x_{j_q}}\right)^0 = 0 \qquad \text{or}$$

$$\left(\frac{\partial^{q-p}\, \theta_{k+1}(\eta_i)\, x_i\, \theta_{k+1}(\eta_i)^{-1}\, x_i^{-1}}{\partial x_{j_{p+1}} \dots \partial x_{j_q}}\right) = 0 , \quad \text{for } 0 \le p \le q-1 . \tag{5.2}$$

Case 1

None of $j_{p+1} ,\dots, j_q$ is i.

Then

$$\left(\frac{\partial^{q-p}[\theta_{k+1}(\eta_i), x_i]}{\partial x_{j_{p+1}} \dots \partial x_{j_q}}\right)^0 = \sum_{a=0}^{q-p}\left(\frac{\partial^a\, \theta_{k+1}(\eta_i)}{\partial x_{j_{p+1}} \dots \partial x_{j_{p+a}}}\right)^0 \left(\frac{\partial^{q-p-a}\, \theta_{k+1}(\eta_i)^{-1}}{\partial x_{j_{p+a+1}} \dots \partial x_{j_q}}\right)^0$$

$$= \left(\frac{\partial^{q-p}\, \theta_{k+1}(\eta_i)\, \theta_{k+1}(\eta_i)^{-1}}{\partial x_{j_{p+1}} \dots \partial x_{j_q}}\right)^0 = 0 .$$

Case 2

Some j_λ is i, $p+1 \le \lambda \le q$.

Then we write

$$\left(\frac{\partial^{q-p}[\theta_{k+}(\eta_i), x_{i,1}]}{\partial x_{j_{p+1}} \cdots \partial x_{j_q}} \right)^0 = \sum_{a+b+c+d=q-p} A_a B_b C_c D_d \,, \tag{5.3}$$

where

$$A_a = \left(\frac{\partial^a \theta_{k+1}(\eta_i)}{\partial x_{j_{p+1}} \cdots \partial x_{j_{p+a}}} \right)^0 , \quad B_b = \left(\frac{\partial^b x_i}{\partial x_{j_{p+a+1}} \cdots \partial x_{j_{p+a+b}}} \right)^0 ,$$

$$C_c = \left(\frac{\partial^c \theta_{k+1}(\eta_i)^{-1}}{\partial x_{j_{p+a+b+1}} \cdots \partial x_{j_{p+a+b+c}}} \right)^0 \quad \text{and}$$

$$D_d = \left(\frac{\partial^d (x_i)^{-1}}{\partial x_{j_{p+a+b+c+1}} \cdots \partial x_{j_q}} \right)^0 , \quad (d = q-a-b-c-p) .$$

Note that b is either 0 or 1.

Now define four subsequences of $\alpha = \{j_{p+1}, \dots, j_q\}$ as follows:

$$\alpha_1 = \{j_{p+1}, \dots, j_{p+a}\}$$
$$\alpha_2 = \{j_{p+a+1}, \dots, j_{p+a+b}\}$$
$$\alpha_3 = \{j_{p+a+b+1}, \dots, j_{p+a+b+c}\}$$
$$\alpha_4 = \{j_{p+a+b+c+1}, \dots, j_q\} .$$

Since $\alpha = \{\alpha_1, \alpha_2, \alpha_3, \alpha_4\}$ and α involves i, at least one of these subsequences involves i. However, some α_i may be empty.

Case 2(a)

α_1 involves i.

Then an actual computation of C_c shows that C_c is a linear combination of $\mu(j_r \cdots j_s i)$, where $\{j_r, \dots, j_s\}$ is α_3 or its subsequence obtained from α_3 by eliminating some members. Therefore, $C_c \equiv 0 \mod m \left(= \bar{\Delta}(j_1 j_2 \cdots j_{k+2})\right)$ unless α_3 is empty. Suppose α_3 is empty. Then B_b and D_d are 0 unless all the integers in α_2 and α_4 are i.

Suppose $\alpha_2 = \{i, i, \dots i\}$ and $\alpha_4 = \{i, \dots, i\}$. Then $A_a = \mu(j_{p+1} \cdots j_{p+a} i) \equiv 0 \mod m$, unless $\alpha_2 = \alpha_4 = \phi$. Therefore, the only possible non-zero integer obtained in this case is

$$A_{q-p} = \mu(j_{p+1} \ldots j_q i) .$$

Case 2(b)

α_2 involves i.

If $\alpha_1 \neq \phi$, then $A_a = \mu(j_{p+1} \ldots j_{p+a} i) \equiv 0 \mod m$.

If $\alpha_1 = \phi$ and $\alpha_3 \neq \phi$, then C_c is a linear combination of $\mu(j_r \ldots j_s i)$, as was seen in Case 2(a), and hence $C_c \equiv 0$, mod m.

If $\alpha_1 = \alpha_3 = \phi$, then $D_d = 0$ unless $\alpha_4 = \{i, i, \ldots, i\}$. Therefore, in this case, only non-zero integer appears when $\alpha_2 = \{i\}$ and $\alpha_4 = \{\underbrace{i, i, \ldots, i}_{q-p-1 \text{ times}}\}$, and then $B_b D_d = (-1)^{q-p-1}$.

Case 2(c)

$\alpha_2 = \phi$ and α_3 involves i.

If $\alpha_1 \neq \phi$, then $A_a = \mu(j_{p+1} \ldots j_{p+a} i) = 0 \mod m$.

Suppose $\alpha_1 = \alpha_2 = \phi$. Then $D_d = 0$ unless $\alpha_4 = \phi$ or $\alpha_4 = \{i i \ldots i\}$. If $\alpha_4 = \{i, i, \ldots, i\}$ $(\neq \phi)$, then again C_c is a linear combination of $\mu(j_r \ldots j_s i)$ and hence $C_c \equiv 0 \mod m$. Therefore, $C_c \neq 0$ only when $\alpha_1 = \alpha_2 = \alpha_4 = \phi$ and then

$$C_{q-p} = \left(\frac{\partial^{q-p} \theta_{k+} (\eta_i)^{-1}}{\partial x_{j_{p+1}} \ldots \partial x_{j_q}} \right)^0 .$$

Case 2(d)

α_4 involves i and $\alpha_2 = \phi$.

Only non-zero integer appears when $\alpha_1 = \alpha_3 = \phi$. Then $\alpha_4 = \{i i \ldots i\}$, and hence, $D_d = (-1)^{q-p}$.

Now combining all the cases, we obtain

$$\sum A_a B_b C_c D_d \equiv \left(\frac{\partial^{q-p} \theta_{k+1} (\eta_i)}{\partial x_{j_{p+1}} \ldots \partial x_{j_q}} \right)^0$$

$$+ (-1)^{q-p-1} + \left\{ \frac{\partial^{q-p} \theta_{k+1} (\eta_i)^{-1}}{\partial x_{j_{p+1}} \cdots \partial x_{j_q}} \right\}^0 + (-1)^{q-p}$$

$$\equiv \left\{ \frac{\partial^{q-p} \theta_{k+1} (\eta_i) \theta_{k+1} (\eta_i)^{-1}}{\partial x_{j_{p+1}} \cdots \partial x_{j_q}} \right\}^0$$

$$\equiv 0 \mod m .$$

The second to the last congruence follows from the fact that the sequence $\{ j_{p+1}, \ldots, j_q \}$ involves i.

This proves Proposition 5.2 .

6 *LINKING FUNCTIONS*

Proposition 5.2 shows that given a sequence $\hat{\xi} = \{ j_1, j_2, \ldots, j_{k+2} \}$, we can define a nilpotent representation $\hat{\phi}_\xi$ of the link group $G(L)$ on $S(\mathbb{Z}_m^k)$ associated with a sub-sequence $\xi = \{ j_1, j_2, \ldots, j_k \}$ of $\hat{\xi}$.

Now consider the covering space M_ξ of $S^3 - L$ associated with $\hat{\phi}_\xi$. To study the linking number between two components of the lift $\tilde{L}$ of L in M_ξ , first we define an appropriate linking function for $G(L)$. (For the definition, see Hartley & Murasugi, 1977).

Let Ω be the orbit of $(0, \ldots, 0) \in \mathbb{Z}_m^k$ under $\phi_\xi(F)$, see § 3 .

For $(a_1, \ldots, a_k) \in \Omega$, let $\mathcal{D}_{\xi,(a_1,\ldots,a_k)}$ denote the Reidemeister-Schreier rewriting function of F . Precisely speaking, $\mathcal{D}_{\xi,(a_1,\ldots,a_k)}$ is a function from F to the free group generated freely by $\{ x_{i,(a_1 \ldots a_k)} \mid i = 1,\ldots,n , (a_1, \ldots, a_k) \in \Omega \}$ such that $\mathcal{D}_{\xi,(a_1 \ldots a_k)}$ is characterized by the following two properties

$$(1) \quad \mathcal{D}_{\xi,(a_1,\ldots,a_k)} x_i = x_{i,(a_1,\ldots,a_k)} \tag{6.1}$$

(2) For $u, v \in F$,

$$\mathcal{D}_{\xi,(a_1,\dots,a_k)}(uv) = \mathcal{D}_{\xi,(a_1,\dots,a_k)}(u)\mathcal{D}_{\xi,\phi_\xi(u)(a_1,\dots,a_k)}(v) .$$

Similarly, let $\hat{\mathcal{D}}_{\xi,(a_1,\dots,a_k)}$ be the Reidemeister-Schreier rewritting function from F^* to the free group freely generated by $\{ x_{i,j,(a_1,\dots,a_k)} \}$.

Furthermore, let ζ_i be a homomorphism

$$\zeta_i : \mathcal{D}_{\xi,(a_1,\dots,a_k)}(F) \to \mathbb{Z}_m \quad \text{(additive group)}$$

defined by

$$\zeta_i (x_{j,(a_1,\dots,a_k)}) = \delta_{i,j}\, a_k .$$

Using ζ_i and $\hat{\mathcal{D}}$, we define a new homomorphism

$$\hat{\zeta}_i : \hat{\mathcal{D}}_{\xi,(a_1,\dots,a_k)}(F^*) \to \mathbb{Z}_m \quad \text{(additive group)}$$

as follows:

$$\hat{\zeta}_i (x_{q,1,(a_1,\dots,a_k)}) = a_k\, \delta_{i,q}$$

$$\hat{\zeta}_i (x_{q,j+1,(a_1,\dots,a_k)}) = \zeta_i\, \mathcal{D}_{\xi,(a_1,\dots,a_k)}(\theta_{k+1}(x_{q,j+1})) ,$$

for $j \geq 0$.

The main purpose of this section is to show that

$$\hat{\zeta}_i (\hat{\mathcal{D}}_{\xi,(a_1,\dots,a_k)}(r_{p,q})) = 0 ,$$

that is, $\hat{\zeta}_i$ is, in fact, a linking homomorphism (Hartley & Murasugi, 1977).

To prove this, first we prove

Proposition 6.1

Let β_r *be the projection from* $\mathbb{Z}_m^r$ *to* $\mathbb{Z}_m$ *defined by* $\beta_r(a_1,\dots, a_r) = a_r$. *Then for any element* g *in* F ,

$$\zeta_i(\mathcal{D}_{\xi,(a_1,\ldots,a_k)}(g)) = \beta_{k+1}(\phi_{\{\xi,i\}}(g)(a_1,\ldots,a_k,0)), \quad k \geq 1, \tag{6.2}$$

where $\phi_{\{\xi,i\}}$ *is an action of* F *on* $\mathbb{Z}_m^{k+1}$ *associated with the extended sequence* $\{\xi, i\} = \{j_1, j_2, \ldots, j_k, i\}$.

Proof. By (3.2), (6.2) is equivalent to

$$\zeta_i(\mathcal{D}_{\xi,(a_1,\ldots,a_k)}(g))$$

$$= a_k \left(\frac{\partial g}{\partial x_i}\right)^0 + a_{k-1}\left(\frac{\partial^2 g}{\partial x_{j_k}\partial x_i}\right)^0 + \ldots + \left(\frac{\partial^{k+1} g}{\partial x_{j_1}\ldots\partial x_{j_k}\partial x_i}\right)^0 \tag{6.3}$$

and hence, we will prove (6.3) by induction on $|g|$, the length of g .

Suppose $|g| = 1$, i.e. $g = x_q$ or x_q^{-1} .

For $g = x_q$,

$$\zeta_i \mathcal{D}_{\xi,(a_1,\ldots,a_k)}(x_q) = \zeta_i\, x_{q,(a_1,\ldots,a_k)}$$

$$= \zeta_{q,i}\, a_k = a_k \left(\frac{\partial x_q}{\partial x_i}\right)^0 .$$

For $g = x_q^{-1}$,

$$\mathcal{D}_{\xi,(a_1,\ldots,a_k)}(x_q^{-1}) = x_{q}^{-1}{}_{,\phi_\xi(x_q^{-1})(a_1,\ldots,a_k)} = x^{-1}_{q,(b_1,\ldots,b_k)},$$

and hence

$$\zeta_i \mathcal{D}_{\xi,(a_1,\ldots,a_k)}(x_q^{-1}) = -\delta_{i,q}\, b_k ,$$

where b_k is, by (3.2),

$$b_k = a_k + a_{k-1}\left(\frac{\partial x_q^{-1}}{\partial x_{j_k}}\right)^0 + a_{k-2}\left(\frac{\partial^2 x_q^{-1}}{\partial x_{j_{k-1}}\partial x_{j_k}}\right)^0 + \dots + \left(\frac{\partial^k x_q^{-1}}{\partial x_{j_1}\dots\partial x_{j_{k-1}}\partial x_{j_k}}\right)^0$$

$$= a_k - \delta_{j_k,q}a_{k-1} + \delta_{j_{k-1},q}\delta_{j_k,q}a_{k-2} + \dots + (-1)^k\delta_{j_1,q}\delta_{j_2,q}\dots\delta_{j_k,q} .$$

Since

$$(-1)^{\lambda+1}\,\delta_{j_k-\lambda+1,q}\dots\delta_{j_k,q}\,\delta_{i,q} = \left(\frac{\partial^{\lambda+1} x_q^{-1}}{\partial x_{j_k-\lambda+1}\dots\partial x_{j_k}\,\partial x_i}\right)^0 ,$$

we have

$$-\delta_{i,q}b_k = a_k\left(\frac{\partial x_q^{-1}}{\partial x_i}\right)^0 + a_{k-1}\left(\frac{\partial^2 x_q^{-1}}{\partial x_{j_k}\partial x_i}\right)^0 + \dots + \left(\frac{\partial^{k+1} x_q^{-1}}{\partial x_{j_1}\dots\partial x_{j_k}\partial x_i}\right)^0 .$$

This proves the first step on the induction.

Now we assume inductively that (6.3) holds for g with $|g| < \ell$ and for all k-tuples $(a_1,\dots,a_k) \in \Omega$.

Then using induction assumption and (3.2), we have

$$\zeta_i\,\mathcal{D}_{\xi,(a_1,\dots,a_k)}(gx_q)$$

$$= \zeta_i\,(\mathcal{D}_{\xi,(a_1,\dots,a_k)}(g)\cdot\mathcal{D}_{\xi,\phi_\zeta(g)(a_1,\dots,a_k)}(x_q))$$

$$= \beta_{k+1}(\phi_{(\xi,i)}(g)(a_1,\dots,a_k,0) + \zeta_i(x_{q,\phi_\xi(g)(a_1,\dots,a_k)})$$

$$= a_k\left(\frac{\partial g}{\partial x_i}\right)^0 + a_{k-1}\left(\frac{\partial^2 g}{\partial x_{j_k}\partial x_i}\right)^0 + \dots + \left(\frac{\partial^{k+1} g}{\partial x_{j_1}\dots\partial x_{j_k}\partial x_i}\right)^0 +$$

$$+ \delta_{i,q}\,\beta_k\,\phi_\xi(g)(a_1,\dots,a_k)$$

$$= \sum_{p=0}^{k} a_{k-p}\left(\frac{\partial^{p+1} g}{\partial x_{j_{k-p+1}}\dots\partial x_{j_k}\partial x_i}\right)^0 + \delta_{i,q}\sum_{p=0}^{k} a_{k-p}\left(\frac{\partial^p g}{x_{j_{k-p+1}}\dots x_{j_k}}\right)^0$$

$$= \sum_{p=0}^{k} a_{k-p} \left\{ \left(\frac{\partial^{p+1} g}{\partial x_{j_{k-p+1}} \cdots \partial x_{j_k} \partial x_i} \right)^0 + \left(\frac{\partial^p g}{\partial x_{j_{k-p+1}} \cdots x_{j_k}} \right)^0 \delta_{i,q} \right\}$$

$$= \sum_{p=0}^{k} a_{k-p} \left(\frac{\partial^{p+1} (gx)_q}{\partial x_{j_{k-p+1}} \cdots \partial x_{j_k} \partial x_i} \right)^0 .$$

This proves (6.3) for gx_q.

A proof of (6.3) for gx_q^{-1} needs a slightly more complicated computation than that for gx_q. However, one can apply the same technique that was used in the proof of (3.2), and hence the details will be omitted.

By using Proposition 6.1, we will be able to prove the following.

Proposition 6.2

For any relator $r_{p,q}$ *in the modified Wirtinger presentation of* G(L), *we have*

$$\hat{\zeta}_i \hat{\mathcal{D}}_{\xi,(a_1,\ldots,a_k)} (r_{p,q}) = 0 ,$$

and hence, $\hat{\zeta}_i$ *is a linking homomorphism.*

Proof. First consider a relator

$$r_{p,q} = u_{p,q} x_{p,1} u_{p,q}^{-1} x_{p,q+1}^{-1} , \quad 1 \le q < \nu_p .$$

From Proposition 6.1, it follows that

$$\hat{\zeta}_i \hat{\mathcal{D}}_{\xi,(a_1,\ldots,a_k)} (r_{p,q}) = \zeta_i \mathcal{D}_{\xi,(a_1,\ldots,a_k)} (\theta_{k+1}(r_{p,q}))$$

$$= \zeta_i \mathcal{D}_{\xi,(a_1,\ldots,a_k)} (\theta_{k+1}(u_{p,q} x_p, u_{p,q}^{-1}) \theta_{k+1}(x_{p,q+1})^{-1})$$

$$= \zeta_i \mathcal{D}_{\xi,(a_1,\ldots,a_k)} (\theta_{k+1}(x_{p,q+1}) \theta_{k+1}(x_{p,q+1})^{-1})$$

$$= \beta_{k+1} \phi_{\{\xi,i\}} (\theta_{k+2}(x_{p,q+1}) \theta_{k+1}(x_{p,q+1})^{-1}) (a_1,\ldots,a_k, 0) .$$

But, since $g = \theta_{k+2}(x_{p,q+1})\,\theta_{k+1}(x_{p,q+1})^{-1} \in F_{k+2}$ by Proposition 5.1, it follows from Proposition 3.1 that

$$\phi_{\{\xi,i\}}(g)(a_1,\dots,a_k,0) = (a_1,\dots,a_k,0)$$

and hence

$$\beta_{k+1}\,\phi_{\{\xi,i\}}(g)(a_1,\dots,a_k,0) = 0 .$$

Next, for a relator $r_{p,\nu_p} = [\,\eta_p,\ x_{p,1}\,]$, we have

$$\hat{\zeta}_i\,\hat{\mathcal{D}}_{\xi,(a_1,\dots,a_k)}(r_{p,\nu_p}) = \zeta_i\,\mathcal{D}_{\xi,(a_1,\dots,a_k)}[\,\theta_{k+1}(\eta_p),x_p\,]$$

$$= \beta_{k+1}\,\phi_{\{\xi,i\}}[\,\theta_{k+1}(\eta_p),x_p\,](a_1,\dots,a_k,0)$$

$$= \sum_{r=0}^{k} a_{k-r}\left(\frac{\partial^{r+1}[\theta_{k+1}(\eta_p),x_p\,]}{\partial x_{j_{k-r}}\cdots\partial x_{j_k}\,\partial x_i}\right)^0 .$$

Therefore, we must show that

$$\left(\frac{\partial^{r+1}[\theta_{k+1}(\eta_p),x_p\,]}{\partial x_{j_{k-r}}\cdots\partial x_{j_k}\,\partial x_i}\right)^0 = 0\,,\quad \text{for } r = 0,\dots,k .$$

However, since the proof is similar to that given in Proposition 5.2, it will be omitted.

Proposition 6.3

For any sequence $\xi = \{j_1,\dots,j_k\}$,

$$\phi_\xi(\theta_{k+1}(\eta_p)) = 1 ,$$

where ϕ_ξ *is an action of* F *on* $S(\mathbb{Z}_m^{\ k})$, *and*

$$m = \overline{\Delta}(j_1 j_2 \cdots j_k\, j_{k+1}\, j_{k+2})\,,\quad p = j_{k+1} \text{ or } j_{k+2} .$$

Proof. By Proposition 3.1, it suffices to show that for r, s, $0 \le s \le k-1$, $1 \le r \le k$, $1 \le r+s \le k$,

$$\left(\frac{\partial^r \theta_{k+1}(\eta_p)}{\partial x_{j_{s+1}} \partial x_{j_{s+2}} \cdots \partial x_{j_{s+r}}}\right)^0 \equiv 0 \mod m .$$

This is immediate, since, by definition and (5.1)

$$\left(\frac{\partial^r \theta_{k+1}(\eta_p)}{\partial x_{j_{s+1}} \cdots \partial x_{j_{s+r}}}\right)^0 = \mu(j_{s+1} j_{s+2} \cdots j_{s+r} p) \equiv 0 \mod m .$$

Note that j_{k+1} or j_{k+2} is p.

Remark 6.1

Proposition 6.3 implies that the lift of a longitude of K_p in the covering space M_ξ is a simple closed curve. In other words, the covering index of the lift of each component of L is one.

7 *MAIN THEOREM*

Let $\hat{\xi} = \{j_1, j_2, \ldots, j_{k+2}\}$ be a sequence and $\xi = \{j_1, j_2, \ldots, j_k\}$ a subsequence. Let M_ξ be the covering space of $S^3 - L$ associated with the transitive nilpotent representation $\hat{\phi}_\xi : \pi_1(S^3 - L) \to \Omega \subset S(\mathbb{Z}_m^{\ k})$, where $m = \overline{\Delta}(j_1 j_2 \cdots j_{k+2})$.

To be more precise, consider an oriented 3-manifold $X = \overline{S^3 - N(L)}$, where $N(L)$ is a tubular neighbourhood of L in S^3. Let $\overline{M}_\xi$ be the covering space of X associated with $\hat{\phi}_\xi$.

Assume $m \ne 0$.

Let $\{\mathcal{O}_0^{(p)}, \mathcal{O}_1^{(p)}, \ldots, \mathcal{O}_\lambda^{(p)}\}$ be the set of all orbits in Ω under the action $\hat{\phi}_\xi(x_{p,1})$ and $\hat{\phi}_\xi(\eta_p)$. It is known (Hartley & Murasugi, 1977), that each orbit $\mathcal{O}_i^{(p)}$ corresponds to one and only one boundary torus of $\overline{M}_\xi$ that covers the boundary of $N(K_p)$.

We say that $O_i^{(p)}$ corresponds to a covering knot $K_{p,i}$.

Now to each orbit $O_i^{(j_{k+1})}$, define

$$q_i = \sum \beta_k \ (b_1, \dots, b_k) \ ,$$

where the summation runs over all $(b_1, \dots, b_k)$ in $O_i^{(j_{k+1})}$, and put

$$\alpha = \sum_{i=1}^{\lambda} q_i \tilde{K}_{j_{k+1},i} \in H_1(\tilde{L} ; \mathbb{Z}) , \qquad (7.1)$$

where $\tilde{L}$ denotes the lift of L in the branched covering space $\hat{M}_\xi$ of L . α will be called the *characteristic link* of L associated with $\hat{\phi}_\xi$.

It is shown in Hartley & Murasugi (1977) that if $i_* : H_1(L ; \mathbb{Z}) \to H_1(\hat{M}_\xi ; \mathbb{Z})$ is the inclusion homomorphism, then $i_*(\alpha) = 0$, and hence α bounds a 2-chain B in $\hat{M}_\xi$.

Now we have the main theorem of this paper.

Theorem 7.1

Given a sequence $\hat{\xi} = \{ j_1, j_2, \dots, j_k, j_{k+1}, j_{k+2} \}$, *let* $\tilde{K}_{j_{k+2},0}$ *be the knot corresponding to the orbit* $O_0^{(j_{k+2})}$ *that contains* $(0, 0, \dots, 0)$.

Assume that $m \neq 0$. *Then*

$$\bar{\mu}(j_1 j_2 \dots j_{k+2}) \equiv \text{int}(\tilde{K}_{j_{k+2},0}, B) \mod m ,$$

where int(,) *stands for the usual intersection number in* $\hat{M}_\Xi$. *Therefore, if* $\tilde{K}_{j_{k+2},2} \sim 0$ *in* M_ξ , *then*

$$\bar{\mu}(j_1 j_2 \dots j_{k+2}) \equiv \ell k(\tilde{K}_{j_{k+2},0}, \alpha) \mod m .$$

Remark 7.1

If $m = \overline{\Delta}(j_1 j_2 \dots j_{k+2}) = 0$, then consider a finite nilpotent representation $\hat{\phi}_\xi : \pi_1(S^3 - L) \to S(\mathbb{Z}_q^k)$ for a sufficiently large $q > 0$. Then "modulo q" will be interpreted as equality.

Remark 7.2

When $j_{k+1} = j_{k+2}$, take a parallel knot $\tilde{K}$ to $\tilde{K}_{j_{k+1}}$, in M_ξ, if necessary, to that $\tilde{K}$ may have only finitely many intersecting points with B. Then we have the same formula with $\mathrm{int}(\tilde{K}, B)$ instead of $\mathrm{int}(\tilde{K}_{j_{k+2},0}, B)$. However, since $\overline{\mu}$ is not changed by a cyclic permutation of the sequence, we may assume, in practice, that $j_{k+1} \neq j_{k+2}$.

Proof of Theorem 7.1 The theorem is a consequence of Corollary 7.3 in Hartley & Murasugi (1977). First, we know that $\hat{\phi}_\xi(\eta_p) = 1$ for $p = j_{k+1}$ or j_{k+2}, by Proposition 6.3. Further, $\hat{\zeta}_{j_{k+1}}$ defined in §6 is a linking homomorphism and $\hat{\zeta}_{j_{k+1}}(x_{j_{k+1}}, (a_1, \dots, a_k)) = a_k$. Therefore, α defined in Hartley & Murasugi (1977), Theorem 7.2 is exactly α defined by (7.1).

Now Proposition 1.1, 7.1 and Corollary 7.3 in Hartley & Murasugi (1977) imply that

$$\mathrm{int}(\tilde{K}_{j_{k+2},0}, B) \equiv \hat{\zeta}_{j_{k+1}} \mathcal{D}_{\xi,(0,\dots,0)}(\eta_{j_{k+2}}) \bmod m. \qquad (7.2)$$

Then (6.2) and Corollary 3.1 show

$$\hat{\zeta}_{j_{k+1}} \hat{\mathcal{D}}_{\xi,(0,\dots,0)}(\eta_{j_{k+2}})$$

$$\equiv \zeta_{j_{k+1}} \hat{\mathcal{D}}_{\xi,(0,\dots,0)} \theta_{k+1}(\eta_{j_{k+2}})$$

$$\equiv \beta_{k+1} \phi_{(j_1 j_2 \dots j_k j_{k+1})} \theta_{k+1}(\eta_{j_{k+2}})(0, \dots, 0)$$

$$\equiv \left(\frac{\partial^{k+1} \theta_{k+1}(\eta_{j_{k+2}})}{\partial x_{j_1} \partial x_{j_2} \cdots \partial x_{j_{k+1}}} \right)^0$$

$$\equiv \mu(j_1 j_2 \cdots j_{k+1} j_{k+2}) \mod m .$$

This proves Theorem 7.1 .

The following corollary is immediate.

Corollary 7.1
If L *is a boundary link in* S^3 , *then all* $\overline{\mu}$ - *invariants vanish.*

8 EXAMPLES

In this section, we will compute $\overline{\mu}$ for some links.

Since $\overline{\mu}(i\,j)$ is simply $\ell k(K_i, K_j)$, the first non-trivial invariant is $\overline{\mu}(i\,j\,k)$.

Let $\hat{\xi} = \{i, j, k\}$ and $\xi = \{i\}$. Let $m = \Delta(i\,j\,k)$. (If $m = 0$, then assume that m is a sufficiently large integer.) Then ϕ_ξ is a cyclic representation of G(L), namely, ϕ_ξ is a homomorphism from G(L) onto $S(\mathbb{Z}_m)$ such that, for $a \in \mathbb{Z}_m$, $\phi_\xi(x_{i,1})(a) = a+1$ and $\phi_\xi(x_{q,1})(a) = a$, for $q \neq i$. Therefore, the branched covering space $\hat{M}_\xi$ associated with ϕ_ξ is exactly the m-fold cyclic covering space of S^3 branched along K_i . Suppose $i \neq j$ and $i \neq k$. Since $\overline{\mu}(i\,j) = \ell k(K_i, K_j)$ and $\overline{\mu}(i, k) = \ell k(K_i, K_k)$ are divisible by m , it follows that each of K_j and K_k is lifted to exactly m knots in $\hat{M}_\xi$. We call them $\tilde{K}_{j,0}, \tilde{K}_{j,1}, \ldots, \tilde{K}_{j,m-1}$ and $\tilde{K}_{k,0}, \tilde{K}_{k,1}, \ldots, \tilde{K}_{k,m-1}$, where a generator of the covering transformation group maps $\tilde{K}_{j,g}$ to $\tilde{K}_{j,q+1}$ and $\tilde{K}_{k,q}$ to $\tilde{K}_{k,q+1}$ $0 \le q \le m-1$.

Note that each knot $\tilde{K}_{j,q}$ corresponds to the orbit $\{q\}$ in $\Omega(=\mathbb{Z}_m)$ under the action $\phi_\xi(x_{j,1})$ and $\phi_\xi(\eta_j)(=1)$. Therefore, if $j \neq i$, then the characteristic link α is

$$\alpha = \sum_{q=0}^{m-1} q \tilde{K}_{j,q} .$$

Theorem 7.1, now, shows that if $\tilde{K}_{k,0} \sim 0$ in $\hat{M}_\xi$, then

$$\bar{\mu}(i j k) \equiv \ell k(\tilde{K}_{k,0}, \alpha) \mod m .$$

If $i = j$ but $i \neq k$, then the characteristic link is

$$\alpha = \sum_{q=0}^{m-1} q K_{i,0} = \binom{m}{2} K_{i,0} .$$

Therefore, $\bar{\mu}(i i k) \equiv \ell k(\tilde{K}_{k,0}, \binom{m}{2} K_{i,0}) \mod m$. Since $m = \bar{\Delta}(i i k) = \bar{\mu}(i k)$ and $\ell k(\tilde{K}_{k,0}, \tilde{K}_{i,0}) = \pm 1$, we have

$$\bar{\mu}(i i k) \equiv \binom{m}{2} \mod m .$$

This is a well-known result. (Murasugi, (1966), Remark p.100 .)

The next case is $\bar{\mu}(i i j k)$. Here, $\hat{\xi} = \{i i j k\}$ and $\xi = \{i i\}$, and ϕ_ξ is again a cyclic representation $\phi_\xi : G(L) \to S(\mathbb{Z}_m \times \mathbb{Z}_m)$ such that

$$\phi_\xi(x_{i,1})(a_1, a_2) = (a_1 + 1, a_2 + a_1) \quad \text{and}$$

$$\phi_\xi(x_{q,1})(a_1, a_2) = (a_1, a_2), \quad q \neq i ,$$

where $m = \bar{\Delta}(i i j k)$. Therefore $\hat{M}_\xi$ is the cyclic covering space of S^3 branched along K_i (if $m \neq 0$). The order d of this covering depends on m. For example, $d = 4$ when $m = 2$, and $d = 3$ when $m = 3$. Even if $\hat{M}_\xi = \hat{M}_{\{i,i\}}$ is the same cyclic covering space as $\hat{M}_{\{i\}}$, $\bar{\mu}(i i j k)$ may be different from $\bar{\mu}(i j k)$. Because, the characteristic links for these cases are usually different. (See Examples 1, 2).

The first interesting case is $\bar{\mu}(i, j, k, \ell)$, $i \neq j$, where $\hat{\xi} = \{i, j, k, \ell\}$ and $\xi = \{i, j\}$. Then ϕ_ξ is an irregular nilpotent representation of $G(L)$. This case will be briefly discussed in § 10 .

Now we compute $\bar{\mu}$ for some links.

Example 1

Torus link of type (6.2).

Figure 1

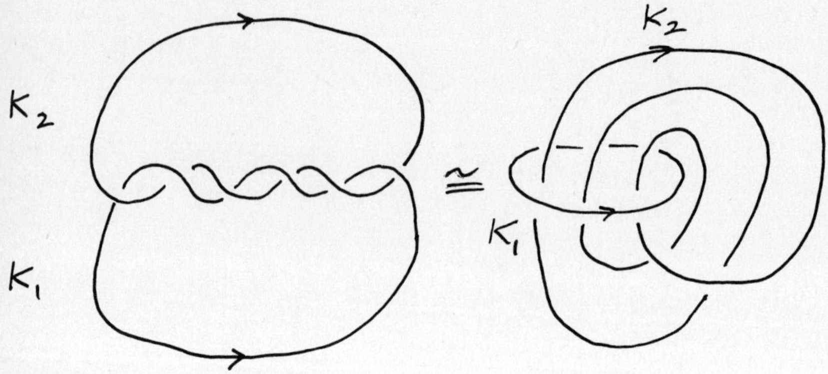

The lift of L in the 3-fold cyclic covering space along K_1 is :

Figure 2

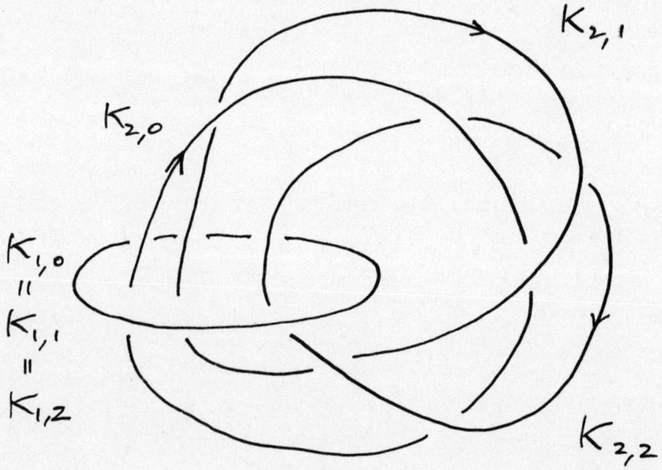

$\bar{\mu}(1\,2) = 3$

$\bar{\mu}(1\,1\,2) \equiv \ell k(K_{2,1}, K_{1,1} + 2K_{1,2}) = 3 \equiv 0 \pmod 3$

$\bar{\mu}(1\,2\,1) \equiv \ell k(K_{1,0}, K_{2,1} + 2K_{2,2}) = 3 \equiv 0 \pmod 3$

$\bar{\mu}(1\,2\,2) \equiv \ell k(K_{2,0}, K_{2,1} + 2K_{2,2}) = 3 \equiv 0 \pmod 3$

Computation of $\bar{\mu}(1\,1\,2\,2)$.

The lift of L in the 3-fold cyclic covering space along K_1 associated with $\phi_{\{1,1\}}$ is exactly the same links given in Figure 2, but components will be indexed differently.

$K_{i,0}$, $K_{i,1}$, $K_{i,2}$ are indexed respectively, by

$$K_{i,(0,0)},\quad K_{i,(1,0)},\quad K_{i,(2,1)},\quad i = 1, 2.$$

Therefore, the characteristic link α is

$$\alpha = 0 \cdot K_{2,(0,0)} + 0 \cdot K_{2,(1,0)} + 1 \cdot K_{2,(2,1)} = K_{2,(2,1)}$$

and hence

$$\bar{\mu}(1\,1\,2\,2) \equiv \ell k(K_{2,(0,0)}, K_{2,(2,1)}) \equiv 1 \mod 3.$$

Example 2

Whitehead link.

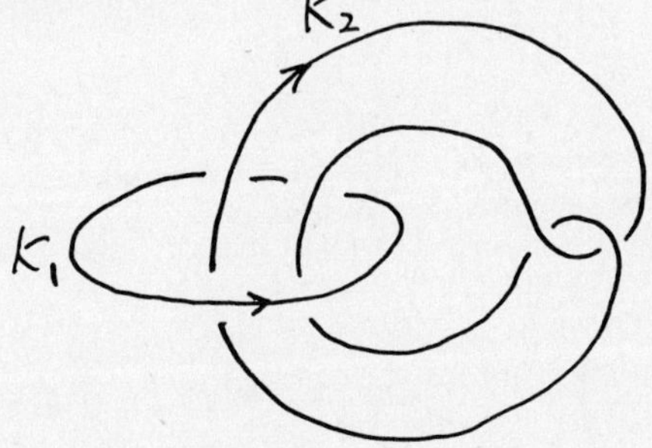

The lift of L in the m-fold cyclic covering space branched along K_1 (with sufficiently large m) :

$$\vdots$$
$$\|$$
$$(K_{1,(2,1)} =)\ K_{1,2}$$
$$\|$$
$$(K_{1,(1,0)} =)\ K_{1,1}$$
$$\|$$
$$(K_{1,(0,0)} =)\ K_{1,0}$$
$$\|$$
$$(K_{1,(-1,1)} =)\ K_{1,-1}$$
$$\|$$
$$(K_{1,(-2,3)} =)\ K_{1,-2}$$
$$\|$$
$$\vdots$$

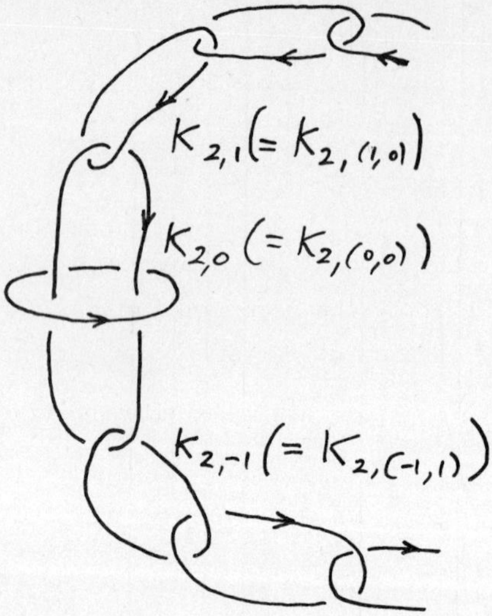

$$\bar{\mu}(1\,1\,2) \equiv \ell k\,(K_{2,0}, \sum_{i=-m+1}^{m-1} i\,K_{1,i}) = 0 .$$

$$\bar{\mu}(1\,2\,2) \equiv \ell k\,(K_{2,0}, \sum_{i=-m+1}^{m-1} i\,K_{2,i}) = \ell k\,(K_{2,0}, K_{2,1} - K_{2,-1}) = 0 .$$

$$\bar{\mu}(1\,1\,2\,2) \equiv \ell k\,(K_{2,(0,0)}, \sum_{j} j\,K_{2,(i,j)})$$
$$= \ell k\,(K_{2,(0,0)}, K_{2,(-1,1)}) = 1 .$$

Example 3

Borromean Ring

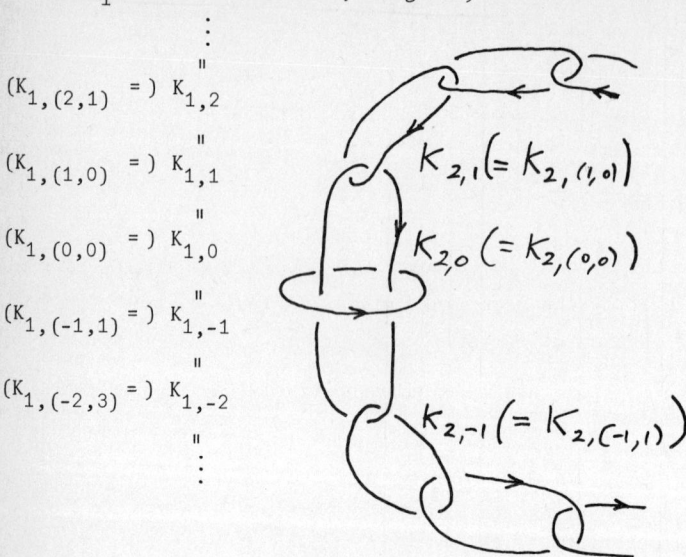

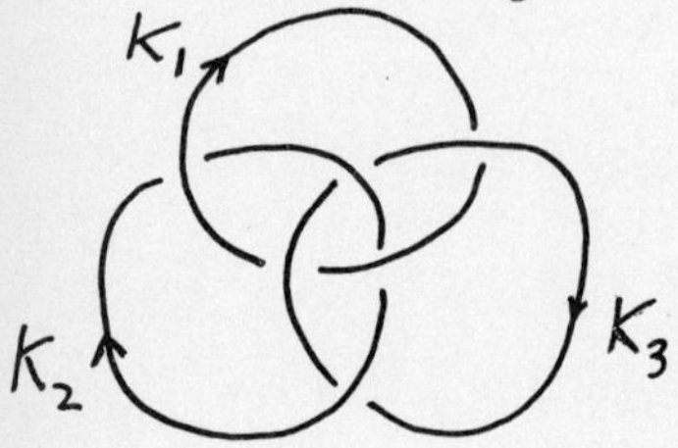

Computation of $\bar{\mu}(1\,2\,3)$.

Since $\bar{\mu}(1\,2) = \bar{\mu}(1\,3) = \bar{\mu}(2\,3) = 0$,

$\bar{\Delta}(1\,2\,3) = 0$.

The lift of the m-fold cyclic covering space branched along K_1 (with a sufficiently large m) :

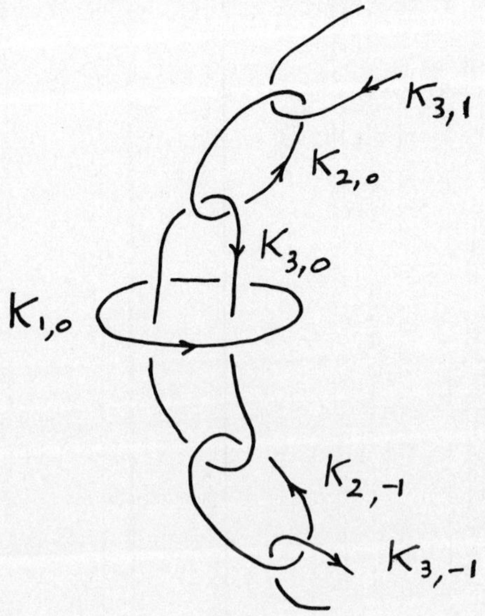

The characteristic link $\alpha = \sum_{i=-m+1}^{m-1} i\, K_{2,i}$

and hence

$$\bar{\mu}(1\,2\,3) \equiv \ell k\,(K_{3,0}, \alpha) = \ell k\,(K_{3,0} - K_{2,-1}) = 1 .$$

Similarly, $\bar{\mu}(1\,3\,2) \equiv \ell k\,(K_{2,0}, \sum_{i=-m+1}^{m-1} i\, K_{3,i}) = \ell k\,(K_{2,0}, K_{3,1}) = -1 .$

Also, $\bar{\mu}(1\,2\,2) \equiv (K_{2,0}, \sum_{i=-m+1}^{m-1} i\, K_{2,i}) = 0$, and,

similarly, $\bar{\mu}(1\,3\,3) = 0$.

9 *SPECIAL NILPOTENT COVERINGS*

In this section, we consider a special type of sequence ξ and the associated coverings of a 2-components link L.

Let ξ be a sequence of the type:

$$\xi = \{\underbrace{1, 1, \ldots, 1}_{r \text{ times}}, \underbrace{2, 2, \ldots, 2}_{s \text{ times}}\}, \quad r, s > 0,$$

which will be denoted by (r, s).

Since L has only two components, there are only two types of $\bar{\mu}$-invariants:

$$\bar{\mu}(1\,1 \ldots 1\,2\,2 \ldots 2\,2\,2) = \bar{\mu}(\xi, 2, 2)$$

or

$$\bar{\mu}(1\,1 \ldots 1\,2\,2 \ldots 2\,1\,2) = \bar{\mu}(\xi, 1, 2).$$

Now, the homomorphism ϕ_ξ associated with ξ does kill not only F_{r+s+1}, but also F'', the second commutator subgroup of F. In fact, we have

Proposition 9.1

Let $\xi = \{1, 1, \ldots, 1, 2, \ldots, 2\} = (r, s)$, $r, s \geq 1$. *Then* $\phi_\xi(F'') = 1$.

Proof. It is sufficient to show that $\phi_\xi[u, v] = 1$ for $u, v \in F'$. Let $g = [u, v]$. Write

$$\phi_\xi(g)(a_1, \ldots, a_r, b_1, \ldots, b_s) = (a_1', \ldots, a_r', b_1', \ldots, b_s').$$

Then

$$a_q' = a_q + a_{q-1}\left(\frac{\partial g}{\partial x}\right)^0 + a_{q-2}\left(\frac{\partial^2 g}{\partial x^2}\right)^0 + \ldots + a_1\left(\frac{\partial^{q-1} g}{\partial x^{q-1}}\right)^0 + \left(\frac{\partial^q g}{\partial x^q}\right)^0,$$

$$1 \leq q \leq r.$$

Since for any $w \in F$ and $\lambda \geq 1$,

$$\left(\frac{\partial^\lambda \omega}{\partial x^\lambda}\right)^0 = \binom{\left(\frac{\partial \omega}{\partial x}\right)^0}{\lambda}, \quad [3, (3.9)],$$

it follows that $\left(\frac{\partial g}{\partial x}\right)^0 = 0$ implies $\left(\frac{\partial^\lambda g}{\partial x^\lambda}\right)^0 = 0$ for $\lambda \geq 1$.

Therefore, $a_q' = a_q$ for $1 \leq q \leq r$.

Next, consider b_q'. Proposition 3.1 shows that

$$b_q' = b_q + b_{q-1}\left(\frac{\partial g}{\partial y}\right)^0 + b_{q-2}\left(\frac{\partial^2 q}{\partial y^2}\right)^0 + \ldots + b_1\left(\frac{\partial^{q-1} g}{\partial y^{q-1}}\right)^0$$

$$+ a_r\left(\frac{\partial^q g}{\partial y^q}\right)^0 + a_{r-1}\left(\frac{\partial^{q+1} g}{\partial x \partial y^q}\right)^0 + \ldots + a_1\left(\frac{\partial^{q+r-1} g}{\partial x^{r-1} \partial y^q}\right)^0$$

$$+ \left(\frac{\partial^{q+r} g}{\partial x^r \partial y^q}\right)^0 .$$

Again, $\left(\frac{\partial^\lambda g}{\partial y^\lambda}\right)^0 = 0$ for $\lambda \geq 1$, since $\left(\frac{\partial g}{\partial y}\right)^0 = 1$.

To show that $\left(\frac{\partial^{p+q} g}{\partial x^p \partial y^q}\right)^0 = 0$, $p \geq 1$, we write

$$\left(\frac{\partial^{q+p} uv u^{-1} v^{-1}}{\partial x^p \partial y^q}\right)^0 = \sum_{a+b+c+d = p+q} A_a B_b C_c D_d , \tag{9.1}$$

where $A_a = \left(\frac{\partial^a u}{\partial \sigma_1 \ldots \partial \sigma_a}\right)^0$, $B_b = \left(\frac{\partial^b v}{\partial \tau_1 \ldots \partial \tau_b}\right)^0$, $C_c = \left(\frac{\partial^c u^{-1}}{\partial \alpha_1 \ldots \partial \alpha_c}\right)^0$,

$D_d = \left(\frac{\partial^d v^{-1}}{\partial \beta_1 \ldots \partial \beta_d}\right)^0$, and the ordered sequence

$\{\sigma_1, \ldots, \sigma_a, \tau_1, \ldots, \tau_b, \alpha_1, \ldots, \alpha_c, \beta_1, \ldots, \beta_d\}$ is identical

to the ordered sequence $\{\underbrace{x, x, \ldots, x}_{p}, \underbrace{y, y, \ldots, y}_{q}\}$.

Since $\left(\frac{\partial u}{\partial x}\right)^0 = \left(\frac{\partial u}{\partial y}\right)^0 = \left(\frac{\partial v}{\partial x}\right)^0 = \left(\frac{\partial v}{\partial y}\right)^0 = 0$, we see that

(1) If $1 \leq a \leq p$, then $A_a = 0$. (9.2)

(2) If $p < a < p+q$, then at least one of b, c, d is not zero, and hence at least one of B_b, C_c and D_d is zero.

A similar argument can be applied on b, c and d, and therefore, only possibly non-zero terms in the summation of (9.1) are A_{p+q}, B_{p+q}, C_{p+q} and D_{p+q}.

However $A_{p+q} + C_{p+q} = \left(\frac{\partial^{p+q}(uu^{-1})}{\partial x^p \partial y^q}\right)^0 = 0$ and

$B_{p+q} + D_{p+q} = \left(\frac{\partial^{p+q}(vv^{-1})}{\partial x^p \partial y^q}\right)^0 = 0$. This proves Proposition 9.1.

Now, ϕ_ξ can be extended to a homomorphism $\hat{\phi}_\xi : G(L) \to S(\mathbb{Z}_m^{r+s})$, $m = \overline{\Delta}(r, s)$, (see §5). Then Proposition 9.1 shows actually that $\hat{\phi}_\xi$ gives a representation of $G(L)/G(L)_{k+1}G''(L)$ into $S(\mathbb{Z}_m^{r+s})$. The quotient group

$$Q_k(L) = G(L)_k G''(L) / G(L)_{k+1} G''(L)$$

is called "Chen group" of L and $Q_k(L)$ is completely determined by Milnor's invariant $\overline{\mu}(\xi)$ for $\xi = (r, s)$, (Murasugi, 1970). This is not surprising in view of the following Proposition 9.2.

Proposition 9.2

Let $\Gamma_{(r,s)}$ *be the kernel of* $\phi_{(r,s)} : F \to S(\mathbb{Z}^{r+s})$. *Then*

$$\bigcap_{\substack{1 \le r+s \le k \\ 0 \le r,s \le k}} \Gamma_{(r,s)} = F_{k+1} F''.$$

Proof. Since $F'' F_{k+1} \subset \Gamma_{(r,s)}$, $1 \le r+s \le k$, by Proposition 9.1, it suffices to show that the reverse inclusion.

Let $g \in \bigcap \Gamma_{(r,s)}$. Since $\{[x, y, \underbrace{x, \ldots, x}_{r \text{ times}}, \underbrace{y, \ldots, y}_{s \text{ times}}],\ r+s = k-2,\ 0 \le r,\ s \le k-2\}$ are free generators of $F'' F_k / F'' F_{k+1}$ [12], we can write

$$g \equiv x^{\alpha_0} y^{\beta_0} \prod_{\substack{0 \le p,q \le k-2 \\ p+q \le k-2}} [\, x, y, \underbrace{x, \dots, x}_{p \text{ times}}, \underbrace{y, \dots, y}_{q \text{ times}} \,]^{\alpha_{p,q}} \mod F'' F_{k+1} .$$

We claim that $\alpha_0 = \beta_0 = \alpha_{p,q} = 0$ for any p, q.

For simplicity, $[\, x, y, \underbrace{x, \dots, x}_{p}, \underbrace{y, \dots, y}_{q} \,]$ will be denoted by $[\, x, y, x^p, y^p \,]$.

First, set $\xi = \{1\}$. Then $\phi_\xi(g)(a_1) = a_1 + \left(\frac{\partial g}{\partial x}\right)^0 = a_1 + \alpha_0$, and hence $\phi_\xi(g) \in \Gamma_{(1,0)}$ iff $\alpha_0 = 0$. Similarly, $\phi_{\{2\}}(g)\,(a_1) = a_1 + \left(\frac{\partial g}{\partial y}\right)^0 = a_1 + \beta_0$ and hence, $\phi_{\{2\}}(g) \in \Gamma_{(0,1)}$ iff $\beta_0 = 0$. Therefore, $g \in F' = F'' F_2$.

Next, to show that $\alpha_{p,q} = 0$, we need the following formula:

$$\left(\frac{\partial^{r+s} \quad x, y, x^{p-1}, y^{q-1}}{\partial x^r \, \partial y^s} \right)^0 = \begin{cases} (-1)^{p-1} & \text{if } p = r \text{ and } q = s \\ 0 & \text{otherwise.} \end{cases} \tag{9.3}$$

Since (9.3) is proved by an easy induction on $p+q$, the details will be omitted.

Now we will return to the proof of Proposition 9.2 .

Let $g \in \bigcap \Gamma_{(r,\, s)}$ and write

$$g \equiv \mathop{}\limits_{\substack{1 \le p,q \le k-2 \\ p+q \le k}} [\, x, y, x^{p-1}, y^{q-1} \,]^{\alpha_{p-1,q-1}} \mod F'' F_{k+1} .$$

Suppose that there is some $\alpha_{p-1,q-1} \ne 0$ with $p + q = \ell$, but $\alpha_{r-1,s-1} = 0$ for all r, s such that $r + s < \ell$. Then for any r, s such that $r + s < p + q$,

$$\left(\frac{\partial^{r+s} g}{\partial x^r \partial y^s} \right)^0 = \alpha_{r-1,s-1} = 0 ,$$

and

$$\left(\frac{\partial^{p+q} g}{\partial x^p \partial y^q}\right)^0 = (-1)^{p-1} \alpha_{p-1,q-1} .$$

Since $g \in \Gamma_{(p,q)}$, we have $\phi_{(p.q)}(g)(a_i) = a_i$ for $1 \le i \le k$, and hence, for $p + q \le k$,

$$\phi_{(p,q)}(g)(a_{p+q}) = a_{p+q} + a_{p+q-1}\left(\frac{\partial g}{\partial y}\right)^0 + \ldots + a_1\left(\frac{\partial^{p+q-1} g}{\partial x^{p-1} y^q}\right)^0 + \left(\frac{\partial^{p+q} g}{\partial x^p \partial y^q}\right)^0$$

$$= a_{p+q} + \left(\frac{\partial^{p+q} g}{\partial x^p \partial y^q}\right)^0$$

$$= a_{p+q} + (-1)^{p-1} \alpha_{p-1,q-1} .$$

Therefore, $g \in \Gamma_{(p,q)}$ implies that $\alpha_{p-1,q-1} = 0$, which contradicts our assumption, and hence, we can conclude that $\alpha_{p-1,q-1} = 0$ for all p, q, with $p + q \le k$, that is, $g \in F'' F_{k+1}$. This proves Proposition 9.2 .

Question

Let Λ_k be the set of all sequences $\xi = \{ j_1, j_2, \ldots, j_k \}$ with $j_i = 1$ or 2. Let Γ_ξ be the kernel of $\phi_\xi : F \to S(\mathbb{Z}^k)$.

Then is it true that $\displaystyle\bigcap_{\substack{\xi \in \Lambda q \\ 1 \le q \le k}} \Gamma_\xi = F_{k+1}$?

10 *p-GROUP REPRESENTATION*

To define $\overline{\mu}(j_1 j_2 \ldots j_{k+2})$, we used a nilpotent representation $\phi_\xi : G(L) \to S(\mathbb{Z}_m^k)$, where $m = \overline{\Delta}(j_1 j_2 \ldots j_{k+2})$. If m is a prime, or more generally, by compounding an obvious reduction $\mathbb{Z}_m \to \mathbb{Z}_p$ for a prime p dividing m, we obtain a new representation $\phi : G(L) \to S(\mathbb{Z}_p^k)$. It is easy to see that $\phi(G(L))$ is a p-group.

In this section, as the simplest example, we consider the case where $\xi = \{1, 2\}$.

Let F be the free group generated by x_1 and x_2. Define a homomorphism

$$\phi_\xi : F \to S(\mathbb{Z}_p \times \mathbb{Z}_p)$$

as before :

$$\phi_\xi(x_1)(a_1, a_2) = (a_1 + 1, a_2)$$

$$\phi_\xi(x_2)(a_1, a_2) = (a_1, a_2 + a_1) .$$

Proposition 10.1

$\phi_\xi(F)$ *is a non-abelian* p*-group of order* p^3 .

Further, $\phi_\xi(F)$ *is a quotient group of* $(p,p \mid p,p)$, *where*

$$(p,p \mid p,p) = \langle X_1, X_2 \mid X_1^p = X_2^p = (X_1X_2)^p = (X_1^{-1}X_2)^p = 1 \rangle .$$

See Coxeter & Moser (1957).

Proof. Let $X_i = \phi_\xi(x_i)$ and let $G = \phi_\xi(F)$. Then X_1 and X_2 satisfy all relations in the presentation of $(p,p \mid p,p) = H$. Therefore, G is the quotient group of H . Further, ${}^G/G' \cong \mathbb{Z}_p \times \mathbb{Z}_p$, since ${}^H/H' \cong \mathbb{Z}_p \times \mathbb{Z}_p$. Now, G' is abelian by Proposition 9.1 and $G_3 = 1$ by Proposition 2.1 . Therefore, G' is a cyclic group generated by $[X_1, X_2]$. Since $[X_1, X_2]$ has order p , G is an extension of $\mathbb{Z}_p$ by $\mathbb{Z}_p \times \mathbb{Z}_p$, and hence the order of G is p^3 .

Corollary 10.1

When $p = 2$, G *is the dihedral group* D_4 *of order* 8 . *When* $p = 3$, $G = (3,3 \mid 3,3)$.

Proof. If $p = 2$, then G has order 8 and hence, G is either D_4 or the quaternion group Q_8 . But G has at least two elements of order 2 . Therefore $G \cong D_4$ (Schenkman, 1965). Further, $(3,3 \mid 3,3)$ has order 27 and hence $G = (3,3 \mid 3,3)$ when $p = 3$.

Proposition 10.2

Let p *be a prime. If* $\ell k(K_1, K_2) \equiv 0 \pmod p$, *then there exists a homomorphism from* $G(K_1 \cup K_2)$ *onto a non-abelian* p*-group of order* p^3 .

Proof. A mapping $\hat{\phi}_{\{1,2\}} : G(K_1 \cup K_2) \to S(\mathbb{Z}_p \times \mathbb{Z}_p)$ defined by $\hat{\phi}_{\{1,2\}}(x_{i,j}) = \phi_{\{1,2\}}\theta_3(x_{i,j})$ is a required homomorphism.

Now the branched covering space of $L = K_1 \cup K_2$ associated with $\hat{\phi}_\xi$ is an *irregular* metabelian covering, and at least we can prove the following proposition. For a proof, see Burde (1971).

Proposition 10.3

Let $L(2p,q)$ *be a 2-bridge link of type* $(2p,q)$. *If* $\ell k(K_1, K_2) \equiv 0 \pmod 2$, *then the irregular* D_4 *- covering space associated with* $\hat{\phi}_\xi$ *is* S^3 .

11 CONCLUDING REMARKS

Remark 1. The nilpotent representations defined in §3 need not be the only nilpotent representation. There are many nilpotent non-equivalent representations of the link group which will be defined in similar fashions.

Remark 2. Proposition 6.3 shows that the longitude element η_i belongs to the kernel of $\hat{\phi}_\xi$. Therefore, our representation $\hat{\phi}_\xi$, in fact, a nilpotent representation of the group of 0-framed link. In particular, $\hat{\phi}_\xi$ gives a nilpotent representation of Fukuhara's framed link group FG(L) of 0-framed link L (Fukuhara), and it will be proved that some μ - invariants are, in fact, invariants of the equivalence classes of FG(L) .

Remark 3. As was seen in §8 , Examples 1, 2, the computation of $\mu(1122)$ or more generally $\mu(11\ldots1ij)$, needs only a representation associated with $\phi_\xi : F \to S(\mathbb{Z}^k)$, where $\xi = \{1, 1, \ldots, 1\}$. ϕ_ξ is, in fact, cyclic, and the corresponding covering space M_ξ is the cyclic covering space of S^3 branched along a knot K_1 . Therefore, ϕ_ξ is always extended to a homomorphism $\phi_\xi : G(L) \to S(\mathbb{Z}_r^k)$, where r need not be $m = \Delta(11\ldots1ij)$.

However, if $\mu(1i)$ is prime to r , then the lift of K_i in M_ξ is a single knot and hence the covering linkage invariants are not very interesting. Therefore, in order to get a non-trivial covering linkage invariants, at least r should be a divisor of $\mu(1i)$. The invariants a_n^*, h_n^* defined by Holmes and Smythe (1966) are presumably interpreted in this manner.

Remark 4. For a fixed prime p, Lanfer (1971), defines the invariant $s\omega(k_1, k_2, \ldots, k_n ; 2, 1)$, called the self-winding number, by taking p-fold cyclic coverings successively, when it is possible.

It is, therefore, an irregular nilpotent branched covering of S^3, but it may not be obtained in the method we discussed in this paper. It is quite unlikely that there are relationships between $\bar{\mu}$ and $s\omega(k_1, k_2, \ldots, k_n ; 2, 1)$ except for a few obvious relations when $n = 1$ or 2.

Remark 5. It is easy to see the homotopy invariance of $\bar{\mu}(1\,2\,3 \ldots n)$ from our construction of covering spaces.

Let $m = \bar{\Delta}(1\,2\,3 \ldots n)$. First consider the m-fold cyclic cover $p_1 : M_1 \to S^3$ branched along K_1. Let $p_1^{-1}(K_2) = K_{2,0} \cup K_{2,1} \cup \ldots \cup K_{2,m-1}$. Since $\tilde{K}_2 = K_{2,1} + 2K_{2,2} + \ldots + (m-1)K_{2,m-1} = 0$ in $H_1(M_1 ; \mathbb{Z}_m)$, (see §7), $\tilde{K}_2$ bounds a 2-chain B_2 and $\bar{\mu}(1\,2\,3) \equiv \mathrm{int}(B_2, K_{3,0})$ (mod m). Then, consider the m-fold cyclic cover $p_2 : M_2 \to M_1$ branched along $\tilde{K}_2$. Since

$p_1^{-1} p_2^{-1}(K_3) = K_{3,(0,0)} \cup K_{3,(0,1)} \cup \ldots \cup K_{3,(m-1,m-1)}$ and

$$\tilde{K}_3 = \sum_{(i,j)=(0,0)}^{(m-1,\, m-1)} j K_{3,(i,j)} = 0 \text{ in } H_1(M_2 ; \mathbb{Z}_m),$$

$\tilde{K}_3$ bounds a 2-chain B_3 in M_2, and $\bar{\mu}(1\,2\,3\,4) \equiv \mathrm{int}(B_3, K_{4,(0,0)})$ (mod m). Repeat these constructions up to M_{n-2}.

If K_i is homotopic to K_i', then K_i' is obtained from K_i by allowing only self-intersections in S^3 at various places. These self-intersections will be lifted to those in the covering spaces, but $\mathrm{int}(B_i, K_{i+1,(0,\ldots,0)})$ (mod m) will be unchanged, and hence $\bar{\mu}(1\,2\,3 \ldots n)$ is a homotopy invariant.

REFERENCES

Burde, G. (1971). On branched coverings of S^3. Can. J. Math. 23 84-89.

Coxeter, H.S.M. & Moser, W.O.J. (1957). Generators and relations for discrete groups. Springer-Verlag.

Fox, R.H. (1953). Free differential calculus, I. Ann. of Math. 57 547-560.

Fukuhara, S. On framed link group. (To appear)

Goldsmith, D.L. (1977). A linking invariant of classical link concordance, Knot Theory, Plans-sur-Bex, Switzerland, 135-170.

Hartley, R. & Murasugi, K. (1977). Covering linkage invariants. Can. J. Math. 29, 1312-1339.

Holmes, R. & Smythe, N. (1966). Algebraic invariants of isotopy of links. Amer. J. Math. 88, 646-654.

Lanfer, H. (1971). Some numerical link invariants. Topology 10, 119-130.

Milnor, J. (1954). Link groups. An.. of Math. 59, 177-195.

Milnor, J. (1957). Isotopy of Links, Algebraic geometry and topology. (Lefschetz Symposium) Princeton Univ. Press, Princeton, 280-306.

Murasugi, K. (1966). On Milnor's invariant for links. Trans. Amer. Math. Soc. 124, 94-110.

Murasugi, K. (1970). On Milnor's invariant for links II. The Chen group. Trans. Amer. Math. Soc. 148, 41-61.

Schenkman, E. (1965). Group Theory. D. van Nostrand Co.

Smythe, N. (1967). Isotopy invariants of links and the Alexander matrix. Amer. J. Math. 89, 693-704.

PRESENTATIONS EN PONTS DES NOEUDS RATIONNELS

J. -P. Otal

Abstract. In this paper, it is shown that any bridge presentation of a national knot is obtained by a sequence of "stabilisations" on a 2-bridge one.

On se fixe une fois pour toutes une fonction de Morse h sur S^3 ayant comme seules singularités un maximum et un minimum; par exemple, on peut prendre pour h la projection de $S^3 \subset \mathbb{R}^4$ sur l'un des axes de coordonnées. Un noeud K (non nécessairement connexe) dans S^3 est dit *présenté en* n *ponts par rapport à* h si $h|K$ est une fonction de Morse à n maxima (et n minima) et s'il existe une sphère de niveau S de h séparant les maxima de $h|K$ de ses minima. Deux présentations en ponts K_0 et K_1 sont isotopes s'il existe une isotopie $\{K_t\}$ de K_0 à K_1 au cours de laquelle K_t reste présenté en ponts.

Etant donnée une présentation à n ponts d'un noeud K , on construit une nouvelle presentation à n+1 ponts par le procédé de stabilisation suivant : Soit B une petite boule dans S^3 , telle que $B \cap K$ soit un arc, et située à un niveau séparant les maxima de $h|K$ de ses minima. On modifie la présentation de K en remplacant $K \cap B$ par le modèle indiqué figure 1 :

Figure 1.

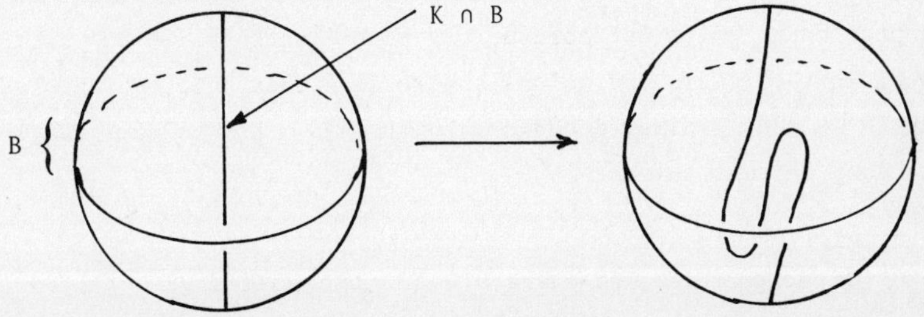

La nouvelle présentation ainsi obtenue est appelée *stabilisation élémentaire* de K. On vérifie sans peine qu'à isotopie près, cette stabilisation élémentaire ne dépend que du choix de la composante de K recontrant B. Plus généralement, une *stabilisation* d'une présentation en ponts est obtenue à partir de celle-ci par une succession de stabilisations élémentaires.

Nous allons ici démontrer le résultat suivant, annoncé dans (Otal, J.-P., 1982).

Theoreme 1.

Toute présentation en ponts d'un noeud rationnel non trivial est obtenue par stabilisations d'une présentation à deux ponts, à isotopie près.

Rappelons que les noeuds rationnels sont précisément les noeuds admettant une présentation à deux ponts. H. Schubert a montré dans (Schubert, H., 1956) que le noeud rationnel $K_{p.q}$ admet exactement 1 ou 2 présentations à deux ponts selon que q est congru à ± 1 modulo p ou non. Nous avons d'autre part déjà démontré dans (Otal, J.-P. 1982) l'analogue du théorème 1 pour le noeud trivial.

Par passage au revêtement double ramifié, le théorème 1 est à rapprocher du théorème sur l'unicité des scindements de Heegaard des espaces lenticulaires (Bonahon, F. et Otal, J.-P. (a & b)) il s'interprète comme une version $\mathbb{Z}/2$ - équivariante de ce résultat. Sa preuve suit les grandes lignes de (Bonahon, F., et Otal, J.-P. (b)) mais les techniques utilisées restent plus proches de celles de H. Schubert (Schubert, H., 1956).

§ 1. *CRITERE DE STABILISATION*

Nous aurons besoin dans la suite d'un critère assurant qu'une présentation en ponts est obtenue par stabilisations. Si K est présenté en ponts, remarquons qu'une sphère de niveau S, séparant les maxima de $h \mid K$ des minima, découpe la paire (S^3, K) en deux tangles triviaux, dans le sens suivant.

Définition.

Un *tangle* est une paire (B, L) de variétés C^∞ telles que B est difféomorphe à la boule B^3 et que L est une réunion d'arcs disjoints proprement plongés dans B. Un tel tangle est dit *trivial* si K est une réunion d'arcs non noués et non enlacés dans B.

On dira par extension qu'une sphère S découpant (S^3, K) en deux tangles triviaux (B_1, K_1) et (B_2, K_2) présente K en ponts. Les compasantes de K_1 (ou de K_2) seront alors appelés *ponts*.

Une *membrane* pour un tangle (B, L) est un disque dont le bord est formé d'un arc de L et d'un arc de ∂B, et dont l'interieur ne rencontre pas $K \cup \partial B$. Le tangle (B, L) à n composantes est trivial si et seulement si il admet un *système de membranes*, c'est-à-dire une collection de n membranes disjointes.

On vérifie maintenant sans peine le critère suivant : étant donnée une présentation en ponts du noeud K, soit S une sphère de niveau de h séparant les maxima des minima de $h|K$. La sphère S découpe (S^3, K) en deux tangles triviaux (B^+, K^+) et (B^-, K^-). Alors, la présentation en ponts est obtenue par stabilisations si et seulement si il existe deux membranes D^+ et D^-, respectivement contenues dans (B^+, K^+) et (B^-, K^-), telles que $D^+ \cap D^-$ soit formée d'un unique point, situé dans $K \cap S$. En effet, lorsque $h|\text{int } D^+$ et $h|\text{int } D^-$ sont du type Morse, sans point critique, on repère directement la stabilisation dans un voisinage de $D^+ \cup D^-$. On peut se ramener à cette situation en utilisant la remarque suivante.

Lemme 2.

Soit M *un système de membranes pour* (B^+, K^+); *après une isotopic ambiante de* B^+, *constante près de* S, *se prolongeant et une isotopie de la présentation en ponts de* K, *on a* : $h|(\text{int } M)$ *est de Morse, sans points critiques.*

Preuve.

Soit M un système de membranes pour le tangle trivial (B, K). On peut obtenir, par une légère isotopie de M, constante sur K, que $h|(\text{int } M)$ est de Morse avec des singularités à des niveaux tous différents. Imposons-nous en outre l'allure de M près des minimums de h K, en exigeant que, près de ces minimums, la normale intérieure à K dans M pointe vers le haut (ainsi, chaque minimum de $h|K$ est aussi un minimum de $h|M$). Par hypothèse, la restriction de h à chaque composante de K a pour seule singularité un minimum. Donc, si

$h|\text{int}(M)$ a une singularité, l'une d'elles est un extremum. En reprenant les notions et le raisonnement de (Otal, J.-P., 1982), proposition 2, on considère une singularité s de type selle apparaissant dans une situation du premier type : l'une des séparatrices issues de s est une courbe C qui borde dans M un disque Δ possédant comme seule singularité un extremum m'.

Si m' est un minimum, le raisonnement décrit dans (Otal, J.-P., 1982) permet d'éliminer la selle s avec l'extremum m'.

Supposons donc que m' est un maximum. Soit S la sphère de niveau de h contenant la selle s : la courbe C borde un unique disque Δ' sur S tel que la sphère $\Delta \cup \Delta'$ borde une boule contenue dans B. Si $\text{int}(\Delta)$ ne rencontre pas les séparatrices issues de s, on peut encore appliquer le raisonnement de (Otal, J.-P., 1982). Dans le cas contraire on remarque tout d'abord que $\Delta' \cap K = \emptyset$, sinon, comme la fonction $h|K$ a pour seules singularités des minimums, l'intersection $K \cap \Delta$ serait non vide. Donc, l'intersection de Δ' avec M est formée de deux courbes fermées issues de s et d'une réunion disjointes de courbes fermées. On peut éliminer ces courbes fermées par la technique suivante : soit ν une telle courbe non singulière et plus intérieure sur Δ'. Sur M, ν borde un unique disque D ; en remplacant sur M le disque D par l'unique disque bordé par ν sur Δ', puis en arrondissant convenablement les coins, on supprime la courbe ν des intersections de M avec Δ' ; cette dernière réduction ne fait que supprimer les selles éventuelles du feuilletage de M contenues dans D.

On peut donc supposer que $\Delta \cap M$ est réduit aux deux séparatrices issues de S. Dans ce cas, si on applique la technique précédente dans un niveau juste inférieur à celui de S, on élimine la selle s et le maximum m' du feuilletage de M. □

Dans la suite, deux membranes D^+ et D^- se rencontrant en un seul point seront dites *en position d'annulation*.

§ 2. *UN SYSTEME DE COORDONNEES POUR LE TANGLE TRIVIAL A DEUX COMPOSANTES.*

Considérons le quotient de la paire $(\mathbb{R}^2, \mathbb{Z}^2)$ par l'action du groupe G engendré par les rotations d'angle π autour des points à coordonnées entières : cette paire est homéomorphe à la paire $(S^2$, 4 points). La projection $p : (\mathbb{R}^2, \mathbb{Z}^2) \to (\mathbb{R}, \mathbb{Z})$, $(x,y) \longmapsto x$ commute avec l'action de G et fournit ainsi une application $\tilde{p}$ de $(S^2$, 4 points) sur le quotient de $(\mathbb{R}, \mathbb{Z})$ par G, qui est homéomorphe à $([0,1], \{0,1\})$. Le cylindre d'application ("mapping cylinder") de $(S^2$, 4 points) et de $\tilde{p}$ est homéomorphe au tangle trivial à 2 composantes (B, L).

On montre assez facilement que toute courbe simple fermée C de $(S^2$, 4 points) qui est *non-triviale*, c'est-à-dire telle que chaque composante de $S^2 - C$ contient exactement 2 des 4 points, est isotope par une isotopie fixant les 4 points à une courbe de pente constante, se relevant dans $\mathbb{R}^2$ en une famille de droites parallèles de pente rationnelle q/p. Le nombre rationnel q/p est par définition la *pente* de la courbe considérée. De même tout arc joignant deux points distincts parmi les 4 points est isotope à un arc de pente rationnelle constante par une isotopie de S^2 fixant les 4 points.

§ 3. *RAPPEL SUR LA CLASSIFICATION DE SCHUBERT DES NOEUDS RATIONNELS.*

Nous avond déjà dit qu'un noeud rationnel est un noeud K admettant une présentation à deux ponts. En particulier, il existe une sphère S decoupant (S^3, K) en deux tangles triviaux à deux composantes $(B_1, K \cap B_1)$ et $(B_2, K \cap B_2)$. Après choix d'une identification de $(B_1, K \cap B_1)$ avec le cylindre d'application de $\tilde{p} : (\mathbb{R}^2, \mathbb{Z}^2)/G \to (\mathbb{R}, \mathbb{Z})/G$ comme ci-dessus, la trace sur $\partial(B_1, K \cap B_1)$ d'une membrane de $(B_2, K \cap B_2)$ est un arc de pente q/p.

Schubert a montré que le couple $(p, q) \in \mathbb{N} \times \mathbb{Z}$, avec $\operatorname{pgcd}(p, q) = 1$, classifie le noeud K modulo la relation $(p, q) \sim (p', q') \Longleftrightarrow p = p'$ et $q' \equiv q^{\pm 1} \pmod p$. On notera $K_{p,q}$ le noeud rationnel ainsi associé au couple $(p, q) \in \mathbb{N} \times \mathbb{Z}$.

§ 4. *MISE EN PLACE*

Nous allons associer, pour $p \neq 0$, au noeud $K_{p.q}$ un objet singulier possédant des propriétés comparables à celles de l'objet Δ singulier construit dans (Bonahon F., et Otal, J.-P.). Considérons le noeud $K_{p.q}$ avec les notations du paragraphe précédent. Dans B_2, il existe, à isotopie fixant K près, un unique disque D proprement plongé, séparant les ponts de $K_{p,q} \cap B$. Après isotopie, ∂D est une courbe de pente constante q/p, $(B, K_{p.q} \cap B_1)$ étant muni des coordonnées précisées ci-dessus.

Notons Δ l'objet singulier réunion de D et du cylindre d'application de $\tilde{p} \mid \partial D$. Abstraitement, Δ est homéomorphe au quotient du disque unité de C par la relation d'équivalence qui identifie sur le bord z_1 et $z_2 \in S^1$ lorsque $z_1^p = z_2^p$ ou $z_1^p = \bar{z}_2^p$. Remarquer que Δ est une variété lisse dans le complementaire de l'arc Σ correspondant à $\mathbb{R}/G$ (et au quotient de S^1 par la relation d'équivalence ci-dessus). Au voisinage d'un point de $\mathrm{int}(\Sigma)$, Δ est formé de 2p feuillets distincts venant se recoller le long de Σ (cf. figure 1).

Figure 2.

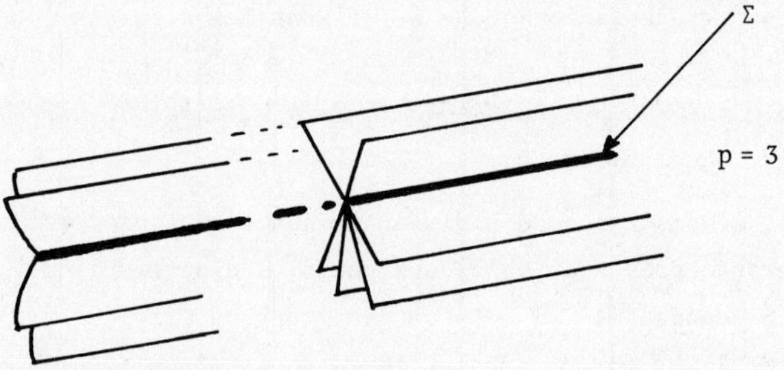

Supposons dorénavant le noeud rationnel $K_{p,q}$ présenté en ponts par rapport à la fonction hauteur h sur S^3. Soit S une sphère de niveau de h présentant $K_{p,q}$ en ponts, et soient B_1 et B_2 les deux boules de S^3 bordées par S.

La preuve du théorème 1 dans le cas du noeud trivial $K_{1,0}$ (cf. Otal, J.-P., 1982) s'adapte exactement au cas de l'enlacement trivial à deux composantes $K_{0,1}$. On peut donc supposes $p \neq 0$ et $q \not\equiv 0$ modulo p. Soit Δ l'objet singulier défini ci-dessus et Σ son ensemble singulier. Horst Schubert demontre dans (Schubert, H., 1956) le résultat suivant dans le cas où $h \mid K_{p,q}$ possède 2 maxima, mais les arguments contenus dans sa preuve ne tiennent pas compte de cette hypothèse :

Proposition 3.

Après une isotopic ambiante de S^3, *induisant une isotopie de la présentation en ponts de* $K_{p,q}$, *on a* $\Sigma \subset S$. □

Nous allons maintenant imposer une *condition de linéarité* pour l'allure de Δ et S près de Σ : pour un petit voisinage U de Σ, la paire $(U, U \cap K)$ est homéomorphe au modèle (B, L) construit au § 2. On réclame alors les deux conditions suivantes, réalisables par isotopie de Δ et de l'homéomorphisme orienté $(U, U \cap K) \cong (B, L)$.

(i) $S \cap U$ est le cylindre d'application d'une courbe horizontale de pente constante 0 sur $\partial(U, U \cap K)$.

(ii) $\Delta \cap U$ est le cylindre d'application d'une courbe de pente constante q'/p (on vérifie alors que $q \equiv q' \mod p$).

Par position générale, on peut en outre imposer que $h|(\Delta - \Sigma)$ soit de type Morse avec des singularités toutes à des niveaux distincts et différents de $h(S)$. Ainsi, l'intersection $S \cap (\Delta - \Sigma)$ est transverse et formée d'arcs ouverts et de courbes fermées.

§ 5. *SIMPLIFICATION DE LA POSITION DE* Δ *PAR RAPPORT A* S

Proposition 4.

Après une isotopie ambiante de S^3 *induisant une isotopie de la présentation en ponts du noeud,* $(\Delta - \Sigma) \cap S$ *ne contient pas de courbes fermées.*

Cette proposition est une conséquence rapide du lemme suivant :

Lemme 5.

Après une isotopie ambiante induisant une isotopie de la présentation en ponts, $h\ (\Delta - \Sigma)$ *n'a pas de singularités de type selle (= points de Morse d'indice* 1*).*

Preuve.

La condition que le niveau $h(\Sigma) = h(S)$ est non critique pour $h|(\Delta - \Sigma)$ entraîne que les courbes de niveau issues d'une singularité du type selle de $h|(\Delta - \Sigma)$ ne rencontrent pas Σ . On peut alors appliquer le raisonnement de la proposition 2 de (Otal, J.-P., 1982). □

Preuve de la proposition 4.

Une composante connexe quelconque D de $\Delta - S$ est maintenant une surface planaire munie de la fonction de Morse $h|D$ dont les seules singularités sont d'indice 0 ou 2 ; D est donc un disque. En particulier, $(\Delta - \Sigma) \cap S$ n contient pas de courbes fermées. □

Remarquons que la condition de linéarité entraîne que le nombre d'arcs de $(\Delta - \Sigma) \cap S$ est non vide. Plus précisément, il est congru à $\pm q$ modulo p . Suivant un raisonnement analogue à celui de (Bonahon, F. et Otal, J.-P.(b)), nous faisons la remarque suivante.

Lemme 6.

Il existe un disque D *proprement plongé dans* B_1 *ou dans* B_2 *, évitant* $K_{p,q}$ *, qui recontre* Σ *en exactement un point.*

Preuve.

Dans la situation permise par la proposition 4, soit k un arc de $(\Delta - \Sigma) \cap S$ "le plus pres du bord" sur $\Delta - \Sigma$, c'est-à-dire tel que l'une des composantes de $(\Delta - \Sigma) - k$ soit un disque D_1 ne recoupant pas S et contenu par exemple dans B_1 .

Soit U un petit voisinage de Σ muni des coordonnées traduisant la linéarité de Δ près de Σ . L'intersection $D_1 \cap \partial U$ est un arc linéaire (de pente constante) sur ∂U et se projette sur $U \cap S$ en un arc rencontrant Σ en un seul point. Cette propriété entraîne que le disque singulier $\overline{D_1}$ peut être approximé par un disque plongé D vérifiant le lemme 6 . □

Lemme 7.

Il existe un système de membranes $M = \bigsqcup_1^n M_i$ *pour* $K_{p,q} \cap B_1$ *tel que* $M \cap (\text{int}\,\Sigma) = \emptyset$.

Preuve.

Il est facile de trouver un système de membranes M pour les ponts de $K \cap B_1$ tel que $M \cap S$ soit transverse à Σ et que $M \cap D = \emptyset$. Parmi ces systèmes choisissons-en qui minimise le nombre de composantes de $M \cap (\text{int}\,\Sigma)$. Supposons $M \cap (\text{int}\,\Sigma) \neq \emptyset$. Il existe alors un arc k contenu dans Σ, joignant le point $D \cap \Sigma$ a un point de $M_i \cap \Sigma$, et dont l'intérieur ne recoupe pas M. On obtient alors une nouvelle membrane M_i' en faisant la somme connexe de M_i et de D par un tunnel le long de k. Le nouveau système de membranes $M' = (M - M_i) \cup M_i'$ contredit la minimalité de M. □

Déjà, ce système de membranes particulier nous permet de faire la remarque suivante :

Affirmation 8.

Les extrémités de Σ *ne sont pas situées sur le même pont de* $K_{p,q} \cap B_1$.

Preuve.

Supposons, en raisonnant par l'absurde, que $\partial\Sigma$ soit contenu dans M_1, où M_1 est une membrane du système M construit au lemme 7. Alors $\Sigma \cup (M_1 \cup S)$ est une courbe fermée tracee sur la sphère S qui rencontre ∂D en un seul point, ce qui est impossible. □

Ainsi, l'arc Σ privilégie deux ponts particuliers p_1 et p_2 de $K_{p,q} \cap B_1$, correspondant par exemple aux membranes M_1 et M_2 de M.

Soit S' une sphère de niveau de h, contenue dans B_1 et voisine de S. Cette sphère découpe S^3 en deux boules B_1' et B_2', avec $B_1' \subset B_1$. En appliquant éventuellement le lemme 2 à M, on peut isotoper $(K_{p,q} \cap B') - p_1 \cup p_2$ à travers S' grâce à M, tout en préservant $K_{p,q}$ présenté en ponts par rapport à h par la sphère S. On a alors :

Lemme 9.

La paire $(B_2', K_{p,q} \cap B_2')$ *est un tangle trivial à deux composantes.*

Preuve.

Il suffit, d'après la construction même de Δ, de vérifier que B_1' est isotope à un voisinage régulier de Σ, par une isotopie qui respecte $K_{p,q}$. Or, B_1' est clairement isotope à un voisinage régulier de $\Sigma \cup M_1 \cup M_2$, lequel est aussi un voisinage régulier de Σ puisque $(M_1 \cup M_2)$ int $\Sigma = \emptyset$ (tout ceci relativement au noeud). □

§ 6. *FIN DE LA PREUVE DU THEOREME 1*

Nous aurons besoin pour conclure d'étendre le théorème sur l'unicité des presentations en ponts du noeud trivial (Otal, J.-P.,1982) au cas du tangle trivial. La définition suivante généralise naturellement celle de présentation en ponts de noeuds.

Définition.

Soient (B, L) un tangle, et h une fonction de Morse sur B, constante sur le bord ∂B et dont la seule singularité est un maximum. Alors, L est présenté en ponts par rapport à h si $h|L$ est de type Morse et s'il existe une sphère de niveau S séparant les maximums des minimums de $h|L$. Le nombre de ponts d'une telle présentation est alors le nombre de maxima de $h|L$.

On définit comme dans l'introduction les notions d'*isotopie de présentations en ponts de tangle et de stabilisation.* De même, la définition de *membranes en position d'annulation* se généralise immédiatement à cette situation.

Proposition 10.

Soit (B, L) *un tangle trivial à* b *composantes, présenté en* n *ponts par rapport à une fonction de Morse* h *par une surface de niveau* S. *Si* $n > b$, *il existe alors deux membranes en position d'annulation (une dans chaque moitié de* (B, L) *délimitée par* S *).*

La démonstration du théorème 1 est clairement achevée par application de la proposition 10 au tangle trivial $(B_2', K_{p,q} \cap B_2')$. En effet, les deux membranes en position d'annulation fournies par cet énoncé permettent d'appliquer le critère du § 1 à la présentation en ponts de $K_{p,q}$ par la sphère S, et donc de conclure.

Signalons au passage que la proposition 10 admet comme corollaire que toute présentation en ponts du tangle trivial à b

composantes est, à isotopie près, obtenue par stabilisations de la présentation en b ponts (qui est clairement unique). On utilise pour cela une extension du lemme 2 au cadre considéré. La démonstration de ce dernier point nécessite toutefois une analyse un petit peu plus compliquée que celle de la proposition 2 de (Otal, J.-P., 1982), et nous ne démontrerons donc pas ces résultats.

Preuve de la proposition 10.

(1) Considérons tout d'abord le cas simple où $b = 1$. En recollant à (B, L) le long de $(\partial B, \partial L)$ un tangle trivial à une composante (B', L') présenté en un seul pont, on obtient un noeud trivial $\tilde{L} = L \cup L'$ dans $S^3 = B \cup B'$, présenté en n ponts par la sphère S. Si S sépare S^3 en B^+ et B^-, où B^+ contient B', il suit de l'unicité des présentations en ponts du noeud trivial (Otal, J.-P., 1982) que chaque composante de $\tilde{L} \cap B^+$ s'appuie sur une membrane en position d'annulation avec une membrance de $(B^-, \tilde{L} \cap B^-)$. Si $n > b = 1$, on peut appliquer ceci à un pont ne rencontrant pas B', et la membrane de $(B^+, \tilde{L} \cap B^+)$ ainsi fournie peut en outre être choisie disjointe de B' (car B' est simplement un voisinage régulier relativement à $\tilde{L}$ d'un point de $\tilde{L}$). On a ainsi deux membranes en position d'annulation pour la présentation en ponts de (B, L) par S.

(2) Quand $b > 1$, on se ramène au cas précédent. En effet, si $n > b$, la restriction de h à une composante L_1 de L possède au moins deux maximums. L'application du cas précédent au tangle trivial à une composante (B, L_1) fournit deux membranes M^+ et M^- en position d'annulation pour la sphère S présentant (B, L_1) en ponts. A priori, M^+ et M^- recoupent $L - L_1$. Pour éviter ceci, on considère un disque D proprement plongé dans le tangle trivial (B, L), évitant L et séparant L_1 de $L - L_1$ (par exemple, D est obtenu par épaississement d'une membrane contnant L_1). On veut fair en sorte que $M^+ \cap D = M^- \cap D = \emptyset$, ce qui assurera bien que M^+ et M^- évitent $L - L_1$.

On utilise d'abord le résultat suivant qui sera démontré dans le paragraph 8.

Lemme 11.

Soit (B, L) *un tangle présenté en ponts par une sphere* S . *Soit* D *un disque proprement plongé dans* B, *évitant* L, *et séparant au moins deux composantes de* L *entre elles. Après une isotopie de* D *dans* B - L , *l'intersection* $D \cap S$ *est réduite à une seule courbe.*

Grâce au lemme 11, on peut donc supposer, dans la situation qui nous intéresse, que $D \cap S$ est réduit à une seule courbe.

Affirmation 12.

Il existe un difféomorphisme ϕ *de* B, *tel que :*

(i) $\phi(S) = S$;

(ii) $(B, \phi(L))$ *est présenté en ponts par rapport à* S ;

(iii) *la fonction* $h : \phi(D)$ *est de Morse avec pour seule singularité un maximum.*

Preuve.

La sphère S découpe B en 2 composantes B^+ et B^- ; supposons $\partial B \subset B^-$. On va simplifier indépendamment les allures de $h | D \cap B^+$ et de $h | D \cap B^-$ par des isotopies respectives de B^+ et de B^- constantes près de S .

On choisit d'abord un système de membranes pour les ponts de $L \cap B^-$; en faisant des découpages, on peut supposer que ces membranes ne rencontrent pas D . Grâce à ces membranes, on peut glisser les ponts de $L \cap B^-$ dans un petit voisinage collier V de S . Si le voisinage V est choisi suffisamment petit, $D \cap V$ est un anneau vertical. Soit V' l'adhérence de $B^- - V$. La paire $(V', L \cap V')$ est difféomorphe à $(S^2, 2k \text{ points}) \times [0,1]$, la fibration au-dessus de l'intervalle $[0,1]$ étant fournie par $h | V'$. D'après un lemme de F. Waldhausen (1968), après une isotopie de V' , l'anneau $D \cap V'$ est vertical. Ceci résoud le cas de $D \cap B^-$. Celui de $D \cap B^+$ se traite de la même facon, quoique beaucoup plus simplement: une fois *poussés* les ponts de $L \cap B^+$ dans un voisinage collier V de S , l'adhérence $V' = B^+ - V$ est une boule; il est clair qu'après une isotopie de $D \cap V'$, la fonction $h | D \cap V'$ sera du type Morse avec pour seule singularité un maximum.

L'isotopie de B ainsi construite nous fournit le difféomorphisme cherché. □

Revenons maintenant à la preuve de la proposition 10.

Le disque D découpe B en deux boules. dont l'une, B', contient $L - L_1$. D'après l'affirmation 12, l'intersection de $\phi(B')$ avec une sphère de niveau de h est un disque ou un point, lorsqu'elle n'est pas vide. On construit alors facilement un difféomorphisme ψ de la boule B tel que :

(i) ψ respecte les sphères de niveau de h ;

(ii) $\psi(\phi(L_1)) = \phi(L_1)$;

(iii) $\psi\,\phi\,(B')$ est un petit voisinage régulier d'un arc vertical disjoint de $\phi(M^+ \cup M^-)$.

Les deux membranes $\phi^{-1}\psi^{-1}\phi\,(M^+)$ et $\phi^{-1}\psi^{-1}\phi\,(M^-)$ sont alors disjointes de D : elles forment donc un couple de membranes en position d'annulation pour le tangle (B, L).

§ 7. *DISQUES DE COMPRESSION ET PRESENTATIONS EN PONTS.*

Ce paragraphe est consacré à la preuve du lemme 11 utilisé au paragraphe précédent. La démonstration consiste à simplifier l'intersection de D avec S pour aboutir à la situation où $D \cap S$ est réduit à une seule courbe. Pour cela, on adapte les arguments du § 4 de (Bonahon, F., et Otal, J.-P.(b)) à la situation des présentations en ponts. Nous renvoyons le lecteur à ce texte pour plus de détails.

Par définition, un *tangle creux* est une paire (B, L) où B est difféomorphe à $S \times [0,1]$ et L est une réunion d'arcs proprement plongés dans B. Soit p la fonction de Morse définie sur B par la projection sur $[0,1]$. Le tangle creux (B, L) est dit *trivial* lorsque $p\,|\,L$ est de Morse avec comme seules singularités des minimums. Nous noterons respectivement $\partial_1 B$ et $\partial_0 B$ les deux composantes $p^{-1}(1)$ et $p^{-1}(0)$ de ∂B.

Les tangles creux triviaux apparaissent naturellement dans les présentations en ponts de tangle : en effet, une sphère de présentation découpe un tangle en deux paires dont l'une est un tangle trivial et l'autre un tangle creux trivial.

La preuve utilise d'une manière essentielle le résultat suivant :

Lemme 13.

Soit T *une surface compacte proprement plongée dans le tangle creux trivial* (B, L), *telle que* $T \cap L = \emptyset$ *et* $T \cap \partial_1 B \neq \emptyset$. *Si* T *n'admet dans* $B - L$ *aucun disque de compression, ni aucun disque de* ∂ *- compression vers* $\partial_1 B$, *alors chaque composante de* T *est ou bien un disque, ou bien un anneau joignant* $\partial_0 B$ *à* $\partial_1 B$.

Preuve.

On verifié que la flèche $\pi_1(\partial_1 B - L) \to \pi_1(B - L)$ est une surjection. Donc, $\pi_1(B - L, \partial_1 B - L) = 0$. D'autre part, si T n'est un disque, ni un anneau joignant $\partial_0 B$ à $\partial_1 B$, $\pi_1(T, T \cap \partial_1 B) \neq \emptyset$, et il suffit d'appliquer le "loop theorem" pour exhiber un disque de compression ou de ∂ - compression vers $\partial_1 B$.

Si D n'est pas dans la situation cherchées, alors $D \cap B^+$ admet, par exemple, un disque de compression ou de ∂ - compression vers $\partial_1 B$. Si $D \cap B^+$ admet un disque de compression, un argument standard, reposant sur l'irréductibilité de $B^+ - L$, permet d'isotoper D sur un nouveau disque D_1^1 rencontrant S en moins de courbes que D.

Si $D \cap B^+$ admet un disque de ∂ - compression δ vers S, on considère l'isotopie de D qui consiste à "tirer" $D \cap \delta$ a travers δ (cf. figure 3).

Figure 3.

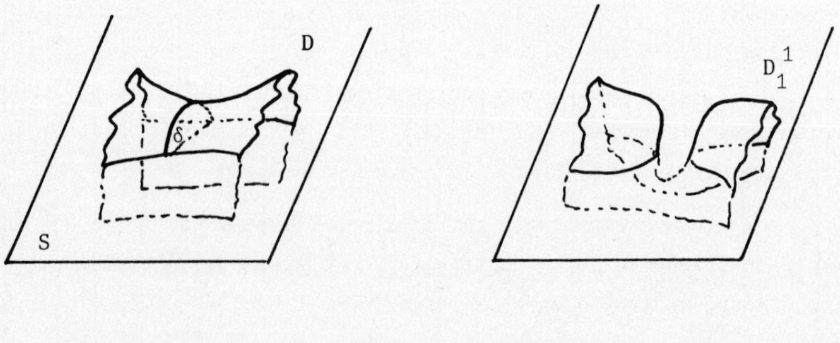

On obtient ainsi un nouveau disque D_1^1, isotope à D. Si $D_1^1 \cap B^+$ admet à nouveau un disque de compression ou de ∂ - compression, on construit un nouveau disque D_1^2. Ainsi de suite jusqu'à ce que l'on obtienne un disque D_1 avec $D_1 \cap B^+$ incompressible et ∂ - incompressible. On recommence alors dans B^- ; on définit ainsi une suite de disques $D_1, D_2, \ldots, D_n$, tous isotopes à D . Les arguments combinatoires du lemme 10 dans (Bonahon, F., et Otal, J.-P.(b)) assurent que le nombre de composantes de $D_{2n} \cap S$ diminue strictement. D'autre part, pour les raisons homologiques, $D_i \cap S \neq \emptyset$. Donc, un disque D_i rencontrera S seulement en une courbe. □

§ *8. REVERSIBILITE DES PRESENTATIONS EN PONTS.*

Pour compléter le théorème 1, nous allons préciser le nombre de présentations à n ponts, distinctes, du noeud rationnel $K_{p,q}$.

Theoreme 14.

Pour $n \geq 3$*, le noeud* $K_{p,q}$ *admet une unique présentation à* n *ponts s'il est connexe (c'est-à-dire si* p *est impair) et exactement* $[\frac{n}{2}]$ *présentations à* n *ponts dans le cas contraire.*

Dans l'enonce ci-dessus, $[\frac{n}{2}]$ designe la partie entiere de $\frac{n}{2}$.

Preuve.

D'après le théorème 1, une présentation à n ponts d'un noeud rationnel est obtenue par stabilisations sur une présentation à 2 ponts . H. Schubert a démontré dans (Schubert, H., 1956) qu'un noeud rationnel admet au plus 2 présentations à 2 ponts reliées entre elles par "échange de haut et bas"; plus précisément, on peut associer à une présentation K en ponts d'un noeud, une autre présentation K' obtenue en modifiant K par un difféomorphisme qui préserve l'orientation de S^3 , respecte les sphères de niveau de h et échange le maximum et le minimum de h . La présentation K est dite *réversible* lorsque K et K' sont isotopes en tant que présentations en ponts.

D'après le théorème de Schubert et l'unicité de l'opération de stabilisation, le noeud rationnel $K_{p,q}$ admet lorsqu'il est connexe au plus 2 présentations en n ponts. S'il n'est pas connexe, rappelons qu'une isotopie (évidente) de la présentation en 2 ponts échange les deux composantes de $K_{p,q}$; donc il admet au plus $2[\frac{n}{2}]$ présentations à n ponts. Le théorème 14 découle maintenant de la proposition suivante.

Proposition 15.

Pour $n \geq 3$, *une présentation à* n *ponts d'un noeud rationnel est réversible.*

La proposition ci-dessus est fausse si $n = 2$; on peut montrer que la présentation a deux ponts du noeud $K_{p,q}$ est réversible si et seulement si $q^2 \equiv 1$ (modulo p) .

Preuve.

En utilisant le théorème 1 et l'unicité de l'opération de stabilisation, on remarque que la proposition 15 se déduit, par récurrence sur n , du cas où $n = 3$. Dans ce cas particulier, nous allons établir la proposition 15 sous une forme légèrement différente.

Rappelons que l'on peut définir une notion de présentation en ponts indépendamment des fonctions de Morse: une sphère S présente un noeud K en ponts si elle découpe la paire (S^3, K) en deux tangles triviaux. De même pour la notion de *stabilisation ;* si la sphère S présente K en ponts, soit B une boule de S^3 telle que $B \cap S$ est un disque et $B \cap K$ un arc non noué dans B . On modifie S à l'intérieur de B par le modèle suivante :

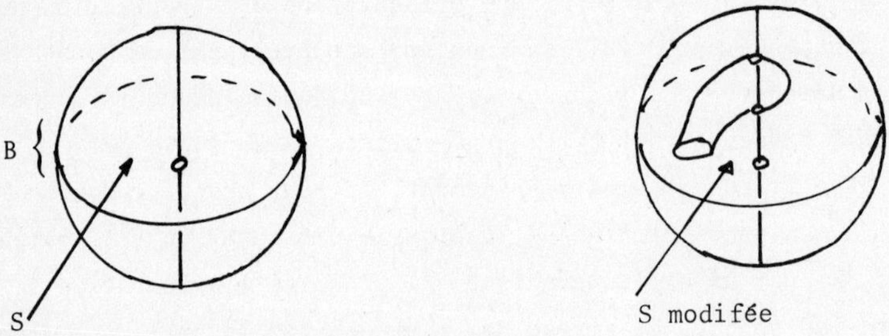

La nouvelle sphère obtenue présente toujours K en ponts, et ne dépend, à isotopie respectant K près, que de la composante de K sur laquelle est effectuée l'opération. On dit qu'elle est obtenue par stabilisation élémentaire sur S .

Le lien entre le résultat suivant et le cadre initial des présentations en ponts sera fait par la suite.

Lemme 16.

Soit S *une sphère orientée présentant* K *avec deux ponts; soit* $(-S)$ *la sphère minuie de l'orientation opposée. On peut isotoper une stabilsation élémentaire de* S *sur la stabilisation élémentaire de* $(-S)$ *effectuée sur la même composante par une isotopie respectant le noeud.*

Preuve.

Notons (B^+, K^+) (respectivement (B^-, K^-)) le tangle trivial dont le bord orienté est $(S, K \cap S)$ (respectivement $((-S), K \cap S)$). Soit S' le complémentaire dans S d'un petit voisinage ouvert du point P de $S \cap K$. Le bord (orienté) d'un petit voisinage régulier V de S' est une sphère orienté S^0 présentant K avec 3 ponts : on vérifie en effet que le complémentaire dans (S^3, K) du tangle trivial $(V, K \cap V)$ a pour adhérence un tangle trivial, soit $(V', K \cap V')$.

Soit D_1 une membrane pour le pont du tangle (B^+, K^+) ne contenant pas P : $D_1 \cap V'$ est alors une membrane pour le tangle $(V', V' \cap K)$. On construit facilement un arc k tracé sur S' joignant $\partial S'$ à un point de $K \cap D_1$ tel que $\mathrm{int}(k) \cap D_1 = \emptyset$. L'arc k fournit alors une membrane pour le tangle $(V, V \cap K)$ qui rencontre la membrane $D_1 \cap V'$ en un seul point. Cette situation présente S^0 comme une stabilisation (orientée) de $(-S)$. Par symétrie, S est aussi une stabilisation (orientée) de S. □

La proposition 15 pour les présentations à 3 ponts se déduit du lemme précédent grâce au résultat général suivant :

Lemme 17.

Soient K *et* K' *deux noeuds présentés en ponts (pour la fonction hauteur* h *) par une même sphère de niveau* S *. Soit* ϕ *un difféomorphisme préservant l'orientation de* S^3 *tel que :*

(i) $\phi(S) = S$ *et* $\phi|S$ *préserve l'orientation ;*

(ii) $\phi(K) = K'$;

alors les présentations K *et* K' *sont isotopes.*

Preuve.

Quitte à modifier K' et ϕ par une petite isotopie dans un voisinage de S respectant les sphères de niveau, on peut supposer que ϕ est l'identité dans un voisinage collier V de S.

Soit M un système de membranes pour le tangle trivial $(B^+, K \cap B^+)$; $\phi(M)$ est alors un système de membranes pour $(B^+, K' \cap B^+)$. Après application du lemme 2 à M, une isotopie de présentations en ponts de K (consistant à glisser le long des nouvelles membranes de M) permet d'amener $K \cap B^+$ dans $V \cap B^+$. De même, en utilisant les membranes de $\phi(M)$ pour $K' \cap B^+$. On peut ainsi obtenir $K \cap B^+ = K' \cap B^+$ car ϕ est l'identité dans V. Le même raisonnement dans la boule B^- termine la preuve.

REFERENCES

Bonahon, F., et Otal, J.-P. (a) Scindements de Heegaard des espaces lenticulaires, C.R. Acad. Sci. Paris série 1, t.294.

Bonahon, F., et Otal, J.-P. (b) Scindements de Heegaard des espaces lenticulaires, Prépublications d'Orsay.

Otal, J.-P., (1982). Présentations en ponts du noeud trivial, C.R. Acad. Sci. Paris série 1, t.294.

Schubert, H., (1954). Uber ein numerische Knoteninvariant, Math. Zeit. 61, 245-288.

Schubert, H., (1956). Knoten mit zwei Brücken, Math. Zeit 65, 133-170.

Waldhausen, F., (1968). On irreducible manifolds which are sufficiently large, Ann. of Math. 87, 56-88.

PIECEWISE-LINEAR I-EQUIVALENCE OF LINKS

Dale Rolfsen

Abstract. Piecewise-linear I-equivalence of codimension two links has been neglected because of the belief (until recently) that it was a trivial equivalence relation, except in classical dimensions. It provides a type of cobordism relation between links which, in a sense, ignores knotting and so constitutes link theory modulo knot theory. Here we discuss its connection with other notions of link equivalence, and outline some algebraic aspects of I-equivalence.

In this report, I take the term *link* to mean a submanifold L^n of a manifold M^m such that L is homeomorphic with a disjoint union of n-spheres :

$$L = L_1 \cup \ldots \cup L_\mu \cong S_1^n \cup \ldots \cup S_\mu^n = \mu S^n .$$

Following Stallings (1965), an *I-equivalence* between such links L and L' is an embedding

$$e : \mu S^n \times I \to M \times I \quad (I = [0, 1])$$

satisfying $e(\mu S^n \times 0) = L \times 0 \subset M \times 0$ and $e(\mu S^n \times 1) = L' \times 1 \subset M \times 1$. One may assume the embedding is proper and transverse to the ends, $M \times 0$ and $M \times 1$, in $M \times I$.

This equivalence relation generalizes two other important notions of equivalence. A *cobordism* is an I-equivalence which is locally flat. Cobordism (sometimes called concordance) was introduced for knots by Fox and Milnor (1966). An *isotopy* (that is, homotopy through embeddings) may be regarded as an I-equivalence which is level preserving, meaning $e(\mu S^n \times t) \subset M \times t$ for all $t \in [0, 1]$, but not necessarily locally flat. We may summarize the implications by the diagram

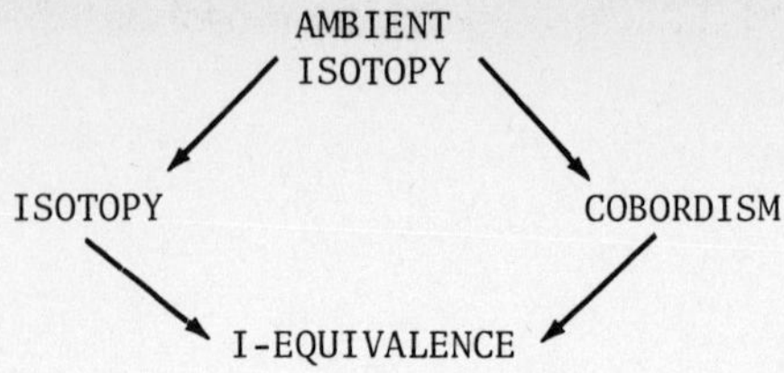

Example 1.

The three 2-component links L, L', L" pictured below are all I-equivalent in S^3.

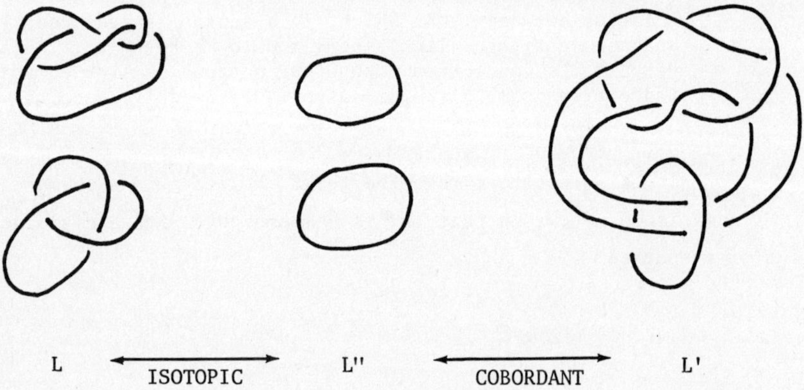

The isotopy from L to L" simply pulls the knots tight until they disappear -- a PL, but not smooth, operation. L' is cobordant with the trivial link L" since its components bound disks with only "ribbon" singularities. The disks may therefore be made to embed in $S^3 \times I$. It can be seen easily, however, that L and L" are not cobordant, nor are L' and L" isotopic. Also, L and L' are neither isotopic nor cobordant with one another.

Restricting discussion to the PL category, one may characterize an ambient isotopy as an I-equivalence which is simultaneously an isotopy and a cobordism, using the isotopy extension theorem of Hudson and Zeeman (1966). Also, by PL unknotting, if $m > n+2$ isotopy coincides with ambient isotopy and I-equivalence coincides with cobordism. (The same holds for the smooth category regardless of codimension). So I am interested here in the more problematical case of codimension 2 in the PL category, in which all four relations are really different. The principal result states that the diagram is a sort of pushout. That is : PL I-equivalence is exactly the relation generated by isotopy plus cobordism.

Theorem 1 (PL) :

Let L *and* L' *be n-dimensional locally flat links in* M^{n+2}. *Then* L *is* I-*equivalent to* L' *if and only if there exists a link* L" *in* M^{n+2} *which is isotopic with* L *and cobordant with* L'.

A related result is that in case $M = S^{n+2}$, the only obstructions to making an I-equivalence into a cobordism lie in the group C_n^{n+2} of cobordism classes of locally flat PL n-knots in S^{n+2} (see Kervaire (1970)).

Following the proofs of these geometric results, the paper considers various algebraic aspects. These include the behaviour under I-equivalence of the Alexander polynomial, the localized Blanchfield pairing of Hillman, the Goldsmith invariant $[\Lambda_{\tilde{M}}(L)]$ and the Murasugi signature. This last algebraic construction must be modified to produce what I call *reduced signature*, which is shown to be invariant under I-equivalence. Further application includes verification that Giffen's wild topological I-equivalence between certain PL links in S^3 cannot be replaced by a PL I-equivalence. Finally, certain links in S^3 are shown, using our results, not to be PL I-equivalent to split links or boundary links.

GEOMETRIC RESULTS.

All that follows is in the PL category. The links are assumed *oriented* and the components *ordered*. I-equivalences are supposed to respect these.

First some elementary observations.

Lemma 1.

Let $N^n \subset M^{n+2}$ *be* PL *manifolds and let* X *denote the set of points of* N^n *at which the embedding fails to be locally flat. Then* X *lies in a subpolyhedron of* N *of dimension at most* $n-2$.

The proof is an easy induction, left to the reader (or see Rolfsen (1972)).

Proposition 1.

Every knot $K^n \subset S^{n+2}$ *(locally flat or not) is isotopic to the trivial knot (hence also* I *- equivalent to the trivial knot).*

Proof: Using lemma 1 we can easily find a ball $B^{n+2} \subset S^{n+2}$ such that $(B, B \cap K)$ is a standard ball-pair. Now the complement $C = S^{n+2} - B$ is also an $(n+2)$-ball with $C \cap K$ and n - ball which has boundary unknotted in ∂C. It is easy to find an n - ball $D^n \subset C$, with $\partial D = \partial(C \cap K)$, such that the n - sphere

$$K' = (B \cap K) \cup D$$

is unknotted in S^{n+2}. By the Alexander cone isotopy trick, $C \cap K$ is isotopic, in C, with D, and hence K is isotopic with K'.

Proposition 2.

Every link L^n *in* S^{n+2} *is isotopic (and hence* I*-equivalent) with a locally flat link.*

Proof: We use a similar trick. By lemma 1 the non-locally flat points may be engulfed by n-balls, one in each component of L. Taking regular neighbourhoods we obtain disjoint $(n+2)$-balls $B_i \subset S^{n+2}$, such that $B_i \cap L_i$ is an n-ball containing, in its interior, all the bad points of L_i. Its boundary is a locally flat $(n-1)$-knot in ∂B_i. Moreover it is null-cobordant since it bounds a locally flat n-ball outside of B_i. So we may replace each $B_i \cap L_i$ by a locally flat n-ball $D_i^n \subset B_i$ having $\partial D_i = (B_i \cap L_i)$. The new link thus formed is locally flat and, by Alexander, isotopic with the original link.

Proof of Theorem 1: One direction is clear. To prove the hard direction we assume that $L = e(\mu S^n \times I) = L_1 \cup \ldots \cup L_\mu$ is an I-equivalence in $M \times I$ between (by abuse of notation) $L \subset M \times 0$ and $L' \subset M \times 1$. We need to produce a link $L'' \subset M$, an isotopy from L to L'' and a cobordism from L' to L''.

If L is locally flat, there is nothing to prove. In any case the set of "bad" points of L may be assumed bounded away from the ends L and L'. I assert that one may engulf the bad points of L in disjoint $(n+3)$-dimensional balls $B_1, \ldots, B_\mu$ satisfying, for each $i = 1, \ldots, \mu$:

(1) $\mathcal{B}_i$ intersects L only in the component L_i. Moreover, if $\mathcal{A}_i$ denotes $\mathcal{B}_i \cap L_i$, then $\mathcal{A}_i$ is an $(n+1)$-ball properly embedded in $\mathcal{B}_i$, transverse at the boundary and containing, in its interior, all the bad points of L_i.

(2) The $(n, n+2)$-sphere pair $\partial\mathcal{A}_i \subset \partial\mathcal{B}_i$ is locally flat, and intersects $M \times 0$ in the locally flat $(n, n+2)$-ball pair $A_i \subset B_i$ where $B_i = \mathcal{B}_i \cap (M \times 0)$ and $A_i = \mathcal{A}_i \cap (M \times 0) = B_i \cap L_i$.

(3) The $\mathcal{B}_i$ do not intersect $M \times 1$.

To construct such $\mathcal{B}_i$, first note that the set of non-locally-flat points of L lies in a subcomplex of dimension at most $n-1$, by Lemma 1. Thus it may be engulfed in L by μ sets of the form $e((n\text{-ball}) \times [0, 1-\varepsilon])$, one for each component. Then just take negular neighbourhoods.

Now we wish to replace $\mathcal{A}_i$, which may be horribly embedded in $\mathcal{B}_i$, by a locally flat $(n+1)$-ball. This will be impossible if the cobordism class of the knot $\partial\mathcal{A}_i$ in $\partial\mathcal{B}_i \cong S^{n+2}$ is nontrivial. But if that occurs, one can take a connected sum, *within* B_i, with a cobordism inverse. This forms a new locally flat ball pair $A_i' \subset B_i$, which guarantees that the knot $(\partial\mathcal{A}_i - A_i) \cup A_i'$ bounds a locally-flat $(n+1)$-ball -- call it $\mathcal{A}_i'$ -- within $\mathcal{B}_i$. Define $L'' = (L - (A_1 \cup \ldots \cup A_\mu)) \cup A_1' \cup \ldots \cup A_\mu'$. Then this link is cobordant to L' via the cobordism $(L - (\mathcal{A}_1 \cup \ldots \cup \mathcal{A}_\mu)) \cup \mathcal{A}_1' \cup \ldots \cup \mathcal{A}_\mu'$. Moreover, L'' agrees with L except within the disjoint $(n+2)$-balls $B_1, \ldots, B_\mu$. And there, one just replaces the n-ball A_i by the n-ball A_i'. It follows by "Alexander's trick" that L'' is isotopic with L. Theorem 1 is proved.

Theorem 2 :

Any PL I*-equivalence between locally flat* n*-dimensional links in an* $(n+2)$*-manifold* M *may be replaced by one which is locally flat except at possibly one point for each component.*

Proof : Using the notation of the above proof, remove $\mathcal{A}_i$ from the I-equivalence and replace by the cone on $\partial\mathcal{A}_i$, using the radial structure of $\mathcal{B}_i$.

A cobordism between L and L' is, in particular, a cobordism of knots $L_i \sim L_i'$ for each i. Therefore a necessary condition for an I-equivalence to be "smoothable" is that the cobordism classes of the components match up. This is the only obstruction, in case $M = S^{n+2}$.

Theorem 3 :

Let $L = L_1 \cup \ldots \cup L_\mu$ *and* $L' = L_1' \cup \ldots \cup L_\mu'$ *be links in* S^{n+2} *which are* I*-equivalent. Then* L *is cobordant with* L' *if and only if the cobordism classes* $[L_i] - [L_i']$ *vanish in* C_n^{n+2}, *for every* $i = 1,\ldots, \mu$.

Proof of Theorem 3 : The problem is to remove the local knots of an I-equivalence $L \subset S^{n+2} \times I$ between L and L' assuming $[L_i] = [L_i']$ for each i. Now $S^{n+2} \times I$ can be made into an $(n+3)$-sphere by attaching cones to the "North" $(S^{n+2} \times 0)$ and the "South" $S^{n+2} \times I$. The cone points are the "poles" N and S. We may assume only one bad point x_i in the component L_i. This determines an element X_i of C_n^{n+2}.

Consider the $(n+1)$-sphere obtained from L_i by attaching cones, toward N and S, along the boundary knots L_i and L_i'. This gives us an $(n+1)$-sphere in S^{n+3} with at most three local knots : at x_i, N and S. These singularities are measured in C_n^{n+2} by the respective elements X_i, $[L_i]$ and $-[L_i']$. By the argument of Fox and Milnor (1966), we have the equation

$$X_i + [L_i] - [L_i'] = 0 \quad \text{in} \quad C_n^{n+2} .$$

From this it follows that $X_i = 0$. Therefore, we may smooth L_i in a small neighbourhood of x_i and Theorem 3 is proved.

The above results are analogous to certain results of Rolfsen (1972), concerning isotopy and ambient isotopy. Isotopic n-dimensional links L and L' in S^{n+2} are ambient isotopic if and only if, for each i, their respective components L_i and L_i' have the same knot type. (This is false in an arbitrary M^{n+2}, see Rolfsen (1974).) Any PL isotopy in M^{n+2} between locally flat links may be replaced by one which has at most one bad point per component. I would like to raise the following.

Question.

Does there exist a manifold M^{n+2} and n-dimensional links L and L' in M such that L and L' are I-equivalent and for each $i = 1,\ldots, \mu$, L_i and L_i' are cobordant knots in M, yet L fails to be cobordant with L' ? In other words, does Theorem 3 fail for arbitrary M^{n+2} replacing S^{n+2} ?

Theorem 4 :

Suppose L *is a locally flat* n *-dimensional link in* S^{n+2} . *Then the following are equivalent:*

(a) L *is* I *- trivial (meaning* I *- equivalent with the trivial link of separated unknotted components).*

(b) L *is* I *- equivalent with a completely splittable link.*

(c) L *is cobordant with a completely splittable link.*

Proof : "Completely splittable" means the components lie in disjoint (n + 2)-balls. By the argument of proposition 1 we see that (b) $\Longrightarrow$ (a), and of course we have (a) $\Longrightarrow$ (b) and (c) $\Longrightarrow$ (b). To show that (b) $\Longrightarrow$ (c), let L be I- equivalent with the splittable L'. By tying small knots in the components of L' we may produce a new splittable link L" isotopic with L' and with components L_i'' determining the same cobordism class in C_n^n as do their counterparts L_i . Applying theorem 3, L and L" are concordant and theorem 4 is proved.

An erroneous result of Gutierrez (1973) stated that for each $n > 1$, every locally flat link L^n in S^{n+2} was cobordant to a splittable link. Thus it was believed for several years that every link was I - trivial. But counterexamples were published independently, by Cappel-Shaneson (1980), ($n \geq 5$) and Kawauchi (1980): for each *odd* $n > 1$ there are links L^n in S^{n+2} which are not "split cobordant". Apparently, Gutierrez' result does hold for *even* n (see Kawauchi (1980)), at least for *boundary* links.

Corollary 1.

For odd $n \geq 1$ *and any integer* $\mu > 0$ *there are links* L^n *of* μ *components in* S^{n+2} *which are not* I *- trivial.*

Another geometric consequence of Theorem 1 concerns link homotopy. The n-dimensional links L and L' in M^m are *link-homotopic* provided there is a homotopy $h_t : S_1^n \cup \ldots \cup S_\mu^n \to M^m$ between $h_0(\mu S^n) = L$ and $h_1(\mu S^n) = L'$ such that for all t, $h_t(S_i^m)$ and $h_t(S_j^m)$ do not intersect when $i \neq j$. In the classical case $n = 1$ it was shown, independently, by Giffen (1979) and Goldsmith (1979) and Fenn (unpublished) that cobordism implies homotopy. Clearly also isotopy implies homotopy. So theorem 1 allows us to conclude (as Giffen (1979) noted):

Corollary 2.

For 1-dimensional links in any 3-manifold M^3, *I-equivalence implies homotopy.*

ALGEBRAIC RESULTS.

Suppose some algebraic object associated with (locally flat) links happens to be invariant under both cobordism and isotopy. Then theorem 1 implies that it is also invariant under I-equivalence. The Milnor $\overline{\mu}$-invariants (Milnor, (1957)) are an example for classical links. This includes as a special case the pairwise linking numbers of the components of the link. A related invariant is the set of lower central quotients G/G_2, G/G_3, ... where $G = \pi_1(S^3 - L)$ and G_i is the ith lower central subgroup. These are invariant under I-equivalence, even topologically, as shown by Stallings (1965) in the paper introducing the term "I-equivalence". Massey has used this to show that the "completion" of the Alexander invariant $H_1(\sim)$ is a topological I-equivalence invariant, see Massey (1980).

GOLDSMITH'S INVARIANTS.

$\Lambda_{\tilde{M}}(L)$. Here is a summary of the construction of Goldsmith (1977). Let $L = K_1 \cup \ldots \cup K_\mu \subset M^3$ be a 1-dimensional link in a 3-manifold. Let A be a specified group, and let $\pi : \tilde{M} \to M$ be a (branched or unbranched) covering space which is 'canonically defined' by L and has branch set a specified sublink of L. Moreover, suppose the group of covering translations of $\tilde{M}$ is (canonically) identified with A. For each component K_i which is *not* in the branch set, let $\tilde{K}_i \subset \pi^{-1}(K_i)$ be an oriented component. Construct the matrix $\Lambda_{\tilde{M}}(L) = [\lambda_{ij}]$ where

$$\lambda_{ij} = \sum_{\sigma \in A} \ell k_{\tilde{M}}(\tilde{K}_i, \sigma \tilde{K}_j) \cdot \sigma$$

regarded as a member of the group algebra $\mathbb{Q}(A)$. The linking number $\ell k_{\tilde{M}}$ in $\tilde{M}^3$ is assumed defined (as a rational number). By convention $\ell k_{\tilde{M}}(x, x) = 0$. The rows and columns correspond to those indices i and j which are *not* those of branch components.

Two matrices Λ and Λ' with entries in $\mathbb{Q}(A)$ will be called *equivalent* iff there is a diagonal matrix

$$B = \begin{bmatrix} \sigma_1 & & 0 \\ & \ddots & \\ 0 & & \sigma_n \end{bmatrix}$$

with entries $\sigma_i \in A$, such that $\Lambda' = B^{-1} \Lambda B$. This corresponds to changing choice of lifting $\tilde{L}_i$ within $\pi^{-1}(L_i)$. Goldsmith's main result is that if L and L' are *cobordant* links in M^3, then their respective matrices $\Lambda_{\tilde{M}}(L)$ and $\Lambda_{\tilde{M}'}(L')$ are equivalent. She also states (without proof, although the proof is easy) that the equivalence class $[\Lambda_M(L)]$ is also invariant under isotopy. So we have, via theorem 1 :

Theorem 5.

The matrix class $[\Lambda_{\tilde{M}}(L)]$ *is invariant under* P.L. I-*equivalence.*

Example 2.

Consider the following link in $M = S^3$.

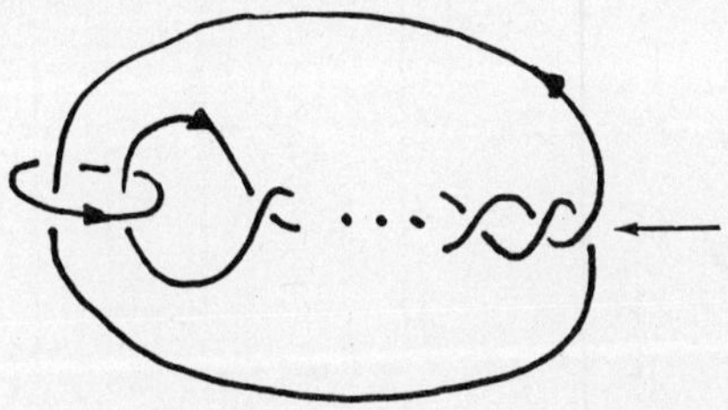

As covering $\tilde{M}$ we take the infinite cyclic covering of the complement of the second component. Since K_2 is unknotted, we have $\tilde{M}$ homeomorphic with $\mathbb{R}^3 = \mathbb{R}^1 \times \mathbb{R}^2$ with covering group $A = \mathbb{Z}$ generated by the translation $\tau : (x, y) \to (x+1, y)$. The liftings of L_1 look like this:

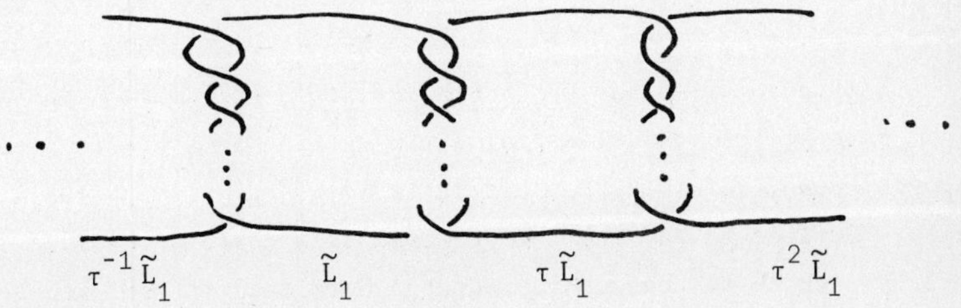

The associated 1×1 Goldsmith matrix is $\Lambda_{\tilde{M}}(L) = [\,q\tau + q\tau^{-1}\,]$. For different q, these are inequivalent.

Proposition 3.

For different values of q, *the links of example 2 are in different* I - *equivalence classes.*

This includes the case $q = 0$ (trivial link) and the two Whitehead links $q = \pm 1$.

Example 3.

Consider links of three components in $M = S^3$. Let $\tilde{M}$ denote 2-fold branched covering over the third component. Its translation group is $A = \mathbb{Z}_2 = \{1, \sigma\}$. Then for the Borromean link

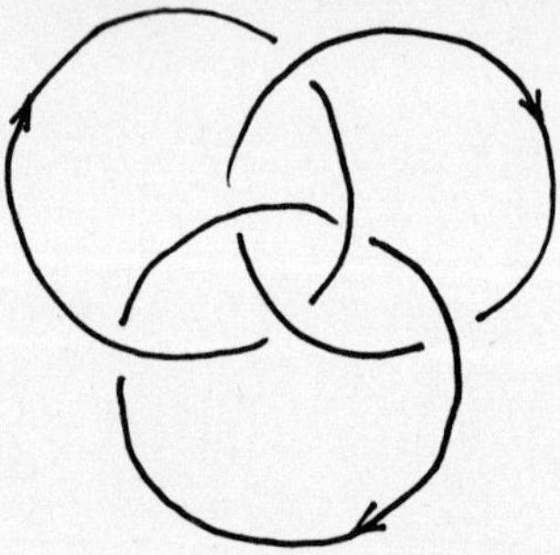

you will enjoy calculating the Goldsmith matrix as

$$\Lambda_{\tilde{M}}(L) = \begin{bmatrix} 0 & \sigma - 1 \\ \sigma - 1 & 0 \end{bmatrix}.$$

This matrix is not equivalent with the zero matrix so we may conclude:

Proposition 4.

The Borromean link is not I - *trivial.*

This is a good place to point out that K. Murasugi has, in his talk at this conference, shown that the Milnor $\bar{\mu}$ - invariants are related to linking and intersection numbers in covering spaces of the type considered by Goldsmith.

HILLMAN'S LOCALIZATION OF BLANCHFIELD'S PAIRING.

Here we consider classical links L in S^3, and to avoid getting bogged down, refer to Blanchfield (1957), and Hillman (1981), for details. Suffice to say that L determines a certain intersection pairing $B(L)$, which may be considered as a Hermitian matrix with entries in K/Λ. Here Λ denotes the ring of Laurent polynomials in variables.

$$\Lambda = \mathbb{Z}[t_1, \ldots, t_\mu, t_1^{-1}, \ldots, t_\mu^{-1}] .$$

"Hermitian" is defined relative to the conjugation $\overline{f}(t_1, \ldots, t_\mu) = f(t_1^{-1}, \ldots, t_\mu^{-1})$, and K is the field of fractions of Λ,

$$K = \mathbb{Q}(t_1, \ldots, t_\mu) .$$

$B(L)$ may be regarded as an element of a certain Witt group, and its class is an invariant of the *ambient* isotopy type of L. But it is *not* invariant under cobordism or isotopy or I-equivalence. To remedy this, Hillman considered localization with respect to the multiplicative set $\Sigma \subset \Lambda$ generated by all nonzero polynomials involving just one variable. (Actually he could have used the finer localization with respect to products of knot polynomials, as in Rolfsen (1975).) This defines an enlarged ring $\Lambda_\Sigma = \{p/q : p \in \Lambda, q \in \Sigma\}$ and a localized Blanchfield pairing $B_\Sigma(L)$ with entries in K/Λ_Σ. Hillman argues that the Witt class $B_\Sigma(L)$ is invariant under both cobordism and isotopy, so we have this result.

Theorem 6.

If L *and* L' *are* I*-equivalent links in* S^3, *then* $B_\Sigma(L) = B_\Sigma(L')$.

Corollary 3.

If L *is* I*-trivial, then* $B_\Sigma(L) = 0$.

Notes.

There seems to be fruitful work yet to be done on Blanchfield pairings in higher dimensions, vis-a-vis I-equivalence.

Hillman's work, and conversations with him, are what prompted me to consider the question: does I-equivalence coincide with the equivalence relation generated by isotopy and cobordism? Since discovering the answer, I've found that some authors (Kawauchi (1978, 1980), and Giffen ( )) apparently already knew the answer, at least in classical dimensions; but it is somewhat buried in their work.

ALEXANDER'S POLYNOMIAL.

Hillman's work, discussed above, implies a somewhat weaker form of the following (see Hillman (1981), p.126). Let $\Delta = \Delta(t_1, \ldots, t_\mu)$ be the (first nonzero) Alexander polynomial of a classical link $L = L_1 \cup \ldots \cup L_\mu \subset S^3$. It is defined only up to multiplication by units of the (Laurent) polynomial ring Λ (i.e. monic monomials). As usual, $\doteq$ denotes "equal up to units".

Theorem 7.

If L *and* L' *are cobordant links with Alexander polynomials* Δ *and* Δ', *then*

$$f\bar{f}\,\Delta \doteq f'\bar{f}'\,\Delta'$$

for some μ-*variable polynomials* f *and* f' *satisfying* $f(1, \ldots, 1) = f'(1, \ldots, 1) = 1$.

The recent result of Kawauchi (1978) and Kakagawa (1978) generalizes a formula of Fox and Milnor for cobordant knots. There is a similar criterion for isotopy, which follows from Rolfsen (1975).

Theorem 8.

If L *and* L' *are isotopic links in* S^3, *then*

$$g_1(t_1) \ldots g_\mu(t_\mu)\,\Delta \mathrel{\overset{\circ}{=}} g_1'(t_1) \ldots g_\mu'(t_\mu)\,\Delta'$$

where $g_i(t)$ *and* $g_i'(t)$ *are knot polynomials, i.e.* $g_i = \bar{g}_i$ *and* $g_i(1) = 1$, *similarly for* g_i'.

These results combine with theorem 1 to show (as noted by Kawauchi) the following.

Theorem 9.

If L *and* L' *are* I-*equivalent links in* S^3, *then*

$$f\bar{f}\,g_1(t_1) \ldots g_\mu(t_\mu)\,\Delta \mathrel{\overset{\circ}{=}} f'\bar{f}'g_1'(t_1) \ldots g_\mu'(t_\mu)\,\Delta'$$

where the polynomials f, f', g_i, g_i' *are as above.*

Example 4.

The following links are all I-equivalent:

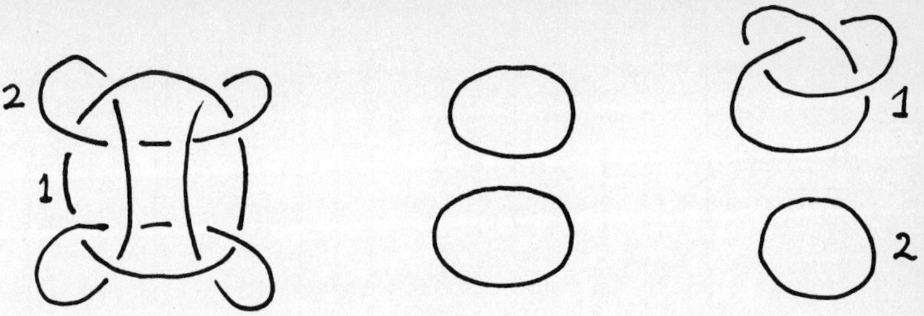

$\Delta = (1 - t_1 + t_1 t_2)(1 - t_2 + t_1 t_2)$ $\quad \Delta = 1$ $\quad \Delta = 1 - t_1 + t_1^2$

Their polynomials are, accordingly, related as in Theorem 9. Note that if $f = 1 - t_1 + t_1 t_2$, then $\bar{f} = 1 - t_1^{-1} + t_1^{-1} t_2^{-1} \doteq t_1 t_2 - t_2 + 1$.

Example 5.

For the following two links $\Delta = 1 + t_1 t_2$ and $\Delta' = 1 + t_1^3 t_2$.

L = 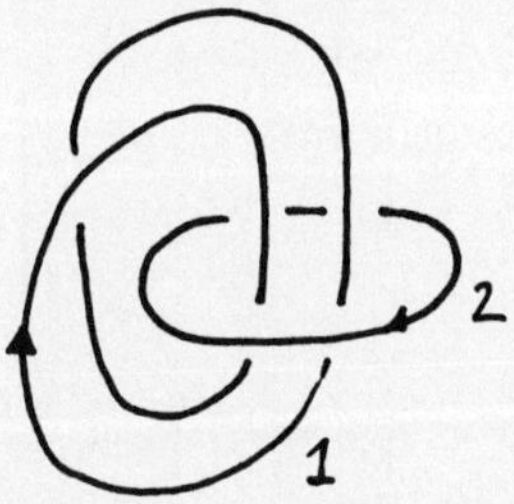

L' = 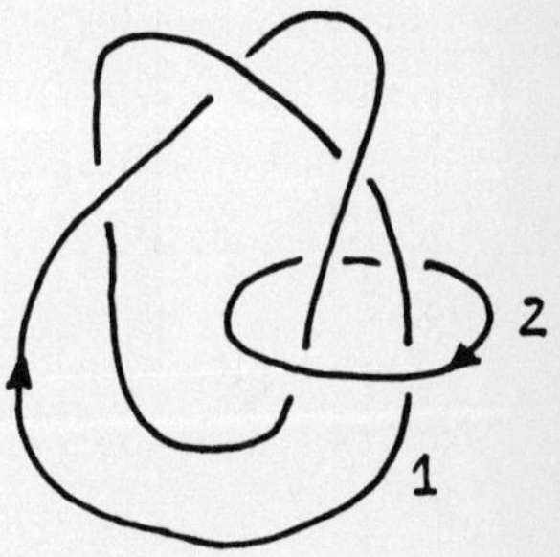

One can show using unique factorization that Δ and Δ' do not satisfy an equation as in theorem 9. Therefore L and L' are not I-equivalent. Notice, however, that they have homeomorphic complements!

REDUCED SIGNATURE.

Introduced first for classical knots by Trotter (1962), signature was defined for oriented links by Murasugi (1965), and shown to be invariant under cobordism. In fact there is an infinite family of signatures, the so-called θ-signatures or p-signatures of Milnor (1967) and Tristram (1969) (see also Litherland's talk (from the Conference at Sussex in 1977)). Signatures of higher-dimensional links have been defined by Viro (1973), who related them to signatures of certain branched covering manifolds (see also Kauffman and Taylor (1976)). For simplicity, I will restrict discussion to classical dimensions and ordinary signature, although the results on reduced signature generalize easily to the other cases.

First we need the notion of a Seifert matrix for an oriented link L^1 in S^3. Let $M^2 \subset S^3$ be a Seifert surface for L, that is M is an oriented surface (not necessarily connected) with L as its oriented boundary. Choose a basis $a_1, \ldots, a_n$ for $H_1(M)$ and define the matrix $V = [v_{ij}]$, where

$$v_{ij} = \ell k\,(a_i^+, a_j)\,.$$

Here a_i^+ is the cycle in $S^3 - M$ obtained by pushing a_i off M in the "positive" normal direction. Then V is called a *Seifert matrix* for L.

The signature $\sigma(L)$ of L is defined to be the signature of the symmetric matrix $V + V^T$. In other words, $V + V^T = XDX^T$ for some diagonal rational matrix D and invertible rational matrix X, and $\sigma(L)$ is the number of positive entries of D minus the number of negative entries. The *nullity* of L is the nullity of $V + V^T$ plus the number of components of M, denoted (L). According to Kauffman and Taylor (1976), we have the following.

Theorem 10.

If L *and* L' *are links in* S^3 *which are* I-*equivalent (or even just topologically* I-*equivalent), then* $\eta(L) = \eta(L')$.

Signature, on the other hand, is certainly not an invariant of I-equivalence. For example, all knots are I-equivalent but any even integer is the signature of some knot. Accordingly, we define *reduced signature* $\bar{\sigma}(L)$ of a link $L = L_1 \cup \ldots \cup L_\mu$ in S^3:

$$\bar{\sigma}(L) = \sigma(L) - \sigma(L_1) - \ldots - (L_\mu).$$

Theorem 11.

If the links L *and* L' *in* S^3 *are* I-*equivalent, then* $\bar{\sigma}(L) = \bar{\sigma}(L')$.

We will see by example that this is *not* true for *topological* I-equivalence of P.L. links!

Proof. For each $i = 1,\ldots, \mu$, let K_i be a knot whose cobordism class satisfies $[K_i] = [L_i'] - [L_i]$. Then revise L by adding (connected sum) to each component, L_i, a knot of type K_i in a small ball. Call the resulting link L''. It is isotopic to L (by Alexander's trick) hence I-equivalent with L'. Moreover, for each i, $[L_i'] = [L_i'']$ and so by theorem 3 we have that L'' and L' are cobordant links. It follows (Kauffman and Taylor (1976), Murasugi (1965)), that $\sigma(L'') = \sigma(L')$. Since a Seifert surface for L'' may be obtained by taking boundary connected sums of a Seifert surface for L with surfaces for the K_i; we also have

$$\sigma(L'') = \sigma(L) + \sigma(K_1) + \ldots + \sigma(K_\mu).$$

But $\sigma(K_i) = \sigma(L_i') - \sigma(L_i)$ and so we conclude that

$$\sigma(L') - \sigma(L_1') - \ldots - \sigma(L_\mu') = \sigma(L) - \sigma(L_1) - \ldots - \sigma(L_\mu),$$

as was to be proved.

Example 6.

The links L and L' pictured below both have zero linking number. Moreover, they have homeomorphic complements and the same signature. Yet they are not I-equivalent because their reduced signatures differ. Furthermore, neither is I-trivial, since the signature (and reduced signature) of the unlink is zero. Note that the unlink has nullity 2, whereas $\eta(L) = \eta(L') = 1$, so they are not even *topologically* I-trivial.

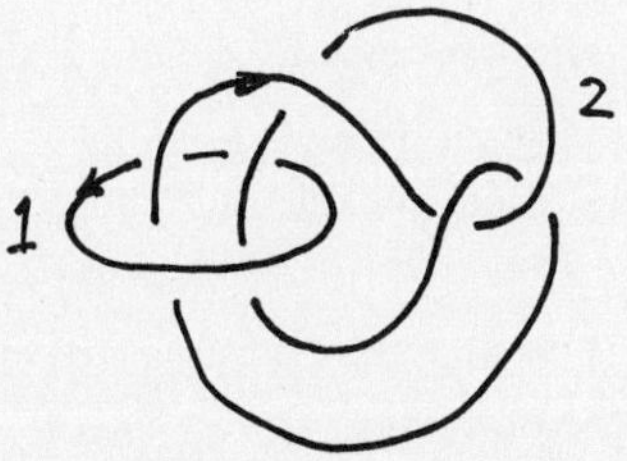

$\sigma(L) = 1$

$\sigma(L_1) = 0$

$\sigma(L_2) = 2$

$\bar{\sigma}(L) = -1$

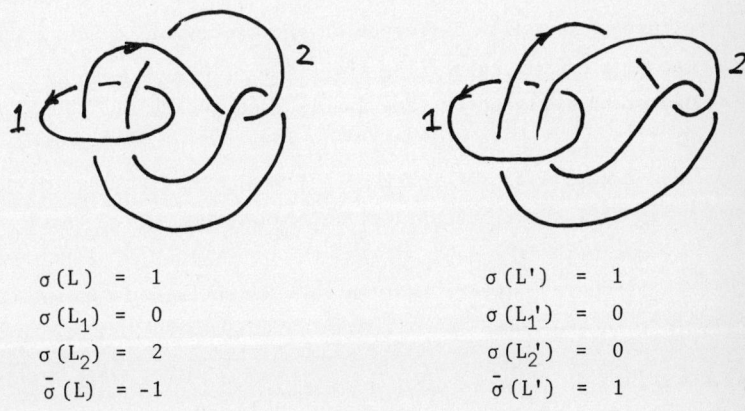

$\sigma(L') = 1$

$\sigma(L_1') = 0$

$\sigma(L_2') = 0$

$\bar{\sigma}(L') = 1$

Finally, neither of these links is I-equivalent with a boundary link, according to Proposition 5.

Proposition 5.

The signature and reduced signature of a classical boundary link are even integers.

Proof: By definition, $L = L_1 \cup \ldots \cup L_\mu$ is a boundary link if it has a Seifert surface $M = M_1 \cup \ldots \cup M_\mu$ which is a disjoint union with $\partial M_i = L_i$. Being an orientable surface with one boundary component, the 1st homology of each M_i has even rank, so (as is well-known) $\sigma(L_i)$ is even for each i. But rank $H_1(M)$ = rank $H_1(M_1) + \ldots +$ rank $H_1(M_\mu)$ is also even, so $\sigma(L)$ is even and so is $\bar{\sigma}(L)$.

Example 7.

There is a remarkable construction of Giffen (1979), called shift-spinning, which shows that the following links are *topologically* I-equivalent.

L =  L' = 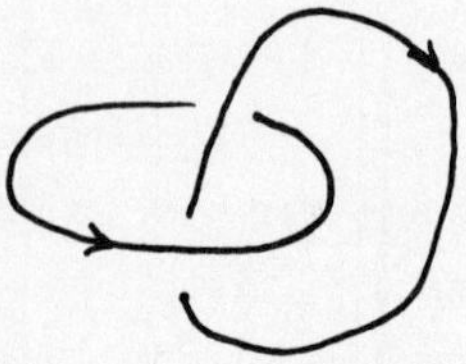

However, they are not P.L. I-equivalent, as calculation shows that $\bar{\sigma}(L) = \sigma(L) = -3$ while $\bar{\sigma}(L') = \sigma(L') = 1$.

This final example illustrates that the PL and topological categories are essentially different in the theory of I-equivalence of links. The wild annuli connecting the (tame) links L and L' in $S^3 \times I$ cannot be replaced by tame annuli even though all components are unknotted. It seems to me that Giffen's ideas point to new directions in which other surprises are in store. I will close with a few questions which, as far as I know, haven't yet been answered.

Question 1.

Is there a (wild) knot in S^3 which is not topologically I-equivalent to the unknot?

Question 2.

Same as 1, but with "isotopic" replacing I-equivalent.

Question 3.

Is there a higher-dimensional (PL) link which is not I-equivalent with a boundary link?

Question 4.

In higher dimensions, does topological I-equivalence (of P.L. links) imply P.L. I-equivalence?

Question 5.

Does P.L. I-equivalence imply homotopy for n-links in S^{n+2}, $n > 1$?

REFERENCES

Blanchfield, R.C. (1957). Intersection theory of manifolds with operators with applications to knot theory. Ann. of Math. 65 34-356.

Cappell, S. & Shaneson, J. (1980). Link cobordism. Comment. Math. Helv. 55 20-49.

Fox, R.H. & Milnor, J. (1966). Singularities of 2-spheres in 4-space and cobordism of knots. Osaka J. Math. 3, 257-267.

Giffen, C. (1979). Link concordance implies link homotopy. Math. Scand. 45, 243-254.

Giffen, C. (1979). F-isotopy and I-equivalence of links. Preprint.

Goldsmith, D. (1979). Concordance implies homotopy for classical links in M^3. Comment Math. Helv. 54, 347-355.

Goldsmith, D. (1977). A linking invariant of classical link concordance. In Knot Theory, Proceedings, Plans-sur-Bex, Switzerland ed. J.-C. Haussman, Springer-Verlag, 135-170.

Gutierrez, M. (1973). Unlinking up to cobordism. Bull. AMS 79, 1299-1302.

Hillman, J. (1981). Alexander Ideals of Links. Springer Verlag Lecture Notes in Mathematics no.895.

Hudson, J.F.P. (1966). Extending piecewise linear isotopies. Proc. Lond. Math. Soc. 16, 651-668.

Kauffman, L. & Taylor, L. (1976). Signature of links. T.A.M.S. 216, 351-365.

Kawauchi, A. (1978). On the Alexander polynomials of cobordant links. Osaka J. Math. 15, 151-159.

Kawauchi, A. (1980). On links not cobordant to split links. Topology 19, 321-334.

Kervaire, M. (1971). Knot cobordism in codimension 2. In Manifolds-Amsterdam, ed. N.H. Kuiper, 85-105.

Litherland, R. (1977). Signatures of iterated torus links. In Topology of Low-Dimensional Manifolds, Proceedings, Sussex 1977, ed. R.A. Fenn, Springer-Verlag, 71-84.

Massey, W.S. (1980). Completion of link modules. Duke Math. J. 47, 399-420.

Milnor, J. (1957). Isotopy of Links. In Algebraic and Geometric Topology, A symposium in honor of S. Lefschetz, ed. R.H. Fox, D.S. Spencer & W. Tucker, Princeton U. Press, 280-306.

Milnor, J. (1967). Infinite cyclic coverings. In Conference on the Topology of Manifolds, Michigan State U. 1967, Prindle, Weber & Schmidt, 115-133.

Murasugi, K. (1965). On a certain numerical invariant of link types. T.A.M.S. 117, 387-422.

Nakagawa, Y. (1978). On the Alexander polynomials of slice links. Osaka J. Math. 15, 161-182.

Rolfsen, D. (1972). Isotopy of links in codimension two. J. Indian Math. Soc. 36, 263-278.

Rolfsen, D. (1974). Some counterexamples in link theory. Canadian J. Math. 26, 978-984.

Rolfsen, D. (1975). Localized Alexander invariants and isotopy of links. Ann. of Math. 101, 1-19.

Stallings, J.R. (1965). Homology and central series of groups. J. Algebra 2, 170-181.

Tristram, S.G. (1969). Some cobordism invariants for links. Proc. Camb. Phil. Soc. 66, 251-264.

Trotter, H.F. (1962). Homology of group systems with applications to knot theory. Ann. of Math 76, 464-498.

Viro, O. Ya. (1973). Branched coverings of manifolds with boundary, and invariants of links. Math. USSR-Izv. 7, 1239-1256.

SOME CLOSED INCOMPRESSIBLE SURFACES IN KNOT COMPLEMENTS WHICH SURVIVE SURGERY

Hamish Short

0 *INTRODUCTION*

I wish to present here two theorems, the first of which is mainly due to Menasco (and is not unrelated to the little known work of Connor). We show that certain closed incompressible surfaces in the complement of a knot remain incompressible after all (p,q)-surgeries ("survive" surgery).

The surfaces which interest us are generalizations of the incompressible, non-boundary parallel torus in the complement of a composite knot, and later, generalizations of the boundary of the solid torus used in the construction of doubled knots. The first class are also present in the complements of Connor's splittable knots (see §2, ii and Connor, A.C., 1969); Connor showed that these knots have property P (amongst other things), and in fact it is usually a straightforward task to generalise the proof of a result for composite knots to splittable knots.

Menasco shows (Menasco, W., 1984) that what I call 2m-surfaces (Definition 1) survive surgery, giving a simple geometric proof. It was in the course of studying Menasco's proof (case (b) of §3) that the generalizations to m-surfaces (Theorem 1) and d-surfaces (Theorem 2) were found. I have since learnt that Menasco was aware of the generalization to m-surfaces. Unfortunately here the result is not as strong, as we cannot deal with the case of integer $(q = \pm 1)$ surgery.

As corollaries we obtain properties P and R for knots with 2m- or d-surfaces, and the proof of the properties for knots with m-surfaces is reduced to the case of ± 1 and (0,1)-surgeries.

In §1 we define m- and 2m-surfaces and present the main theorem (Theorem 1). Before giving the proof, in §2 we give some examples of knots with these surfaces, including composite and splittable knots. In §3 the proof of Theorem 1 is given, following Menasco's proof, which essentially occurs as case b). In §4 we give the definition of d-surfaces and prove the analogous theorem for knots with d-surfaces. In Corollaries 1, 2 and 4, several results concerning properties P and R are given; for a list of known results and references on these two properties, see Kirby's list 1978 , 1.15 and 1.16. The methods given here to establish these properties are, I believe, relatively straightforward geometric ones, and do not "involve detailed and difficult analysis of the knot group", as Dale Rolfsen says of some of the methods of the papers listed by Kirby ("Knots and Links", page 283).

1 *DEFINITIONS AND STATEMENTS OF RESULTS*

By knot we mean a tame embedding of S^1 in S^3 , and we use N(k) (or just N when no confusion is possible) to denote a regular neighbourhood of k in S^3 , its boundary is dN(k) . A meridian and longitude for k (on dN(k)) are denoted by m and ℓ . Note that here a link has more than one component; for complete terminology see Rolfsen (loc. cit.).

The manifold obtained by (p,q)-surgery on the knot k we shall denote *S(k;p,q)* we shall always suppose that p and q are coprime, and that $(p,q) \neq (\pm 1,0)$. We obtain S(k;p,q) from S^3 by removing the interior of N(k) from S^3 and gluing on the solid torus $D^2 \times S^1$ such that $D^2 \times \{point\}$ is identified with $m^p \ell^q$.

A closed connected surface embedded in a 3-manifold is *incompressible* if it is not a 2-sphere and the inclusion map induces an injection on the fundamental group, or it is an essential 2-sphere. (We are only dealing with orientable manifolds here.)

Definition 1

A simple closed curve s in S^3 - N(k) is called an *m-loop* if there is a non-singular annulus A in S^3 - k such that $dA = s \cup m$, where m is a meridian on dN(k) .

A closed, orientable, incompressible, connected, non-boundary-parallel surface embedded in S^3 - N(k) is called an *m-surface* for k if it carries an m-loop.

An m-surface is called a *2m-surface* if it carries two m-loops which are non-isotopic on the surface, and which have annuli (as above) not meeting the surface in their interiors.

Recall that a 3-manifold which is compact, orientable,

irreducible and contains an incompressible 2-sided surface is said to be *Haken*.

Recently Menasco (1984) has given a beautiful geometric proof of :

Menasco's theorem

Let k be an alternating knot (or alternating non-splittable link).

Any closed, incompressible, non-boundary-parallel surface in $S^3 - k$ is an m-surface.

Later in the same paper, he proves the second half of the main theorem of this section (case (b) of the proof of Theorem 1). We here extend his proof to cover nearly all m-surfaces. (I have since discovered that Menasco was aware of this generalization.)

The main results of this first section are :

Proposition 1

Let k be a knot with an m-surface S, p, q coprime integers. If either i) $|q| > 1$;

or ii) S is a 2m-surface ;

then S is incompressible in S(k;p,q) .

A small extension of the method of proof gives :

Theorem 1

Under the same conditions as above, S(k;p,q) is Haken.

Proofs of these results are given in §3.

Recall that a knot is said to have property P (resp. property R , homotopy property R) if no non-trivial surgery gives a manifold which is simply connected, (resp. is $S^1 \times S^2$, has a fundamental group $\mathbb{Z}$) . We call the last "property R' ".

Corollary 1

Let k be a knot with a 2m-surface. Then k has properties P , R and R' .

Note that property R implies property R' ; for a list of known results concerning these properties, see Kirby, R. (1978) 1.15, 1.16.

To establish property P for knots with m-surfaces which are not 2m-surfaces, it remains to check $(\pm 1,1)$-surgery; note that composite knots have m-tori. (Other examples of knots with m- and 2m-surfaces are given in §2.)

Menasco in fact shows that the only m-surface in the complement of an alternating knot which is not a 2m-surface, is an m-torus, in which case the knot is not prime, and so we have :

Corollary 2 (Menasco)

Let k be a prime alternating knot whose complement contains a closed incompressible non-boundary parallel surface. Then k has properties P, R and R' , and S(k;p,q) is Haken for all coprime p,q .

Recall that a (p,q)-cable knot with core k is obtained as a simple closed curve on dN(k) representing $m^p \ell^q$ (when the core is trivial the knot so obtained is a (p,q)-torus knot.). It is known that (pq,1)-surgery on a (p,q)-cable knot gives a non-prime manifold (see Gordon, C.McA., 1983; for torus knots Moser, L., 1971) and it has been conjectured (see Gonzalez Acuna, F. & Short, H., preprint) that these are the only examples of knot surgery giving non-prime manifolds. This gives:

Corollary 3

A (p,q)-cable knot with |p| and |q| greater than 1 has no 2m-surface.

2 *EXAMPLES OF KNOTS WITH m-SURFACES*

i) Composite knots: In the complement of a composite knot there are two non-boundary-parallel incompressible tori, each of which carries an m-loop; in fact a knot with an m-torus is composite.

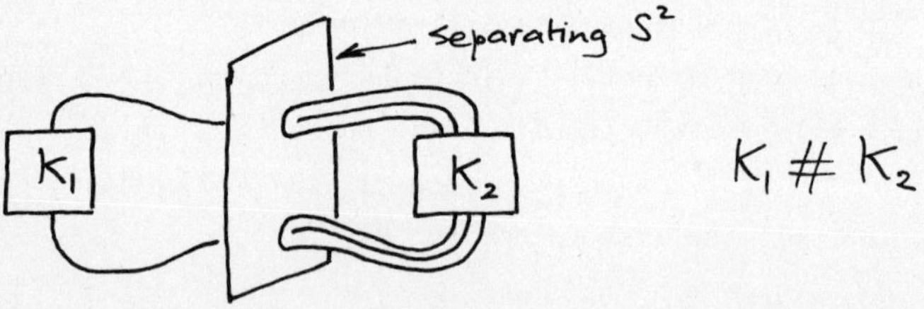

ii) Splittable knots: In his thesis and a number of unpublished papers, Connor, A.C. (1969), introduces the ideas of splittable, semi-splittable, and weakly splittable knots. A knot is said to be *splittable* if there is a closed surface S in S^3 meeting the knot in 2n points (with n greater than zero) such that S splits S^3 - N(k) into two boundary incompressible manifolds. When n = 1 and S is a two-sphere, then we have the definition of a composite knot. The two surfaces obtained from S by tubing up the boundary components of $S \cap \partial N(k)$ by annuli in $\partial N(k)$ are m-surfaces for the knot, and in fact 2m-surfaces when the knot meets S more than twice.

The definitions of semi- and weakly splittable knots are a bit too long to give here; Connor shows in any case that weakly splittable knots are splittable, and it can be shown that semi-splittable knots have 2m-surfaces. Amongst other things, Connor shows that splittable knots have Property P , essentially by generalizing the proofs for composite knots.

iii) Some Sums of Prime Tangles: A *tangle* on n strings is a pair (B,t) , where B is a 3-ball, and t is a set of n properly embedded arcs in B (see Lickorish, W.B.R., 1981). A *sum* of two tangles (both on the same number of strings) (B_1,t_1) and (B_2,t_2) is obtained by identifying ∂B_1 with ∂B_2 by a homeomorphism identifying ∂t_1 with ∂t_2 . We say that the link (possibly a knot) $t_1 \cup t_2$ in S^3 so obtained is a sum of the two tangles.

A tangle is said to be *separable* if there is a 2-disc D properly embedded in B such that $D \cap t = \phi$ and both components of B - D meet t .

A tangle is said to be *incompressible* if $(\overline{B - N(t)})$ has incompressible boundary; note that incompressible implies non-separable, but that the converse is false.

A tangle is said to be *prime* if it is non-separable, and any two-sphere in B meeting t in two points bounds a three-ball B' in B such that B' meets t in a single unknotted arc. Note that there are some prime tangles which are not incompressible, for example the first three examples of prime tangles given by Lickorish in (1981):

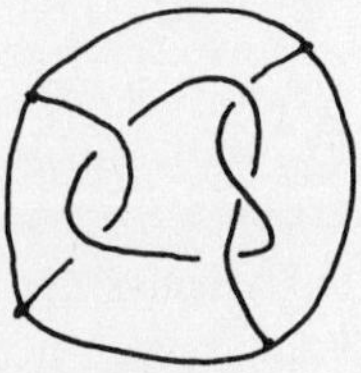

(a)

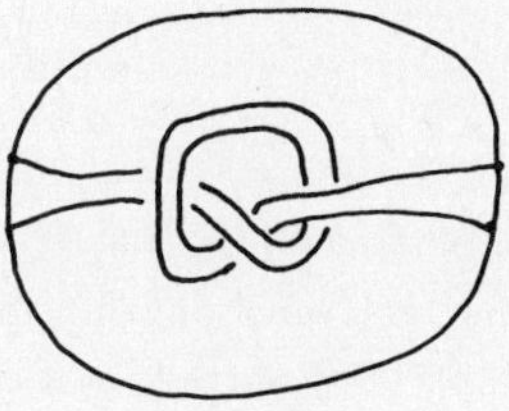

(b)

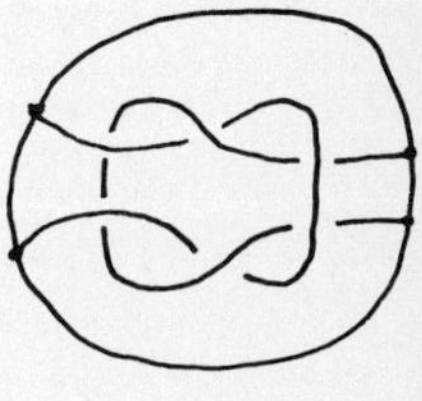

(c)

Lickorish shows that sums of prime tangles give prime links (or prime knots); we note here that the sum of an incompressible tangle and a non-separable tangle gives a link (or knot) with a 2m-surface (when each has more than one string). For instance H.R.Morton has pointed out that the following tangle is incompressible (in Menasco, W., 1984; it appears as example (d) in Lickorish, W.B.R., 1981): Thus any knot formed as the sum of this and a prime tangle has a 2m-surface; for instance:

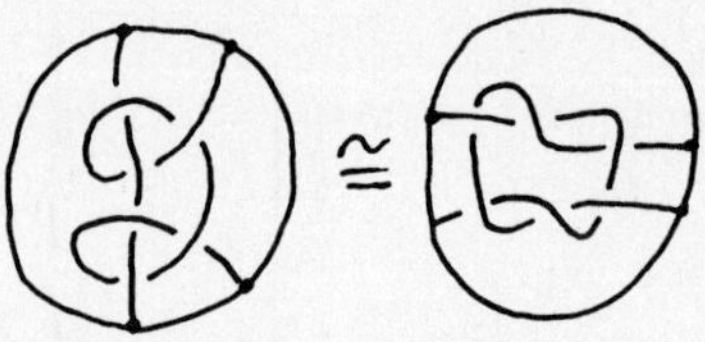

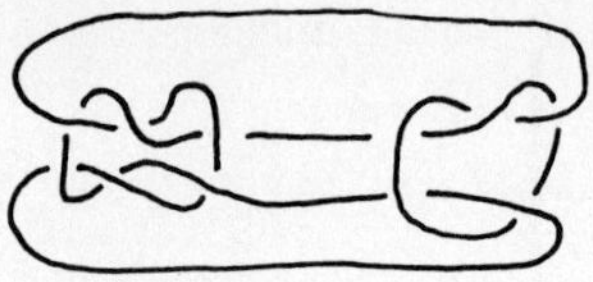

3 PROOF OF PROPOSITION 1

We can regard $\overline{S^3 - N(k)}$ as $M_1 \cup {}_S M_2$, where $dN(k) \subset M_1$; then $S(k;p,q) = M' \cup_S M_2$, where $M' = M_1 \cup {}_{dN}(S^1 \times D^2)$.

We know that S is not a 2-sphere as knot-complements are irreducible; so suppose that the theorem fails, and that S is a closed surface, not a 2-sphere, incompressible in $\overline{S^3 - N(k)}$, and S is compressible in S(k;p,q) . By Dehn's Lemma, there is a properly embedded disc D in S(k;p,q) such that $dD = D \cap S$; by the incompressibility condition, D lies in M' .

As S is an m-surface, let A be the annulus bounded by the m-loop s on S and the meridian m on dN(k) . After making A transverse to S (in $\overline{S^3 - N(k)}$) we can suppose that $A \cap S = s$, because: let $A \cap S = Z$; loops in Z which bound discs in A can be removed by disc exchange, and if Z contains a loop which is non-trivial on A , then we can replace A by the closure of the component of A - Z which contains m .

$D \cap dN(k)$

Make D transverse to dN ; in $D \cap dN$ there are simple closed curves, which are either trivial on dN and can be removed by disc exchange (innermost ones on dN first), or are $(m^p \ell^q)^{\pm 1}$-loops which are capped off by surgery, and thus bound discs in $\overline{(D - dN)}$ - these latter we call discs of D .

$D \cap A$

Make D transverse to A ; in $D \cap A$ there are closed loops, and arcs properly embedded in $\overline{D - (\text{discs of } D)}$. The closed loops which are innermost on D can be removed by disc exchange, as they must bound discs in A (m is non-trivial in $S^3 - k$). The components of $D \cap A$ can be transversely oriented, by choosing an orientation for k , which gives a sign to L and a positive side to A .

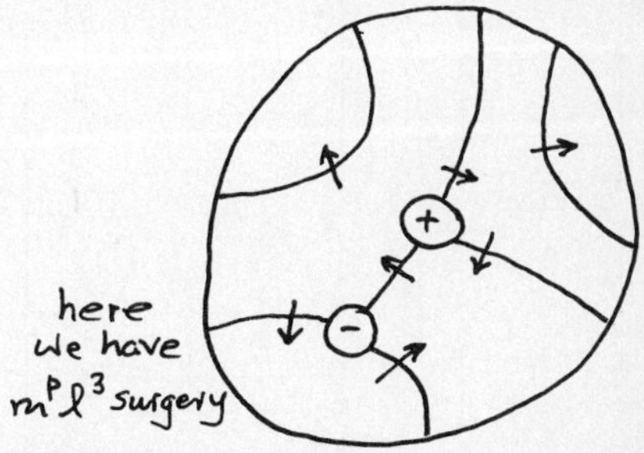

Note that the discs of D can be given a + or a - label, according to whether the disc oriented clockwise in D corresponds to $m^p L^q$ or to $m^{-p} L^{-q}$. Note also that each disc of D meets A in $|q|$ points, and that by assumption $q \neq 0$.

For our proof, we are going to study the *graph of* D , $G = dD \cup (D \cap (A \cup dN))$. The method of proof is to show how to remove all components of G except for dD , thus leaving a compression disc for S in M_1 , contrary to the assumed incompressibility of S in $S^3 - N(k)$.

Reduction I Remove an extremal arc from G . Suppose that G contains an arc a such that $D - a$ contains a component whose closure R is a 2-ball, and $R \cap G = dR = a \cup b$, where b is a segment of $dD - a$.

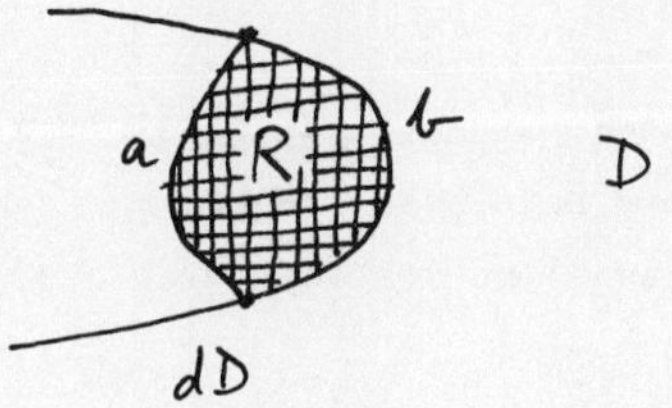

On A there is a 2-ball region R' such that $dR' = a \cup b'$, where b' is a segment of $s - a$.

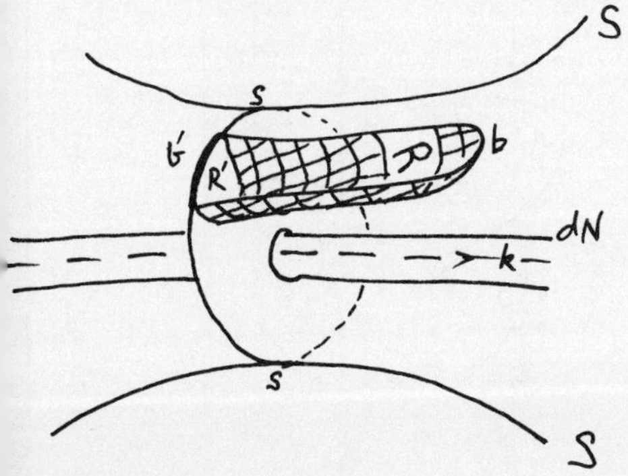

Then $b \cap b'$ is a simple closed curve on S which bounds the 2-cell $R \cup_a R'$ in M_1 , and hence by the incompressibility of S in M_1 , $b \cap b'$ bounds a 2-cell on S , so that we can replace the annulus A by A'

where $A' = (A - R') \cup_a R''$, where R'' is a copy of R pushed off slightly ; this new annulus has the same properties as A , and is still transverse to D (as $(\text{int}R) \cap D = \phi$) so the reduction is possible.

Reduction II Remove a pair of discs of D joined by an arc.

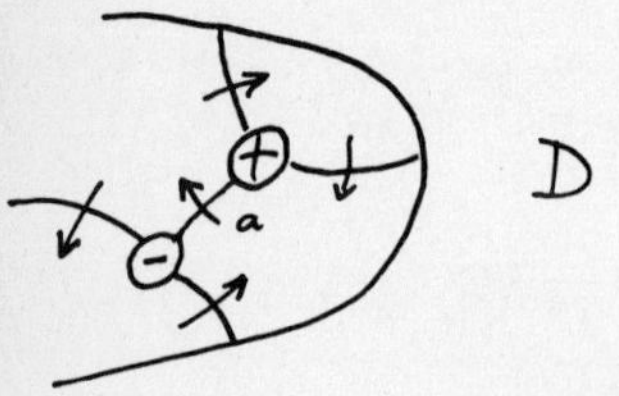

Note that the transverse orientation on the arcs assures us that only discs of opposite sign can be joined by an arc.

In A this corresponds to an arc with endpoints on the meridian m , and pushing an innermost (on A) such arc down onto dN , we see that a neighbourhood of (arc (two discs)) can be exchanged for a disc on dN , without affecting dD in any way.

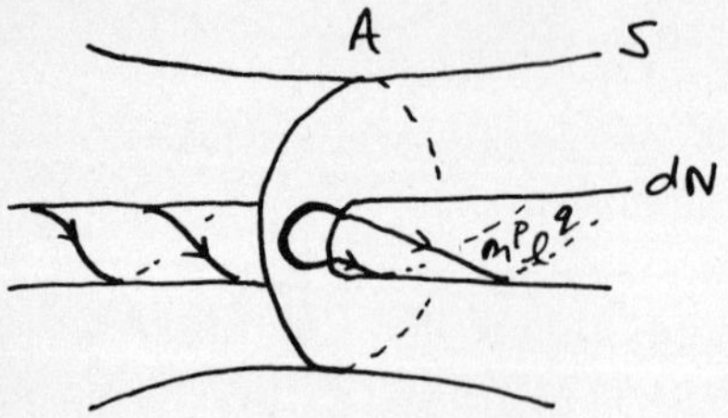

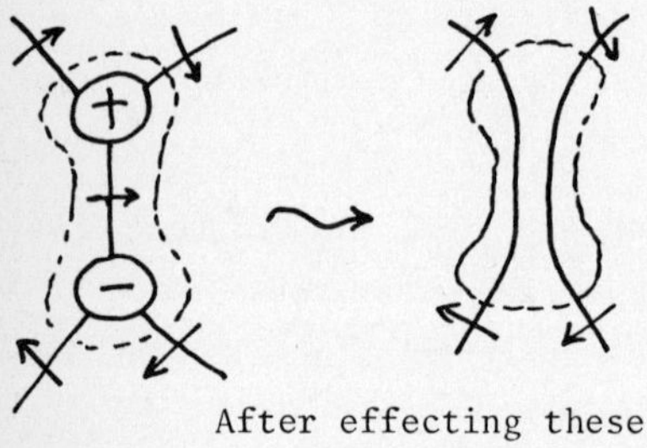

After effecting these two reductions as often as possible, and removing all closed loops in $A \cap D$ which may have been created by reductions of type II (innermost first, as usual) we are left with a graph G where all discs are joined by arcs to dD :

e.g. for $|q| = 3$ or if $|q| = 1$

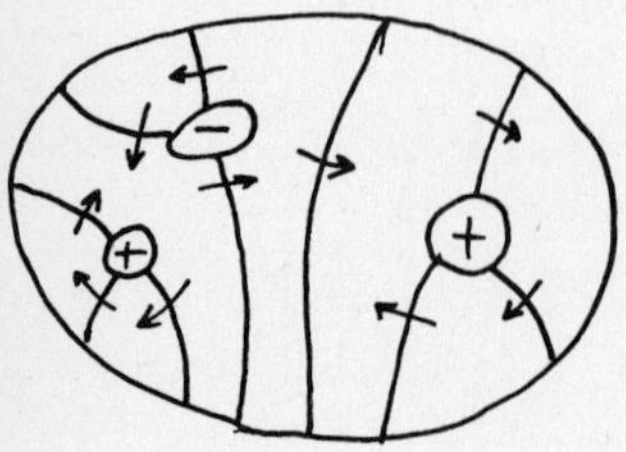

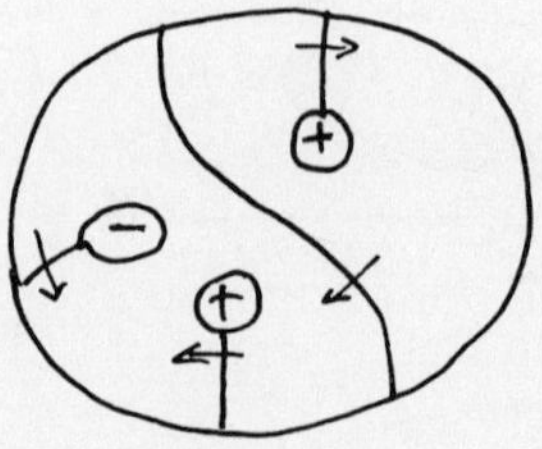

Case a) *If* $|q| > 1$ *then* S *is incompressible in* S(k;p,q)

In this case, each disc of D has at least two radiating arcs, so that there is some component of D - G whose closure R is a 2-cell such that $dR = a_1 \cup b \cup a_2 \cup c$, where a_1 and a_2 are arcs of $A \cap D$, b is a segment of $dD - \{a_1, a_2\}$, and c is a segment of some disc of D. Note that c "goes once along k", that is c lies on dN, and meets a meridian just twice, 'once from each side'.

Now let b' be a component of $s - \{a_1, a_2\}$; then the closed curve $b' \cup a_1 \cup c \cup a_2$ represents the element $m^x L$ in $\pi_1(S^3 - k)$ for some integer x. Also $b \cup b'$ represents an element $z \in \pi_1(S)$ which is non-trivial in $\pi_1(S)$ as $b \cup b'$ meets an m-loop (s moved slightly) just once.

The region R represents a null-homotopy in M_1, showing that $z(m^x L)^{-1} = 1$ holds in $\pi_1(M_1)$. As z and m are elements of $\pi_1(S)$, and $\pi_1(S)$ injects into the group $\pi_1(M_1)$, we have that L too must be in $\pi_1(S)$. The orientable surface S therefore carries commuting elements which are not powers of a common element, and is hence a torus; also S carries a meridian and a longitude, and is therefore boundary parallel, contrary to the original assumptions.

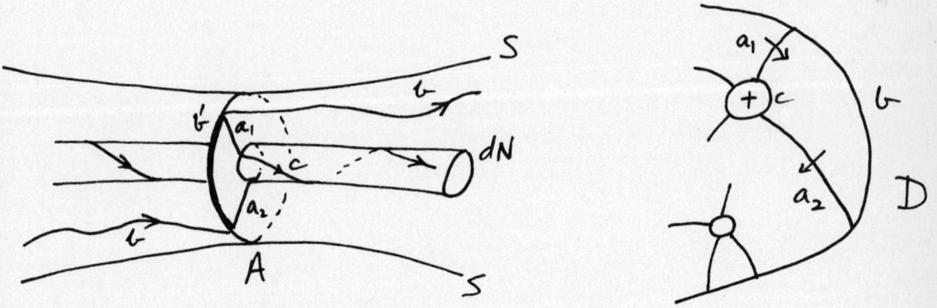

Case b) *If* S *is a 2m-surface then* S *is incompressible in* S(k;p,q) (Menasco, W., 1984).

Let A and A' be the annuli for the two m-loops s , s' ; by the definition of a 2m-surface, we have that A and A' can be taken to be disjoint and such that $S \cap (A \cup A') = \{s, s'\}$.

After reducing the graph G , add in the arcs due to $D \cap A'$;

as in case a) there must be an extremal region R ; that is, there is a component of $D - G$ whose closure R is a 2-cell, and $dR = a_1 \cup b \cup a_2 \cup c$, where a_1 is an arc of $D \cap A$, a_2 an arc of $D \cap A'$, b is a segment of $dD - (A \cup A')$, and c is a segment of a disc of D . We can think of this as a collar from an arc b on S to an arc c on dN .

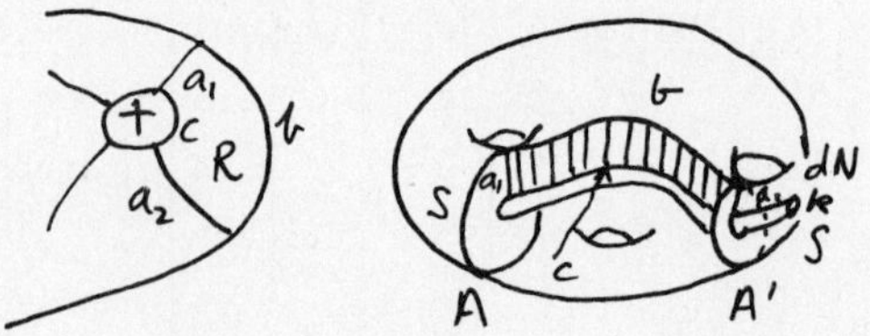

Now take two copies of R glued onto the annulus on dN enclosed by the two meridians m , m' ; this is a null-homotopy of the loop $bs'b^{-1}s^{-1}$. But this means that s and s' are conjugate in $\pi_1(S)$, as B is an arc on S , contrary to the definition of a 2m-surface.

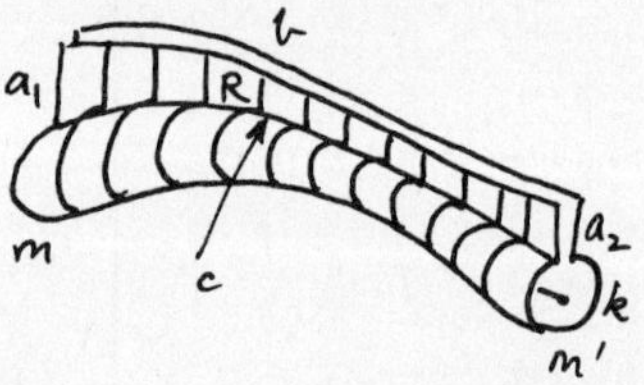

Thus S is incompressible in $S(k;p,q)$.

To complete the proof of Theorem 1, it remains to show that the surgered manifold is irreducible.

Lemma 1

Suppose that k is a knot with an m-surface which is incompressible in $S(k;p,q)$. Then $S(k;p,q)$ is irreducible.

Proof

Suppose that $S(k;p,q)$ is not irreducible, and let Z be an essential embedded 2-sphere in $S(k;p,q)$. Make Z transverse to S , so that the intersection is a set of closed curves; as S is incompressible, these are all trivial curves on S , so that innermost ones can be removed by disc exchange and we can suppose that Z does not meet S .

As before, we examine $Z \cap (A \cup dN)$; we know that Z does not meet $dA \cap S$, so that if there are any loops in $Z \cap dN$ corresponding to surgery discs, they must be joined to others of opposite sign, and can therefore be removed by reductions of type II. Thus we are left with closed loops of $Z \cap \text{int}(A)$, which can all be cut and pasted away, as before.

This moves our essential 2-sphere into $S^3 - N(k)$, which is an irreducible space, so that it bounds a 3-ball contradicting its assumed essential nature.

4 *GENERALIZED DOUBLES AND d-SURFACES*

We now adapt the method of §3 to obtain a similar result for a large class of knots which includes doubled knots (in much the same way that the class of knots with m-surfaces includes composite knots). Recall that for a doubled knot there is an incompressible torus embedded in S - k such that a section of S looks like:

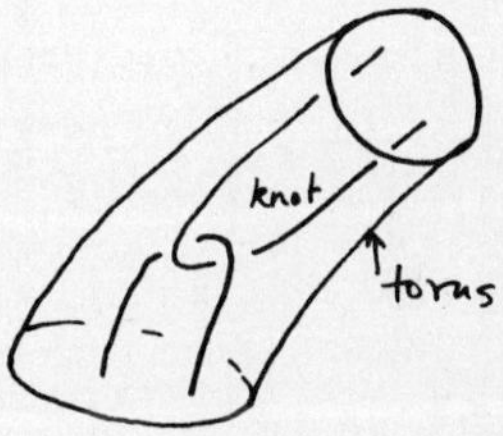

We want to generalize this contruction, retaining the essential points - that there is a section of S "inside" an incompressible surface, which meets the knot in two arcs, such that the arcs cannot be pushed out of this section. Formally:

Definition 2

A closed, orientable, incompressible, non-boundary parallel surface S embedded in $S^3 - N(k)$ is called a *d-surface* for k ("d" = double) if there is a homeomorphism h from $B^2 \times I$ into S^3 such that:

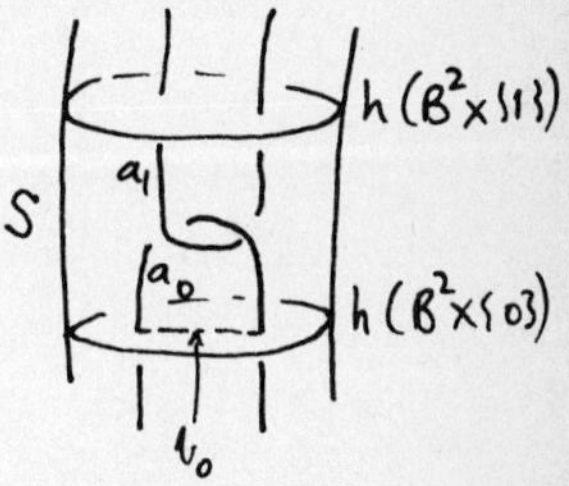

i) $h(B^2 \times I) \cap S = h(S^1 \times I)$

ii) $h(B^2 \times I) \cap k = \{\text{two arcs } a_0 , a_1\}$

iii) $h(B^2 \times \{i\}) \cap k = h(B^2 \times \{i\}) \cap a_i$ $= \text{two points, for } i = 0,1$.

iv) if b_i is an arc in $h(B^2 \times \{i\})$ such that $a_i \cup b_i$ is a loop, then $a_i \cup b_i$ does not bound a disc in $h(B^2 \times I) - \{a_{1-i}\}$ (a_i cannot be "pushed onto" $h(B^2 \times \{i\})$)

Doubles of non-trivial knots have d-surfaces, and replacing the full twist by any number of full twists we also get a knot with a d-surface, in fact a d-torus. This construction is studied in greater

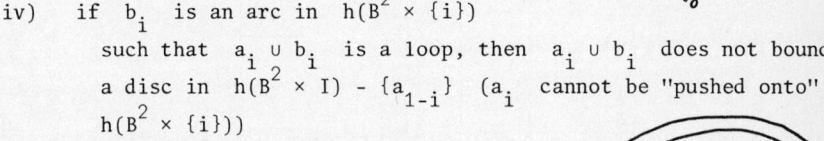

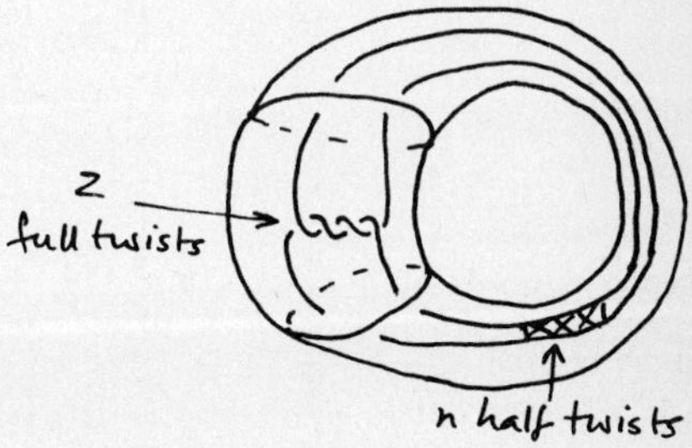

generality (allowing half twists) by Neuzil (Trans. AMS, 1975, vol. 204); he shows that these knots have property P .

Proposition 2

Let k be a knot with a d-surface S . Then S is incompressible in S(k;p,q) .

Proof

The proof follows that of §3, with the two "discs with two holes" A_0 and A_1 corresponding to $h(B^2 \times \{0,1\}) - N(k)$ playing the rôle of the annulus; we examine the graph $G = dD \cup D \cap (A_0 \cup A_1 \cup dN)$, where D is again an embedded disc compressing S .

In Lemma 2 we show that reduction I is possible, or S is a 2m-surface, in which case proposition 1 applies. Loops of $D \cap A_i$ which are innermost on D can be removed as in §3, and in Lemma 3 we show that after reductions of type II, there are no regions bounded by discs of the same sign, thus no discs.

Hence we show that D is a disc whose boundary represents a non-trivial loop on S , and D does not meet dN , contradicting the assumed incompressibility of S in $S^3 - k$.

Lemma 2

If S is a d-surface reduction I can be carried out. (That is, extremal arcs of G can be removed.)

Proof

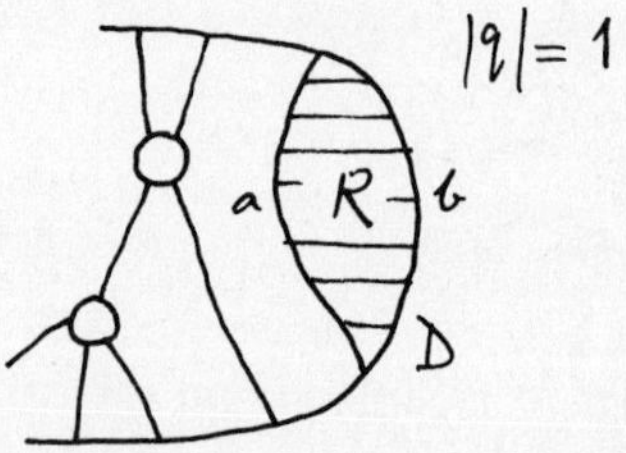

Let a be an extremal arc of G - that is there is a component of D - G whose closure is a 2-ball R such that $dR = a \cup b$ where b is a segment of $dD - \{A_0, A_1\}$. We can suppose that a is in $D \cap A_0$.

On A_0 there are two possible forms which a can have, depending upon whether a separates the boundary components or not.

Case i) a does not separate the boundary components of A_0.

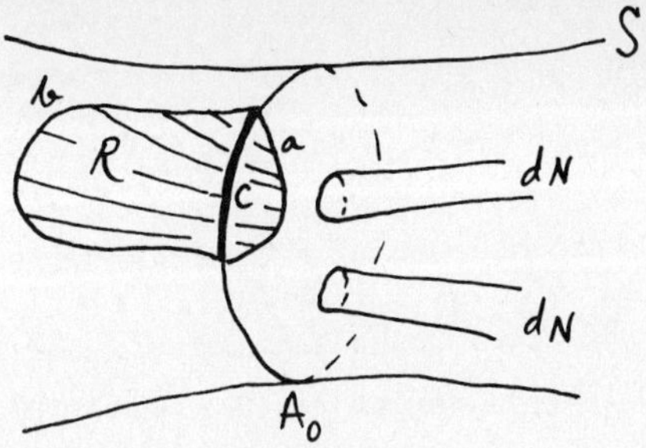

that is, there is a segment c on $A_0 \cap S$ such that $a \cup c$ bounds a two ball R' in A_0. As in lemma 1, we can alter A_0 so that this intersection is removed.

Case ii) a separates the boundary components of A_0.

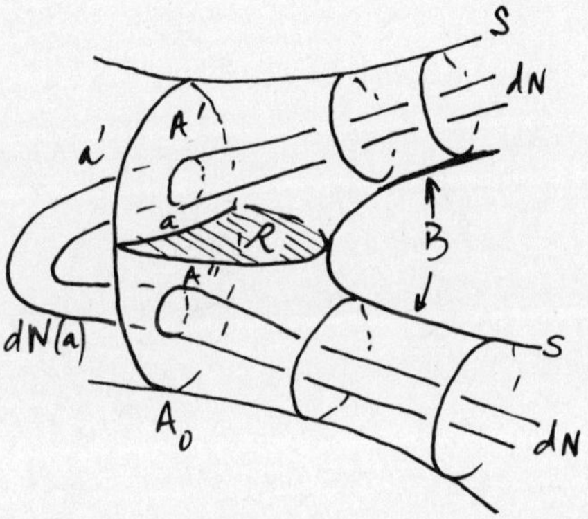

Let a' be a segment of $(dA_0 \cap S) - a$; then $a \cup a'$ together with a meridian on dN cobound an annulus A' in A_0. Gluing on the region R to A' along the arc a, shows that $a' \cup b$ is an m-loop on S. Similarly there is another m-loop on S, $a'' \cup b$, where a'' is the other segment of $(S \cap dA) - a$. These two m-loops can be perturbed slightly so that they bound disjoint annuli; to show that S is a 2m-surface, it remains to show that these m-loops are non-isotopic on S; call them s and s'.

Suppose that B is an annulus on S such that $\partial B = s \cup s'$; without loss of generality, we can suppose that B does not meet either A_0 or A_1. First extend B to $\partial A_0 \cap S$, giving B', a torus with a hole; now glue on A_0, giving a torus with two holes; now to this glue on an annulus $\partial N(a_0) \cap \partial N(k)$, to give a genus two closed surface which contains a_0 on one side, and a_1 on the other, contradicting the fact that k is a knot.

Thus in this case S is in fact a 2m-surface, and the proposition follows from Proposition 1.

Lemma 3

All discs can be removed from G.

Proof

As before, we can put a transverse orientation on the arcs of G, induced by the normals on A_0 and A_1 which point into $h(B^2 \times I)$. With this orientation, the arrows on the arcs radiating from a disc in G alternate in direction, and at least four arcs leave each disc.

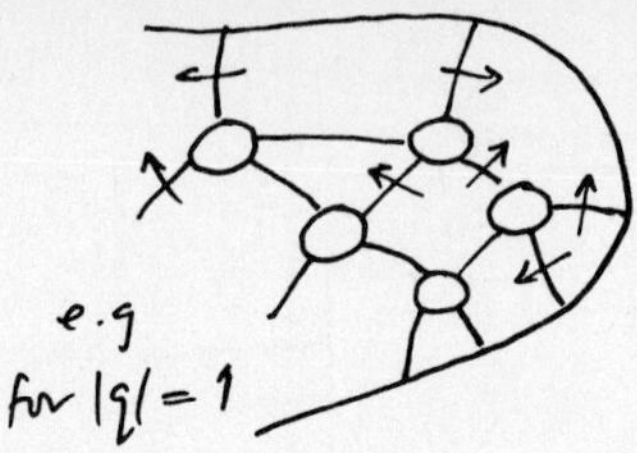

There is now a 2-ball region of $D - G$ whose closure R has the property that all arrows on the arcs point inwards and either $\partial R \cap \partial D = \phi$ or $\partial R \cap \partial D =$ a single connected segment of $D -$ (arcs). Note: $\partial R - S$ lies on $A_i \cup (N \cap h(B^2 \times I))$, as all the arcs in ∂R meet the inward normal side of A_i, and so we can suppose that $i = 0$, and that $\partial R \cap A_1 = \phi$.

An arc joins discs of opposite sign if and only if it has both endpoints on the same boundary component of $\partial A_0 \cup \partial A_1$, so that reduction II is applicable, as before.

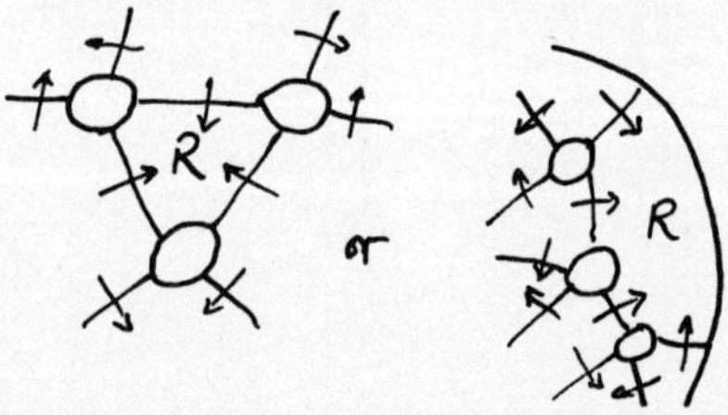

The only remaining arcs join discs with the same sign; again, by the orientations, the endpoints of such arcs must lie on different components of ∂A_0 (or of ∂A_1).

We first show that there is no interior region R as above; that is, one which does not meet ∂D. Here ∂R can be thought of as a cable about the loop $(a_0 \cup b_0)$ for some arc b_0 on $h(B^2 \times \{0\})$. In $h(B^2 \times I) - \{a_1\} = H$, the cable becomes $(a_0 \cup b_0)^k$ in $\pi_1(H)$, where k is the number of discs meeting ∂R. But this is non-trivial in $\pi_1(H)$, except when $k = 0$, so that there can be no such region (see illustration on following page).

Similarly, when $\partial R \cap \partial D = b$, we see that b is an arc on $h(B^2 \times I)$ so that as above, ∂R represents an element $(a_0 \cup b)^k$ in $\pi_1(H)$, which is non-trivial unless $k = 0$.

Hence all discs can be removed by reductions of type II.

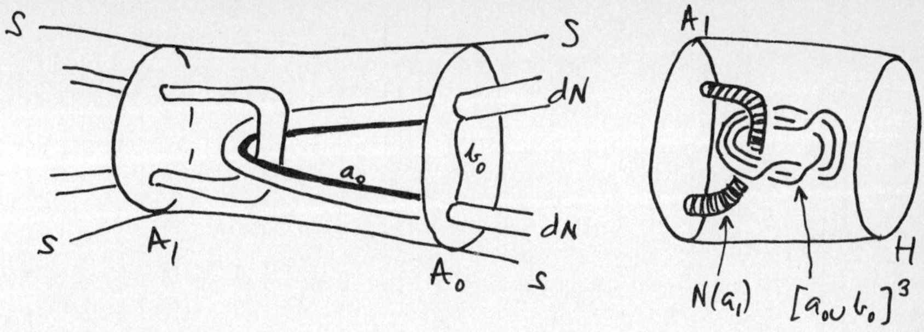

Exactly as in Lemma 1, irreducibility is established to give:

Theorem 2

Let k be a knot with a d-surface, p and q coprime integers. Then S(k;p,q) is Haken.

As in the corollaries in §1 , concerning properties P , R , and R' , we have:

Corollary 4

Let k be a knot with a d-surface S , p and q coprime integers. Then S is incompressible in S(k;p,q) ; in particular k has properties P , R , and R' .

Note that this corollary applies to all doubles of non-trivial knots, although not to twist-knots (doubles of the unknot), a known result (for references see [K] 1.15, 1.16) .

Note also that this implies that surgery on a double of a non-trivial knot never gives a manifold with a hyperbolic structure, as there is an incompressible torus.

Also, just as in Corollary 3, we have:

Corollary 5

A (p,q)-cable knot with $|p|$ and $|q|$ greater than 1 has no d-surface.

REFERENCES

Connor, A.C. (1969). Thesis, University of Georgia. (See also various unpublished articles, Splittable knots, Semi-splittable knots and Concerning splittable knots.)

Gordon, C. McA. (1983). Dehn surgery and satellite knots. Trans. AMS 275, 687-708.

Gonzalez Acuna, F. & Short, H. (Preprint) Knot surgery and primeness.

Kirby, R. (1978). Problems in low-dimensional manifold theory. Proc. Symposia in Pure Math. 32.

Lickorish, W.B.R. (1981). Prime knots and tangles. Trans. AMS 267, 321-332.

Menasco, W. (1984). Closed incompressible surfaces in alternating knot and link complements. Topology 23, 37-44.

Moser, L. (1971). Elementary surgery along a torus knot. Pacific J. of Math. 38, 737-745.

Morton, H.R. (1977). A criterion for an embedded surface in R^3 to be unknotted. In Topology of Low-Dimensional Manifolds, Proceedings, Sussex 1977. Lecture Notes in Maths., 722, Springer-Verlag. Ed. Roger Fenn.

SIMPLE ELEMENTS OF $\pi_2(M^3, x_0)$

Bärbel Wicha-Krause

1. *INTRODUCTION*

The problem here considered is to decide for a given element $S \in \pi_2(M, x_0)$ whether it is *simple*, i.e. can be represented by an embedded sphere. Here M is assumed to be a closed connected orientable PL 3-manifold whose fundamental group is a proper free product of indecomposable non-cyclic infinite factors. We shall indicate where these conditions are used in the argument.

If there are no infinite cyclic factors every embedded sphere (containing x_0) decomposes the manifold into two components and thus induces a free product splitting of the fundamental group of M:

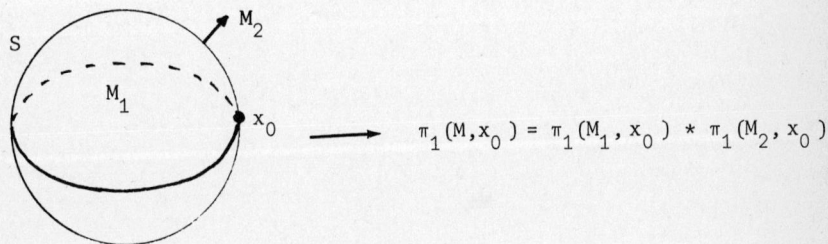

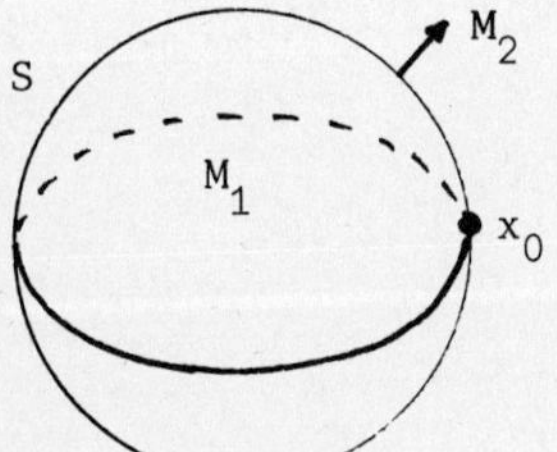

$$\longrightarrow \quad \pi_1(M,x_0) = \pi_1(M_1, x_0) * \pi_1(M_2, x_0)$$

The order of the factors is determined by the condition that the positive normal of S - indicating its orientation - point into M_2.

For $\pi_1 = G_1 * \ldots * G_n$ there are (up to x_0) disjoint embedded spheres $S_1, \ldots, S_n$ realizing the splitting $G_i * (G_1 * .. * \check{G}_i * .. * G_n)$. By a theorem of G.A. Swarup (1973) the homotopy classes of these spheres (for which we use the same names) generate π_2 as a $\mathbb{Z}\pi_1$-module and $S_1 + \ldots + S_n = 0$ is the only relation. (If π_1 also is allowed to have finite and infinite cyclic factors there are two more types of relations which are difficult to handle.) The elements S_i are uniquely determined by the splitting and we call the S_i *canonical generators* for this splitting.

Every element S of π_2 can be written as $S = S_1 z_1 + \ldots + S_n z_n$ with $z_i \in \mathbb{Z}\pi_1$, and we first derive conditions for the z_i provided that S is simple. For the case $n = 3$ we then describe an algorithm to detect simple elements. (In the case $n = 2$ every simple element of π_2 can be found immediately as an easy consequence of Lemma 4.3 .)

For $M = \#_q S^1 \times S^2$ such an algorithm has been given by M. Henry (Dijon)(1980).

2. *THE COEFFICIENTS FOR SIMPLE ELEMENTS*

An embedded sphere S induces a splitting of π_1 into two factors A and B which can be further decomposed into indecomposable factors: $A * B = (A_1 * \ldots * A_\ell) * (B_{\ell+1} * \ldots * B_m)$. By Kurosh's theorem we have $m = n$ and after reordering isomorphisms $G_i \xrightarrow{\cong} A_i$, $i = 1, \ldots, \ell$; $G_j \xrightarrow{\cong} B_j$, $j = \ell+1, \ldots, n$. Together these isomorphisms form an automorphism ϕ of π_1. Let S_i^ϕ denote the canonical generators for the new splitting of π_1. Then we have the following diagram:

$$\begin{array}{ccc} S_1^\phi, \ldots, S_n^\phi & \longleftrightarrow & A_1 * \ldots * A_\ell * B_{\ell+1} * \ldots * B_n \\ \uparrow ? & & \uparrow \phi \\ S_1, \ldots, S_n & \longleftrightarrow & G_1 * \ldots\ldots\ldots * G_n \end{array}$$

Now $S = S_1^\phi + \ldots + S_\ell^\phi$ and we have to describe S_i^ϕ with respect to $S_1, \ldots, S_n$ given $\phi \in \operatorname{Aut}\pi_1$.

2.1 *Aut* π_1 :

By a theorem of D.I. Fouxe-Rabinowitsch (1940), $\operatorname{Aut}\pi_1$ is generated by:

2.1.1 Automorphisms within one factor: $G_i \to G_i$, G_j fixed for $i \neq j$.

2.1.2 Isomorphisms between isomorphic factors: $G_i \underset{\omega}{\to} G_j$, $G_j \underset{\omega^{-1}}{\to} G_i$, G_ℓ fixed for $\ell \neq i,j$.

2.1.3 Fouxe-Rabinowitsch-Conjugations (FRCs), $g_j \in G_j$, $j \neq i$: $c_{G_i}(g_j) : g_i \quad g_j^{-1} g_i g_j \in g_i \quad G_i$, G_ℓ fixed for $\ell \neq i$.

Every automorphism ϕ has the form $c_1 \circ \ldots \circ c_\ell \circ \tilde{\phi}$. Here $\tilde{\phi}$ is a product of the first two types of automorphisms and each c_i is an FRC. (For 2.1 it is necessary that there be no infinite cyclic factors in π_1.)

Since the application of $\tilde{\phi}$ leaves the corresponding generators (up to the order) fixed, we only consider the effect of the FRCs on the generators.

2.2 Proposition

Let $M(\phi)$ be the $n \times n$-matrix whose columns are the coefficients of $S_1^\phi, \ldots, S_n^\phi$ with respect to $S_1, \ldots, S_n$. Then:

2.2.1 $\qquad M(c_{G_i}(g_i)) = \begin{pmatrix} 1 & & & & 0 \\ & \ddots & & & \\ & & g_j & \cdots & 1-g_j & \\ & & \vdots & & \vdots & \\ & & 0 & & 1 & \\ & & & & & \ddots \\ 0 & & & & & 1 \end{pmatrix} \begin{matrix} \\ \leftarrow i \\ \\ \\ \leftarrow j \\ \\ \end{matrix} \qquad \begin{matrix} \text{for } i<j \text{ and} \\ \text{analogously for} \\ i>j. \end{matrix}$

$\qquad\qquad\qquad\qquad\qquad \uparrow \qquad\quad \uparrow$
$\qquad\qquad\qquad\qquad\qquad i \qquad\quad j$

2.2.2 $\qquad M(c \circ \phi) = M(c) \cdot [cM(\phi)]$ for any FRC c and any product ϕ of FRCs. (Here $cM(\phi)$ denotes the result of applying c to each matrix-entry.)

Proof. 2.2.1 : Old generators:

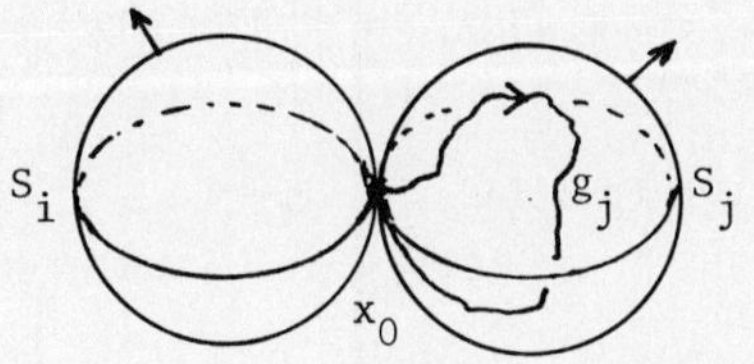

New generators:

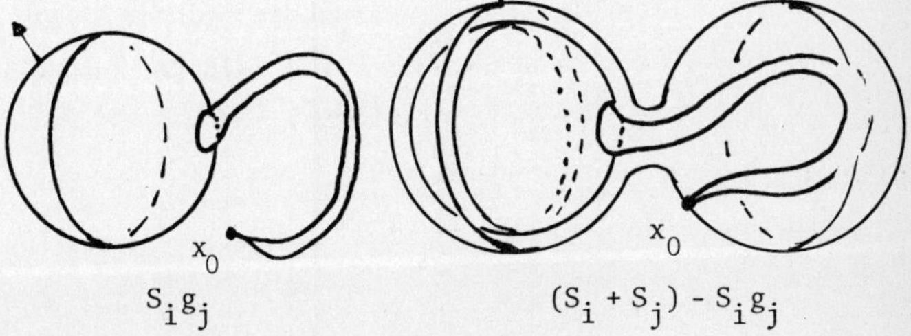

It is easy to verify that $S_i g_i$ and $S_j + S_i(1-g_j)$ are disjoint and disjoint from S_k $(k=i,j)$ and realize the splittings $g_j^{-1} G_i g_j * (G_1 * .. * \check{G}_i * .. * G_n)$ and $G_j * (G_1 * .. * g_j^{-1} G_i g_j * .. * \check{G}_j * .. * G_n)$ respectively.

2.2.2 is proved by induction on the number of FRCs in the product ϕ, after a suitable rewriting of the product (comp. Wicha-Krause, B., 1983). □

Computing the matrices for all products of FRCs we can theoretically get all systems of coefficients of embedded spheres as columns or sums of columns of these matrices, but it seems to be difficult to decide for a *given* matrix, whether it can be constructed in the above way.

Nevertheless, we get a necessary condition for the coefficients of a simple element using the fact that augumentation yields the unit matrix for any of the above matrices.

2.3 Corollary.

Let $\alpha : \mathbb{Z}\pi_1 \to \mathbb{Z}$ *be the augumentation and let* $S = S_1 z_1 + \ldots + S_n z_n$ $(z_i \in \mathbb{Z}\pi_1)$ *be a simple element. Then* $\alpha(z_1, \ldots, z_n) = (\varepsilon_1, \ldots, \varepsilon_n) + (m, \ldots, m)$ *for* $\varepsilon_i \in \{0,1\}$ *and for some* $m \in \mathbb{Z}$.

The summand $(m, \ldots, m)$ is due to the relation $m(S_1 + \ldots + S_n) = 0$. □

To get an algorithm for the case $n=3$ we introduce a tree which describes elements of π_2 better than a column of a matrix of 2.2.

3. *THE TREE FOR AN ELEMENT OF* π_2

3.1 Definition

A *tree* T is a simply connected finite graph whose edges are labelled with syllables from π_1 (i.e. elements of a factor) and are directed away from an exceptional vertex, the *basepoint* BP. The vertices may or may not carry *weights* from the abelian group generated by $S_1, \ldots, S_n$. The *length* of T is the number of edges in T.

3.2 Example

$\pi_1 = A * B * C$, $a \in A$, $b, b', b'' \in B$, $c \in C$:

T :

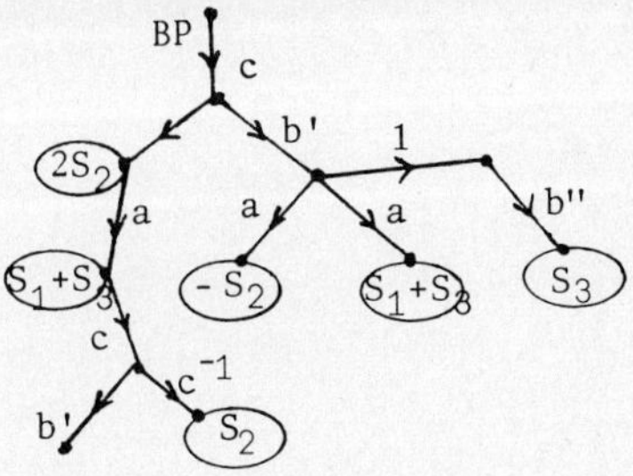

3.3 To a tree there corresponds a unique element S of π_2 which is computed as follows:

$$S = S_1 z_1 + \ldots + S_n z_n \quad \text{where} \quad z_i = \sum_{j=1}^{k_i} m_{ij} w_j \quad \text{and} \quad w_j \quad \text{is the product of}$$

the labels of the edges along the path in T from the weight in which $m_{ij} S_i$ occurs to BP .

3.4 In the example we get:

$$S = S_1(abc + ab'c) + S_2(c^{-1}\,cabc + 2bc - ab'c) + S_3(abc + ab'c + b''1b'c) .$$

It is clear that one can construct a tree T(S) describing any particular element S of π_2 . From 3.4 it is clear that several reductions (with respect to the length) can be applied to the tree without changing the corresponding element of π_2 :

3.5.1 Dropping edges with label 1 .

3.5.2 Dropping weights of the form $m(S_1 + \ldots + S_n)$, $m \in \mathbb{Z}$.

3.5.3 Dropping extremal edges which carry no weight at the end.

3.5.4 Fusing two edges beginning at the same vertex which have same direction and label.

3.5.5 Replacing two consecutive edges bearing labels from the same factor by a branch.

3.5.6 Adding a suitable multiple of $S_1 + \ldots + S_n$ to a weight, to achieve that in one weight all signs are positive or all signs are negative and at least one element S_i does not occur.

3.6 Example

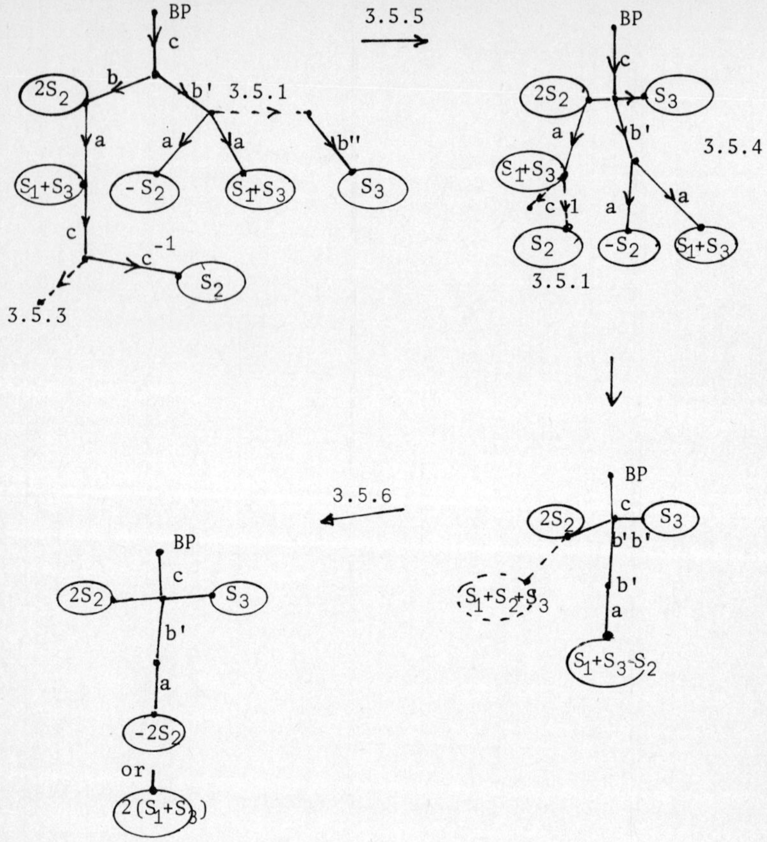

Simple combinatorial arguments show (comp. Wicha-Krause, B., 1983):

3.7 Proposition

The reduced tree (i.e. no reduction other than 3.5.6 is possible) for one element of π_2 *is uniquely determined up to two possibilities for each weight (comp. 3.5.6) and has minimal length among all trees describing this element.* □

As all members of a π_1 - *class* ($= S\cdot\pi_1$) of elements of π_2 are simple provided one member is simple, we now look for the shortest tree in one π_1 - class.

3.8 Definition

A tree is called *completely reduced* if it is reduced and

3.8.1 there is a non-zero weight at BP or

3.8.2 at least two edges from different factors emanate from BP .

3.9 Example

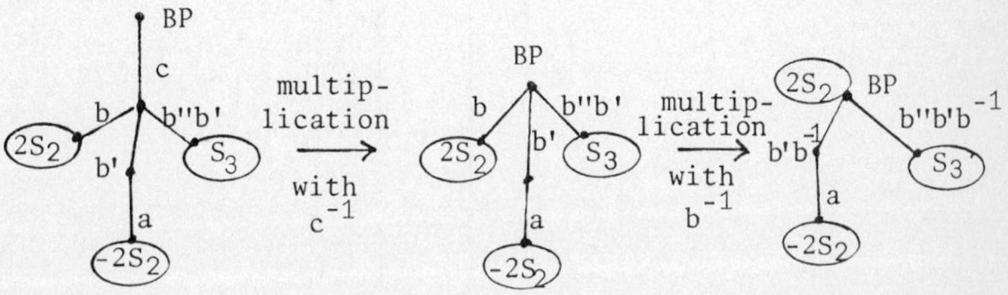

Again by combinatorial arguments it follows:

3.10 Proposition

The completely reduced tree for a π_1 *- class of elements of* π_2 *is uniquely determined up to two possibilities for each weight (comp. 3.7) and the position of BP . It has minimal length among all trees describing this class.* □

"Position of BP" means that a multiplication of T with an element of π_1 (in the example say with $(b'b^{-1})^{-1}$) can shift BP to another position, but the new completely reduced tree has as many edges as the old. Since there are only finitely many positions for BP there are only finitely many completely reduced trees for one π_1 - class.

The effect of an application of a product of FRCs to any simple element of π_2 (its $\mathbb{Z}\,\pi_1$ - coefficients are a column or a sum of columns of an $n \times n$ - matrix as in 2.2) can be derived from 2.2 which in turn gives rise to operations on a tree describing this element.

3.11 Operations on the tree

Let T be a tree for a simple element S of π_2. Then we get a tree for $c_{G_i}(g_j)(S)$ by applying the following operations on T :

3.11.1 Edges:

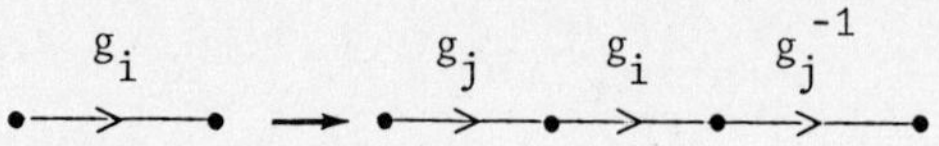

3.11.2 Weights:

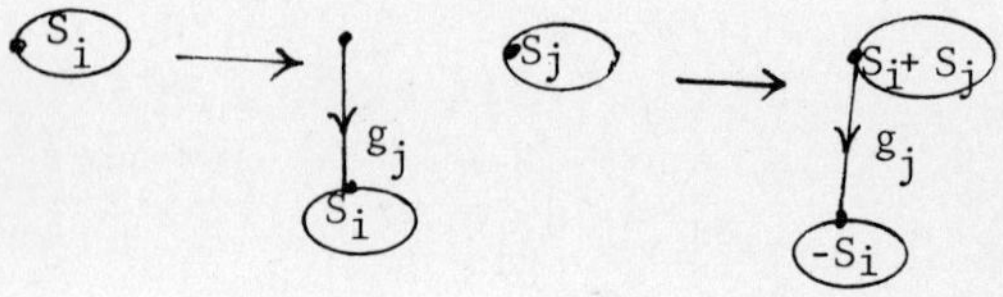

Weights and edges with other indices remain unchanged and 3.11.2 extends canonically to sums.

3.11.2 corresponds to the geometrical situation described in the proof of 2.2.1, whereas 3.11.1 reflects the application of c to each matrix-entry in 2.2.2 .

From now on we will only consider completely reduced trees and assume therefore that after 3.11 a complete reduction is always carried out.

4. *THE ALGORITHM IN THE CASE* $\eta = 3$

From now on we suppose π_1 to be $G_1 * G_2 * G_3$ and use the letters X, Y and Z for arbitrary but different groups G_i. Small letters denote elements of the corresponding groups.

The main proposition is that for any simple element S of π_2 there is a product of FRCs which builds a completely reduced tree for its π_1-class out of BP ($\pm S_i$) $i = 1,2,3$ with a proper increase of length at each step:

$$\text{BP}\ (\pm S_i) \xrightarrow[<]{c_1} T^1 \xrightarrow[<]{} \cdots \xrightarrow[<]{c_n} T^n = T(Sw)\ .$$

(We may start with $\pm S_i$ because for $n = 3$ any sum of generators is minus the third, but this is not essential for the proof, we could also start with $S_{i_1} + \ldots + S_{i_k}$.)

There are two problems to solve:

4.1 Show that the stated product $c_n \circ \ldots \circ c_1$ exists.

4.2 Determine c_n by looking at a completely reduced tree $T(S\tilde{w})$ for S.

For the proof of 4.1 we need a lemma which can only be proved in the case $n = 3$.

4.3 *Lemma*

Up to global conjugation $(\pi_1 \to g^{-1} \pi_1 g)$ *every product of FRCs can be rewritten so that each FRC* $c_X(y)$ *is followed by an FRC of one of the two types* $c_X(z)$ *or* $c_Z(x)$ *:*

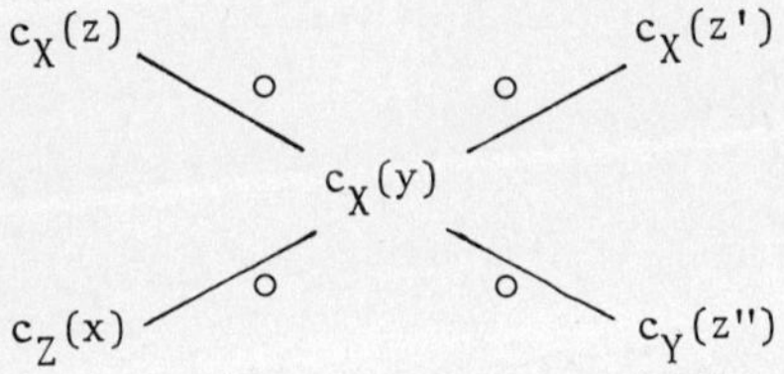

Proof. The lemma follows from the observation that $A * a^{-1} Ba * a^{-1} Ca = a^{-1}(A * B * C)a$ up to an automorphism of A and thus $c_B(a)$ can be replaced by $c_C(a^{-1})$ up to global conjugation (which leads to a multiplication of the corresponding sphere and of the corresponding tree). □

The proof of 4.3 shows that the case $n = 2$ is trivial because there are only global conjugations and the Sw for $S = S_1 = -S_2$, $w \in \pi_1$ are the only simple elements of π_2.

4.4 *Theorem*

For a product of FRCs as in 4.3 it holds that

4.4.1 *The tree constructed by this product out of* $BP(\pm S_i)$ *gets longer at each step.*

4.4.2 *If* $c_X(y)$ *was the last FRC to be applied the tree has the following properties:*

(a) *The only weights are* $\pm S_X$ *and* $\pm S_Z$ *(here* $S_X = S_i$ *for that* i *for which* $X = G_i$ *).*

(b) *There is at least one terminal branch of the form* 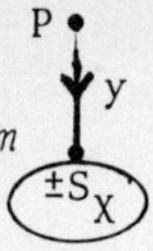

where the vertex P *is not a branch point.*

(c) *Any extremal part of the tree has the form* 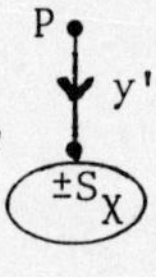

and if P *is not a branch point we have* $y' = y$.

Proof. The proof of both parts is by induction on the number of FRCs already applied. It makes use of the fact that after the application of $c_X(y)$ each edge labelled with a syllable from X is surrounded by syllables from Y. When we now apply $c_X(z)$ or $c_Z(x)$ no cancellation is possible. (Compare Wicha-Krause, B., 1983, 4.1 and 4.3.1.) □

The uniqueness-proposition for completely reduced trees together with 4.4.2 now allows us to read c_n from the completely reduced tree for a given element S of π_2. If the application of c_n^{-1} yields a shorter tree, we may repeat this procedure, if not S is not simple.

4.5 Example.

$\pi_1 = A * B * C$, $a \in A$, $b \in B$, $c \in C$:

$S = S_1\,(ca - cba + cbc^{-1}aca) + S_2\,(-ba + bc^{-1}aca)$.

4.5.1 Augmentation : (1, 0, 0)

4.5.2

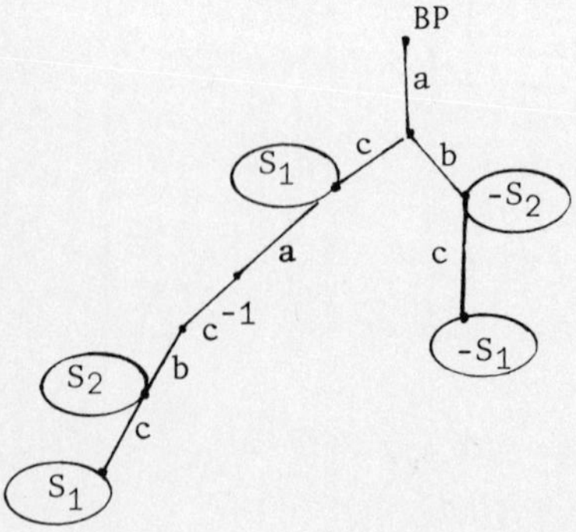

4.5.3 Complete reduction : $T(Sa^{-1})$

4.5.4 Extremal parts : 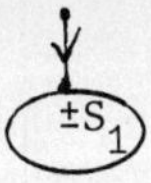

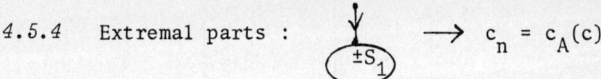

$\longrightarrow c_n = c_A(c)$

4.5.5 Apply $c_A(c^{-1})$: 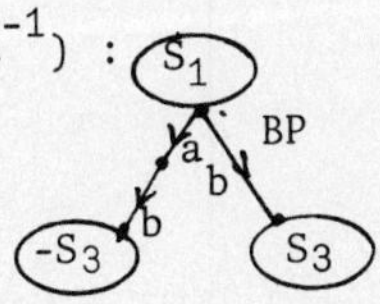

shorter length!

4.5.6 Extremal parts : 

b $\longrightarrow c_{n-1} = c_C(b)$

4.5.7 Apply $c_C(b^{-1})$: 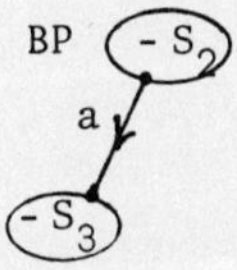

4.5.8 Apply $c_C(a^{-1})$: 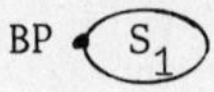

ready!

If S is simple we can also determine the splitting of π_1 which it realizes by applying the inverse of the product of FRCs we used to shorten the tree in the algorithm together with the global conjugations used to reduce the trees completely if necessary.

4.5.9 $k(a) \circ c_A(c) \circ c_C(b) \circ c_C(a) [A * (B * C)] =$

$$a^{-1}c^{-1}Aca * (a^{-1}Ba * a^{-1}c^{-1}a^{-1}cb^{-1}Cbc^{-1}aca).$$

REFERENCES

Fouxe-Rabinowitsch, D.I. (1940). On the Automorphism Groups of a Free Product I (Russian), Matm. Sbornik 8 (50), 265-276.

Henry, M. (1980). Classe d'Homotopie d'une Sphère Plongée dans $\#_q S^1 \times S^2$, Preprint.

Swarup, G.A. (1973). On Embedded Spheres in 3-Manifolds, Math. Ann. 203 89-102.

Wicha-Krause, B. (1983). On Embedded 2-Spheres in 3-Manifolds, Mathematische Zeitschrift.

A NOTE ON THE MAPPING CLASS GROUPS OF SURFACES AND PLANAR DISCONTINUOUS GROUPS

Heiner Zieschang

Abstract. Let $\hat{\mathfrak{G}}$ be the free group on a canonical set of generators for the fundamental group of a closed surface S. In this paper, the automorphisms of $\hat{\mathfrak{G}}$ which preserve the kernel of the natural map from $\hat{\mathfrak{G}}$ to the fundamental group of S are discussed. This enables one to give a slightly different description of the mapping class group.

The fundamental group $\pi_1 S$ of a closed surface S is closely related to the fundamental group of the punctured surface $\hat{S}$, that is, to the free group $\hat{\mathfrak{G}}$ in the canonical generators of $\pi_1 S$. This has been exploited in some of the proofs of the Nielsen theorem that the automorphisms of $\pi_1 S$ are induced by homeomorphisms of S and its generalization to planar groups by showing that each minimal system of generators is obtained from a given one by simple steps corresponding to bifurcations.

Let $\mathrm{Aut}_* \hat{\mathfrak{G}}$ denote the automorphisms of $\hat{\mathfrak{G}}$ that preserve the kernel of $p : \hat{\mathfrak{G}} \to \pi_1 S$. We will determine the kernel of $\mathrm{Aut}_* \hat{\mathfrak{G}} \to \mathrm{Aut}_* \pi_1 S$ and show that it is generated by the inner automorphisms with the defining relation as conjugation factor, see Proposition 1.2. This gives a slightly different description of the mapping class group. In particular the result reduces the number of steps in applying the Whitehead algorithm to find a presentation of mapping class group, following McCool (1975).

1. *The kernel of* $\mathrm{Aut}_* \hat{\mathfrak{G}} \to \mathrm{Aut}\, \pi_1 S$.

Let

$$\mathfrak{G} = \langle s_1, \ldots, s_m, t_1, u_1, \ldots, t_g, u_g \mid s_1^{-a_1}, \ldots, s_m^{-a_m}, \prod_{i=1}^{m} s_i \prod_{j=1}^{g} [t_j, u_j] \rangle$$

or

$$\mathfrak{G} = \langle s_1, \ldots, s_m, v_1, \ldots v_g \mid s_1^{-a_1}, \ldots, s_m^{-a_m}, \prod_{i=1}^{m} s_i \prod_{j=1}^{g} v_j^2 \rangle$$

be a canonical presentation of a Fuchsian group or an NEC-group without reflections, respectively, where $a_i \geq 2$ if $a_i \neq 0$. In either case if $a_i \geq 2$, then s_i represents a rotation, otherwise corresponds to a hole on the surface $S = \mathcal{H}/\mathcal{G}$. The images of the rotation centres are marked on S, see Zieschang, H., (1981), 31.2. (For small m, h, a_i either presentation above belongs to a group acting on $\mathbb{R}^2$ or S^2; the following arguments are applicable to crystallographic groups of $\mathbb{R}^2$ also.) Let

$$\hat{\mathcal{G}} = \langle S_1, \ldots, S_m, T_1, U_1, \ldots, T_g, U_g \mid \rangle \quad \text{or}$$

$$\hat{\mathcal{G}} = \langle S_1, \ldots, S_m, V_1, \ldots, V_g \mid \rangle, \qquad \text{respectively,}$$

be the free group in the generators and let $p : \hat{\mathcal{G}} \to \mathcal{G}$ be the "standard projection" defined by $S_i \mapsto s_i$, $T_j \mapsto t_j$, $U_j \mapsto u_j$, $V_j \mapsto v_j$. Let $\varepsilon : \hat{\mathcal{G}} \to \mathbb{Z}_2 = \{1, -1\}$ denote the orientation marking homomorphism $S_i, T_j, U_j \mapsto 1$, $V_j \mapsto -1$; the induced homomorphism $\mathcal{G} \to \mathbb{Z}_2$ is also denoted by ε.

Let $\Pi_* = \prod_{i=1}^{m} S_i \prod_{j=1}^{g} [T_j, U_j]$ in the Fuchsian case or $\Pi_* = \prod_{i=1}^{m} S_i \prod_{j=1}^{g} V_j^2$, if G is an NEC-group. Denote by $\hat{\mathcal{A}}$ the group of automorphisms $\hat{\alpha} : \hat{\mathcal{G}} \to \hat{\mathcal{G}}$ with $\hat{\alpha}(\Pi_*) = \Pi_*^{\pm 1}$ and $\hat{\alpha}(S_i) = L_i S_{k(i)}^{\varepsilon(L_i)} L_i^{-1}$ for any i such that $a_{k(i)} = a_i$. Let $\mathcal{A}$ denote the group of automorphisms α of $\mathcal{G}$ such that $\alpha(s_i) = l_i s_{k(i)}^{\nu_i} l_i^{-1}$ for $a_i = 0$. Thus, if $a_i = 0$ then $a_{k(i)} = 0$ and $\nu_i = \pm 1$. Moreover, for the other s_j: $\alpha(s_j) = l_j s_{k(j)}^{\nu_j} l_j^{-1}$ with $a_{k(j)} = a_j$ and $\nu_j = \pm 1$. Let $\Phi : \hat{\mathcal{A}} \to \mathcal{A}$ be the homomorphism $\hat{\alpha} \mapsto p \circ \hat{\alpha} \circ p^{-1}$.

By Zieschang, H., (1980), 5.8, each automorphism $\alpha \in \mathcal{A}$ is induced by an automorphism of $\hat{\mathcal{G}}$ which maps Π_* to a conjugate of $\Pi_*^{\pm 1}$ and by a homeomorphism f of the plane $\mathcal{H}$: $\alpha(x) = f^{-1} x f$. (The geometric statement is an easy consequence of the algebraic result.) We will show that, in fact, there is an $\hat{\alpha} \in \hat{\mathcal{A}}$ which induces α. For this we describe some special automorphisms of $\hat{\mathcal{G}}$ mapping Π_* to Π_* or Π_*^{-1}.

1.1 "Inner automorphisms of $\hat{\mathfrak{G}}$ *preserving* $\Pi_*^{\pm 1}$.

Denote by ι_x the inner automorphism $g \mapsto x^{-1}gx$.

(a)

$$\hat{\sigma}_k \ : \begin{cases} S_i \mapsto S_k\Pi_*^{-1}\cdot S_i\cdot\Pi_* S_k^{-1}, \ 1 \le i < k, \\ S_k \mapsto \qquad S_k \\ X \ \mapsto \Pi_*^{-1}S_k \cdot X \cdot S_k^{-1}\Pi_*, \ X \in \{S_i | k < i \le m\} \cup \{T_j, U_j V_j | 1 \le j \le g\} ; \end{cases} \quad (1 \le k \le m)$$

$$\hat{\sigma}_k(\Pi_*) = \Pi_*, \ \Phi(\hat{\sigma}_k) = \iota_{s_k}^{-1} .$$

(b) For $\Pi_* = \prod_{i=1}^{m} S_i \prod_{j=1}^{g} [T_j, U_j]$, $1 \le k \le g$:

$$\hat{\tau}_k \ : \begin{cases} X \ \mapsto T_k \ \Pi_*^{-1}\cdot X\cdot\Pi_* T_k^{-1}, \ X \in \{S_i | 1 \le i \le m\} \cup \{T_j, U_j | 1 \le j \le k\}, \\ T_k \mapsto \qquad T_k \\ U_k \mapsto \Pi_*^{-1}T_k \cdot U_k \cdot T_k^{-1} \\ X \ \mapsto \Pi^{-1}T_k \cdot X \cdot T_k^{-1}\Pi_*, \ X \in \{T_j, U_j | k < j \le g\} ; \end{cases}$$

$$\hat{\tau}_k(\Pi_*) = \Pi_*, \ \Phi(\hat{\tau}_k) = \iota_{t_k}^{-1} ;$$

$$\hat{\mu}_k \ : \begin{cases} X \ \mapsto \Pi_* U_k\cdot X\cdot U_k^{-1}\Pi_*^{-1}, \ X \in \{S_i | 1 \le i \le m\} \cup \{T_j, U_j | 1 \le j < k\}, \\ T_k \mapsto \Pi_* U_k \cdot T_k \cdot U_k^{-1} \\ U_k \mapsto \qquad U_k \\ X \ \mapsto U_k\Pi_* \cdot X \cdot \Pi_*^{-1}U_k^{-1}, \ X \in \{T_j, U_j | k < j \le g\} ; \end{cases}$$

$$\hat{\mu}_k(\Pi_*) = \Pi_*, \ \Phi(\hat{\mu}_k) = \iota_{u_k}^{-1} .$$

(c) For $\Pi_* = \prod_{i=1}^{m} S_i \prod_{j=1}^{g} V_j^2$, $1 \le k \le g$:

$$\hat{\nu}_k \ : \begin{cases} X \ \mapsto V_k\Pi_*^{-1}\cdot X\cdot\Pi_* V_k^{-1}, \ X \in \{S_i | 1 \le i \le m\} \cup \{V_j | 1 \le j < k\}, \\ V_k \mapsto \qquad V_k \cdot \Pi_*^{-1} \\ V_j \mapsto \Pi_* V_k \cdot V_j \cdot V_k^{-1}\Pi_*^{-1}, \qquad k < j \le g ; \end{cases}$$

$$\hat{\nu}_k(\Pi_*) = \Pi_*^{-1}, \quad \Phi(\hat{\nu}_k) = \iota_{V_k}^{-1} .$$

(d) For $\Pi_* = \prod_{i=1}^{m} S_i \prod_{j=1}^{g} [T_j, U_j]$:

$$\rho : \begin{cases} S_i & S_{m+1-i} , \quad 1 \le i \le m \\ T_j & \prod_{i=1}^{m} S_i \cdot U_{g+1-j} \cdot \left(\prod_{i=1}^{m} S_i\right)^{-1} , \\ & \qquad\qquad 1 \le j \le g \\ U_j & \prod_{i=1}^{m} S_i \cdot T_{g+1-j} \cdot \left(\prod_{i=1}^{m} S_i\right)^{-1} ; \end{cases}$$

for $\Pi_* = \prod_{i=1}^{m} S_i \prod_{j=1}^{g} V_j^2$:

$$\rho : \begin{cases} S_i & S_{m+1-i} , \quad 1 \le i \le m \\ V_j & \prod_{i=1}^{m} S_i \cdot V_{g+1-j}^{-1} \cdot \left(\prod_{i=1}^{m} S_i\right)^{-1} , \quad 1 \le j \le g ; \end{cases}$$

$$\rho(\Pi_*) = \Pi_*^{-1} .$$

Since the automorphisms $\hat{\sigma}_i$, $\hat{\tau}_k$, $\hat{\mu}_k$, $\hat{\nu}_k$, ρ belong to $\hat{\mathcal{A}}$ the following proposition follows from Zieschang, H., 1980, Theorem 5.8.2.

1.2 Proposition.

$\Phi : \hat{\mathcal{A}} \to \mathcal{A}$ is an epimorphism. □

1.3 Proposition

Let $\hat{\alpha} \in \hat{\mathcal{A}}$ induce the identity on $\mathfrak{G}$, that is $\hat{\alpha}(\Pi_*) = \Pi_*^{\pm 1}$ and $p \circ \hat{\alpha} \circ p^{-1} = \text{id}$. Then $\hat{\alpha}$ is an inner automorphism of $\hat{\mathfrak{G}}$. In other words, $\ker \Phi = \langle \iota_{\Pi_*} \rangle$

Proof. We assume that all $a_i = 0$. Let S be a compact surface of genus g and m boundary components, orientable, if we consider the Fuchsian case, and non-orientable otherwise. Choose a basepoint $*$ in the interior of S and a disc D which is a regular neighbourhood of $*$. Define $\hat{S} = \overline{S - D}$ and let $\hat{*} \in \partial D$ and λ be an arc in D from $*$ to $\hat{*}$. By $\hat{\gamma} \mapsto \lambda^{-1} \hat{\gamma}^{\lambda}$, if γ is a closed curve in $\hat{S}$ we obtain the canonical projection

$$\hat{\mathfrak{G}} = \pi_1 \hat{S} \xrightarrow{p} \pi_1 S = \mathfrak{G} \quad .$$

If we realize the generators of $\hat{\mathfrak{G}}$: S_i, T_j, U_j, V_j, by a canonical system of curves starting at $\hat{*}$ then Π_* represents ∂D.

By the Dehn-Nielsen Theorem, see Zieschang, H., et al. 1980, 5.7.1, 5.7.2, $\hat{\alpha}$ is induced by a homeomorphism $\hat{h} : \hat{S} \to \hat{S}$ with $\hat{h} \mid \partial D = id_{\partial D}$, $\hat{h}_* = \hat{\alpha}$. We extend $\hat{h}$ by the identity to D and obtain a homeomorphism $h : S \to S$ with $h(*) = *$ and $h_* = id_{\pi_1 S} = id_{\mathfrak{G}}$, Epstein's refined version of the Baer Theorem, see Epstein, D.B.A., 1966, Zieschang, H., et al, 1980; 5.13.2, guarantees an isotopy

$$H: S \times [0, 1] \to S \times [0, 1], \; H(x,t) = (h_t(x), t)$$

such that

$$h_0 = h, \; h_1 = id_S \text{ and } h_t(*) = * \text{ for } 0 \le t \le 1 .$$

We may even assume that $h_t(D) = D$ for all t as follows easily from the combinatorial construction of the isotopy described in Zieschang, H., et al, 1980; 15.13.

A consequence is that the point $\hat{*}$ glides along ∂D and its path gives the conjugating factor for the effect on $\pi_1 \hat{S}$. This proves that $\hat{h}$ is obtained from the identity $id_{\hat{S}}$ by an isotopy moving $\hat{*}$ along ∂D. Hence, the automorphisms induced by $\hat{h}$ and $id_{\hat{S}}$ differ by an inner automorphism with a power of the class of ∂D, that is, of Π_* as conjugation factor. The case where some $a_i \neq 0$ can be reduced to the above as in Zieschang, H., et al, 1980; 5.14. □

From the theorems of Baer-Epstein and Nielsen, see Zieschang, H., et al, 1980; 5.15.3, it follows that the mapping class group of the pair $(S, *)$ is isomorphic to $\mathcal{A}$, and we obtain

1.4 Corollary. $\mathcal{M}(S, *) \cong \mathcal{A} \cong \hat{\mathcal{A}} / \langle \iota_{\Pi_*} \rangle$,

$$\mathcal{M}(S) \cong \mathcal{A} / \operatorname{Inn}(\pi_1 S) \cong \begin{cases} \hat{\mathcal{A}} / \langle\langle \iota_{\Pi_*}, \hat{\sigma}_1, \ldots, \hat{\sigma}_m, \hat{\tau}_1, \hat{\mu}_1, \ldots, \hat{\tau}_g, \hat{\mu}_g \rangle\rangle . \\ \hat{\mathcal{A}} / \langle\langle \iota_{\Pi_*}, \hat{\sigma}_1, \ldots, \hat{\sigma}_m, \hat{\nu}_1, \ldots, \hat{\nu}_g \rangle\rangle \quad . \end{cases}$$

where $\mathcal{M}$ denotes mapping class groups, << >> denotes the normal subgroup generated by the quoted elements and Inn is the group of inner automorhpisms. □

1.5 Remark.

Unfortunately, there is no section $s : \mathcal{A} \to \hat{\mathcal{A}}$, that is, no homomorphism s with $\Phi \circ s = \mathrm{id}_{\mathcal{A}}$. To see this consider the case where $g = 1$, $m = 0$ and the automorphism $\gamma : \mathbb{Z}^2 \to \mathbb{Z}^2 = \mathfrak{G}(t \mapsto u, u \mapsto t^{-1})$ is of order 4 induced by $\hat{\gamma} : \hat{\mathfrak{G}} \to \hat{\mathfrak{G}}$ $(T \mapsto TUT^{-1}, U \mapsto T^{-1})$. Thus, $\hat{\gamma}([T,U]) = [T,U]$ and $\hat{\gamma}^4 = \iota^{-1}_{[T,U]}$; hence, $\hat{\gamma} \in \hat{\mathcal{A}}$. Any other automorphism from $\hat{\mathcal{A}}$ which induces γ has the form $\hat{\delta} = \hat{\gamma} \cdot \iota^{c}_{[T,U]}$. However, since $\hat{\delta}^4 = \iota^{4c-1}_{[T,U]} \neq 1$, γ is not the image of an element of $\hat{\mathcal{A}}$ of order 4.

Proposition 1.3 can also be proved using only combinatorial group-theoretical arguments. However, the proof becomes rather long and does not clear the situation; it consists of many case considerations.

2. *SOME APPLICATIONS*

2.1 Theorem.

The mapping class group $\mathcal{M}(\mathfrak{G})$ of a finitely generated Fuchsian $\mathfrak{G}$ or a non-euclidean crystallographic group $\mathfrak{G}$ without reflections is finitely presentable.

By the Nielsen theorem, see e.g. Zieschang, H., et al., 1980; 5.8.2, 6.6.11, and the Baer theorem, Zieschang, H., et al, 1980; 5.14.1, in either case $\mathcal{M}(\mathfrak{G})$ is isomorphic to $\mathcal{A}/\mathrm{Inn}\,\mathfrak{G}$. By 1.4, $\mathcal{A}/\mathrm{Inn}\,\mathfrak{G} \cong \hat{\mathcal{A}} / \langle\langle \iota_{\Pi_*}, \hat{\sigma}_1, \ldots, \hat{\sigma}_m, \hat{\tau}_1, \hat{\mu}_1, \ldots, \hat{\tau}_g, \hat{\mu}_g \rangle\rangle$ or $\cong \hat{\mathcal{A}} / \langle\langle \iota_{\Pi_*}, \hat{\sigma}_1, \ldots, \hat{\sigma}_m, \hat{\nu}_1, \ldots, \hat{\nu}_g \rangle\rangle$. The group $\hat{\mathfrak{S}}$ is the subgroup of the automorphism of $\hat{\mathfrak{G}}$ which stabilize $\{\Pi_*, \Pi_*^{-1}\}$ and permute the conjugacy classes of the elements from $\{S_1, S_1^{-1}, \ldots, S_m, S_m^{-1}\}$ which have the same exponent a_i. By Whitehead's theorem $\hat{\mathcal{A}}$ is finitely generated (Zieschang, H., 1981; Higgins, P.J. & Lyndon, R.C., 1974) and McCool's result (McCool, J., 1975) ensures that a finite number of relators suffices. Therefore $\hat{\mathcal{A}}$ and $\mathcal{A}/\mathrm{Inn}\,\mathfrak{G}$ are finitely presentable. □

Since the algebraic version of Nielsen's theorem (see Zieschang, H., et al, 1980; 5.8.2) was known the above result became evident when McCool's result (McCool, J., 1975) was available. Here one allows the map Π_* to a conjugate of $\Pi^{\pm 1}$. The number of generators and relators obtained by the Whitehead-McCool-method is very large. Our restriction of the action on Π_* diminishes this number slightly. It would be of interest to find a general reduction for the Whitehead algorithm when applied for genus g to the cases of smaller genus, for instance, to the algorithm for genus $g-1$.

Geometric approaches lead to smaller presentations in the case of orientable surfaces and give much more insight into the situation, see (Hatcher, A., & Thurston, W., 1980), (Harer, J., 1983), (Wajuryb, B., 1983).

Corollary 1.4 enables one to solve the Nielsen realization problem for some types of finite groups of mapping classes by using the construction invented by Nielsen (Nielsen, J., 1942).

2.2 *Proposition.*

Assume that $\hat{\Gamma} \subset \hat{\mathcal{A}}$ and $\hat{\Gamma} / \hat{\Gamma} \cap \langle \iota_{\Pi_*} \rangle$ is finite. Then $\Gamma = \Phi(\hat{\Gamma})$ can be realized by a finite group $\mathcal{H}$ of homeomorphisms of $S = \mathcal{H}/\mathcal{G}$ or, in other words, by a finite extension of $\mathcal{G}$ acting on $\mathcal{H}$. It is $\mathcal{H} \cong \Gamma / \Gamma \cap \operatorname{Inn} \mathcal{G} \cong \hat{\Gamma} / \hat{\Gamma} \cap \langle \iota_{\Pi_*} \rangle$, where the isomorphisms are the natural ones.

Proof. We use the notation of Proposition 1.3. We assume that $\hat{S}$ has an analytic structure such that ∂D lifts to a hyperbolic line in $\mathcal{H}$. Γ can be interpreted as a group of automorphisms of $\pi_1 \hat{S}$ each of which can be realized by a homeomorphism of $\pi_1 \hat{S}$. By assumption, $\hat{\Gamma}$ defines a finite group of mapping classes of $\hat{S}$ and all mappings map ∂D onto ∂D. Thus ∂D defines a $\hat{\Gamma}$-simple axis, for the definition see (Zieschang, H., 1981; 43.1), and the Nielsen construction gives a finite group $\hat{\mathcal{H}}$ of homeomorphism of $\hat{S}$ which induces the group $\hat{\Gamma} / \hat{\Gamma} \cap \langle \iota_{\Pi_*} \rangle$ on $\pi_1 \hat{S}$, see (Zieschang, H., 1981; 47.5). Extending the homeomorphism $\hat{h} \in \hat{\mathcal{H}}$ radially to D gives a homeomorphism $h : S \to S$. Altogether we obtain a finite group $\mathcal{H}$ of auto-homeomorphisms of S and $\mathcal{H}$ induces Γ. □

The assumptions in 2.2 are very restrictive, as shown by the following corollary.

2.3 Corollary.

$\Gamma / \Gamma \cap \mathrm{Inn}\, \mathfrak{G}$ is a finite cyclic or dihedral group. This follows from the fact that the elements of $\hat{\mathcal{H}}$ map ∂D onto itself and, by 1.3, that only the neutral element induces the identity on ∂D. □

Of course, Proposition 2.2 is a consequence of the general Realization Theorem of Kerckhoff (Kerckhoff, S.P., 1983) or, under the assumption that Γ is cyclic, of Fenchel's result ((Fenchel, W., 1948), (Fenchel, W., 150), (Macbeath, A.M., 1962)). The proofs in these papers, however, are based on much deeper results than are used in Nielsen's approach.

REFERENCES

Epstein, D.B.A., (1966). Curves on 2-manifolds and isotopics. Acta Math. 115, 83-107.

Fenchel, W., (1948). Estensioni gruppi descontinui e transformazioni periodiche delle surficie. Rend. Acc. Naz. Lincei (Se. fis., math. e nat.), 5, 326-329.

Fenchel, W., (1950). Bemarkungen om endelige gruppen af abbildungsklassen. Mat. Tidsskrift B, 90-95.

Harer, J., (1983). The second homology group of the mapping class group of an orientable surface. Invent. math., 72, 221-239.

Hatcher, A., & Thurston, W., (1980). A presentation of the mapping class group of a closed orientable surface. Topology, 19, 221-237.

Higgins, P.J., & Lyndon, R.C., (1974). Equivalence of elements under automorphisms of a free group. J. London Math. Soc. 8, 254-258.

Kerckhoff, S.P., (1983). The Nielsen realization problem. Ann. of Math., 117, 235-265.

Macbeath, A.M., (1962). On a theorem by J. Nielsen. Quart. J. Math. Oxford, (2) 13, 235-236.

McCool, J. (1975). Some finitely presented subgroups of the automorphism group of a free group. J. Algebra, 35, 205-213.

Nielsen, J., (1942). Abbildungsklassen endlicher Ordnung. Acta Math., 75, 23-115.

Wajnryb, B., (1983). A simple presentation for the mapping class group of an orientable surface. Israel J. Math., 45, 157-174.

Whitehead, J.H.C., (1936). On equivalent sets of elements in a free group. Ann. of Math., 37, 782-800.

Zieschang, H., (1981). Finite groups of mapping classes of surfaces. Lecture Notes in Math. 875. Berlin-Heidelberg-New York: Springer-Verlag.

Zieschang, H., Vogt, E., & Coldewey, H.-D., (1980). Surfaces and planar discontinuous groups. Lecture Notes in Math. 835. Berlin-Heidelberg-New York: Springer-Verlag.

ZUR KLASSIFIKATION HÖHERDIMENSIONALER SEIFERTSCHER FASERRÄUME

Bruno Zimmermann

Abstract. This paper deals with Seifert fibre spaces of dimension 4 where the fibres are 2-dimensional tori. H. Zieschang has described a set of algebraic invariants for these spaces. These invariants distinguish the fundamental groups up to isomorphism. In this paper, a method is given to decide whether two fundamental groups of 4-dimensional Seifert fibre spaces are isomorphic if the structure group is a subgroup of $SL(2, \mathbb{Z})$.

(n+2)-dimensionale Seifertsche Faserräume seien in Analogie zu den klassischen 3-dimensionalen Seifertschen Faserräumen als Faserungen über Flächen mit gewissen Singularitäten definiert, wobei als Faser der n-dimensionale Torus $T^n = (S^1)^n$ an Stelle von $T^1 = S^1$ im 3-dimensionalen Fall auftritt. Nach (Zieschang, H., 1969), und (Conner, P.E. & Raymond, F., 1969 & 1977) sind diese Mannigfaltigkeiten bzgl. fasertreuer Homöomorphie durch ihre Fundamentalgruppen bestimmt (abgesehen von einigen Ausnahmefällen, in denen die Basis das Geschlecht 0 oder 1 hat). Hauptresultat der vorliegenden Arbeit ist eine Isomorphieklassifikation der Fundamentalgruppen 4-dimensionaler transversal orientierbarer Seifertscher Faserräume (abgesehen von den Ausnahmefällen), und in Spezialfällen auch in höheren Dimensionen. Genauer geben wir einen Algorithmus an, der in endlich vielen einfachen Schritten zu entscheiden gestattet, ob zwei gegebene Fundamentalgruppen 4-dimensionaler transversal orientierbarer Seifertscher Faserräume isomorph sind oder nicht. Unsere Ergebnisse gestatten eine weitgehende Normierung der Fundamentalgruppen, insbesondere kann man Repräsentanten der Isomorphieklassen zu fest vorgegebener Strukturgruppe angeben. Für die Klasse der Fundamentalgruppen (n+2)-dimensionaler Seifertscher Faserräume ist für $n > 3$ das Isomorphieproblem in dieser Allgemeinheit unlösbar.

EINLEITUNG

In der Arbeit (Seifert, H., 1933) hat Seifert 3-dimensionale Mannigfaltigkeiten studiert, die durch Kreislinien S^1 gefasert sind. Die Faserung ist lokaltrivial ausser in endlich vielen Ausnahmefasern, die gefaserte Volltorusumgebungen der Form $(D^2 \times S^1)/h$ besitzen; dabei ist $D^2 := \{z \in \mathbb{C} : |z| \leq 1\}$ die Kreisscheibe und h eine Drehung endlicher Ordnung von D^2 und S^1. Die Basis der Faserung ("Zerlegungsfläche") ist eine Fläche F. Die Fundamentalgruppe $\pi_1 M$ eines solchen "Seifertschen Faserraumes" besitzt eine Gruppe F isomorph zu einer diskontinuierlichen Bewegungsgruppe der Sphäre, der euklidischen oder hyperbolischen Ebene als Faktorgruppe. Diese Mannigfaltigkeiten wurden von Seifert durch ein System numerischer Invarianten bzgl. fasertreuer Homöomorphie klassifiziert. Dass diese Grössen im wesentlichen auch Homöomorphie-invarianten sind, wurde erst 1967 von Waldhausen (Waldhausen, F., 1967) und danach mit anderen Methoden von Orlik-Vogt-Zieschang (Orlik, P., et.al., 1967) gezeigt; Waldhausen zeigte darüber hinaus, dass jeder Homöomorphismus isotop ist zu einem fasertreuen. Abgesehen von einigen Ausnahmefällen bestimmt dann die Fundamentalgruppe $\pi_1 M$ den (fasertreuen) Homöomorphietyp von M, insbesondere falls F eine Fuchssche Gruppe ist, d.h. eine Gruppe von Bewegungen der hyperbolischen Ebene $\mathbf{H}^2$ (vgl. auch Orlik, P., 1970).

Der Begriff des Seifertschen Faserraumes ist später von vielen Autoren verallgemeinert worden, indem man allgemeinere Mannigfaltigkeiten für Basis und Faser zulässt, vgl. (Holmann, H., 1964), (Thornton, M.C., 1967 & 1968), (Zieschang, H., 1969), (Vogt, E., 1973 & 1977), (Conner, P.E. & Raymond, F., 1977). Diese allgemeine Konstruktionsmethode liefert interessante Klassen höherdimensionaler Mannigfaltigkeiten, insbesondere auch asphärischer Mannigfaltigkeiten. Besonders weitreichende Resultate erhält man, wenn man als Faser höherdimensionale Tori $T^n = (S^1)^n$ betrachtet. In dieser Situation kann man die Kohomologietheorie von Gruppen anwenden und viele Aussagen über die Basis auf den Faserraum übertragen; dieses ist sehr allgemein in (Conner, P.E. & Raymond, F., 1970 & 1977) durchgeführt. In der vorliegenden Arbeit betrachten wir den Kodimension-2-Fall, d.h. Mannigfaltigkeiten, die über Flächen gefasert sind durch Tori; dieses ist die nächstliegende Verallgemeinerung der klassischen 3-dimensionalen Seifertschen Faserräume.

0.1 Definition.

Unter einem Seifertschen Faserraum verstehen wir eine geschlossene (n+2)-dimensionale Mannigfaltigkeit M, die durch Tori (als reguläre Fasern) über einer geschlossenen Fläche gefasert ist. Die Faserung ist lokaltrivial ausser in endlich vielen Ausnahmefasern; diese besitzen gefaserte Umgebungen der Form $(D^2 \times T^n)/G$, wobei G eine endlich-zyklische Gruppe von Drehungen von D^2 ist und frei auf T^n operiert als Gruppe affiner Homöomorphismen (operiert G trivial auf $\pi_1 T^n$, so ist die entsprechende Ausnahmefaser $(\{0\} \times T^n)/G$ wieder ein Torus, andernfalls eine allgemeinere flache Mannigfaltigkeit).

Die Fundamentalgruppe $\pi_1 M$ besitzt wieder eine Gruppe F als Faktorgruppe, die isomorph ist zu einer 2-dimensionalen Bewegungsgruppe. Ist F isomorph zu einer Fuchsschen Gruppe, so sagen wir, M bzw. $\pi_1 M$ ist *von hyperbolischem Typ* (unter einer Fuchsschen Gruppe verstehen wir im folgenden eine diskrete Gruppe von Bewegungen der hyperbolischen Ebene $\mathbf{H}^2$ mit kompaktem Fundamentalbereich ohne Spiegelungen); in diesem Fall induziert die Inklusion $T^n \hookrightarrow M$ der regulären Faser eine Einbettung $\mathbb{Z}^n = \pi_1 T^n \hookrightarrow \pi_1 M$, vgl. (Thornton, M.C., 1967), (Zieschang, H., 1969), (Vogt., E., 1977). Wir haben daher eine exakte Sequenz

$$1 \to \mathbb{Z}^n \hookrightarrow \pi_1 M \to F \to 1 .$$

Hieraus erhalten wir einen Homomorphismus $\phi : F \to \mathrm{Aut}\,\mathbb{Z}^n = GL_n(\mathbb{Z})$. Wir nennen $\phi(F) \subset Gl_n(\mathbb{Z})$ die Strukturgruppe von M bzw. $\pi_1 M$. Aus den bekannten Beschreibungen Fuchsscher Gruppen (vgl. z.B. (Zieschang, H. et al, 1980) erhalten wir folgende Beschreibung für $\pi_1 M$ (ausgehend von dieser Beschreibung lässt sich der Seifertsche Faserraum wie im klassischen 3-dimensionalen Fall rekonstruieren, vgl. (Zieschang, H., 1969), (Vogt, E., 1977), (Thornton, M.C., 1967 & 1968).

0.2 $\pi_1 M = <$*Erzeugende* $e_1, \ldots, e_n$ *von* $\mathbb{Z}^n$; $a_1, b_1, \ldots, a_g, b_g$ *bzw.*

$v_1, \ldots, v_g$; $s_1, \ldots, s_q$ | Relationen von $\mathbb{Z}^n$ (d.h. $[e_i, e_j] = 1$, $1 \leq i, j \leq n$),

$$\prod_{i=1}^{g} [a_i, b_i] \prod_{j=1}^{q} s_j = w \in \mathbb{Z}^n \text{ bzw. } \prod_{i=1}^{g} v_i^2 \prod_{j=1}^{q} s_j = w \in \mathbb{Z}^n ,$$

$$s_j^{\alpha_j} = w_j \in \mathbb{Z}^n ,$$

$$a_i x a_i^{-1} = \phi(a_i)(x), \quad b_i x b_i^{-1} = \phi(b_i)(x) \text{ bzw.}$$

$$v_i x v_i^{-1} = \phi(v_i)(x), \quad i = 1, \ldots, g,$$

$$s_j x s_j^{-1} = \phi(s_j)(x), \quad j = 1, \ldots, q, \quad x \in \mathbb{Z}^n > ,$$

mit Faktorgruppe $\pi_1 M / \mathbb{Z}^n$ isomorph zu einer Fuchsschen Gruppe.

In Verallgemeinerung von (Orlik, P. et. al., 1967), zeigte Zieschang (1969), dass 2 Seifertsche Faserräume von hyperbolischem Typ genau dann fasertreu homöomorph sind, wenn ihre Fundamentalgruppen isomorph sind, vgl. auch (Vogt, E., 1977) und insbesondere (Conner, P.E. & Raymond, F., 1977), wo dieses Resultat im Rahmen einer allgemeineren Theorie gezeigt wird (dabei sei stets vorausgesetzt, dass in Dimensionen $n+2 > 5$ die "geometrische Strukturgruppe" eines Seifertschen Faserraumes aus affinen Homöomorphismen des Torus T^n besteht: für $n > 3$ ist nicht bekannt, ob ein beliebiger Homöomorphismus des Torus T^n isotop zu einem affinen ist, vgl. (Thornton, M.C., 1967), (Zieschang, H., 1969). Für Seifertsche Faserräume von hyperbolischem Typ sind daher Homöomorphie, fasertreue Homöomorphie und Isomorphie der Fundamentalgruppe äquivalente Aussagen. Eine Klassifikation der verschiedenen (fasertreuen) Homöomorphie- bzw. Isomorphietypen wurde bisher nur für die klassischen 3-dimensionalen Seifertschen Faserräume bzw. für 4-dimensionale Seifertsche Faserräume mit trivialer Strukturgruppe durchgeführt (Seifert, H., 1933, Orlik, P. et al., 1967) bzw. (Orlik, P. & Raymond, F., 1970 & 1974). Dieser Klassifikation im Falle 4-dimensionaler Seifertscher Faserräume von hyperbolischem Typ mit beliegiger Strukturgruppe (wobei wir uns allerdings auf Untergruppen von $Sl_2(\mathbb{Z})$ an Stelle von $GL_2(\mathbb{Z})$ beschränken) und in Spezialfällen für beliebige Dimensionen ist die vorliegende Arbeit gewidmet. Im Falle trivialer Strukturgruppe lässt sich die Klassifikation bzgl. fasertreuer Homöomorphie weitgehend analog zum 3-dimensionalen Fall (Seifert, H., 1933) durchführen, vgl. (Orlik, P., & Raymond, F., 1970 & 1974), insbesondere lassen sich die numerischen Invarianten vollständig normieren (bis auf einen Basiswechsel in $\mathbb{Z}^n$). Im Falle nichttrivialer Strukturgruppe lässt sich eine Normierung nicht so einfach durchführen. Das eigentliche Problem liegt hier in einem Vergleich bzw. einer Klassifikation der Operationen $\phi : F \to Gl_n(\mathbb{Z})$; ausserdem werden die numerischen Invarianten wesentlich von der Strukturgruppe beeinflusst.

Bei unserer Klassifikation gehen wir rein algebraisch vor und beschäftigen uns ausschiesslich mit den Gruppen der Form 0.2 . Eine Isomorphieklassifikation der Gruppen 0.2 erfordert einen Vergleich der Operationen $\phi : F \to Gl_n(\mathbb{Z})$ sowie der numerischen Grössen w und w_j . Für $n=1$ ist $Gl_1(\mathbb{Z}) = \mathrm{Aut}\,\mathbb{Z} \cong \mathbb{Z}_2$ und eine Klassifikation der Operationen schnell durchgeführt (Orlik, P., et al, 1967). Die Klassifikation für $n=2$ ist der Inhalt der ersten drei Paragraphen dieser Arbeit. Dabei beschränken wir uns auf Homomorphismen $\phi : F \to Sl_2(\mathbb{Z}) \cong \mathbb{Z}_4 *_{\mathbb{Z}_2} \mathbb{Z}_6$ und werwenden wesentlich die Struktur von $PSl_2(\mathbb{Z}) \cong \mathbb{Z}_2 * \mathbb{Z}_3$ als freies Produkt (ausserdem operieren die s_j dann trivial auf $\mathbb{Z}^2$, $\phi(s_j) = 1$, und man kann sich daher auf Flächengruppen F beschränken). Nach (Serre, J.-P., 1980) besitzt bereits $Sl_3(\mathbb{Z})$ bzw. $Gl_3(\mathbb{Z})$ nicht mehr die Struktur eines freien Produkts (mit Amalgam), und für $n > 3$ gibt es in $Sl_n(\mathbb{Z})$ bzw. $Gl_n(\mathbb{Z})$ unlösbare Probleme: man kann z.B. allgemein nich entscheiden, ob zwei endliche Mengen von Elementen dieselbe Untergruppe erzeugen (Miller III, C.F., 1971, Chapter III.C). Als einfache Folgerung hieraus ergibt sich, dass das Isomorphieproblem für die Klasse der Fundamentalgruppen $(n+2)$-dimensionaler Seifertscher Faserräume für $n > 3$ unlösbar ist, vgl. Satz 4.9 .

Wir gehen im folgenden kurz auf den Inhalt der einzelnen Paragraphen ein.

Wir beginnen in § 1 mit einem vorbereitenden Satz (1.5) über quadratische Gleichungen in freien Produkten, d.h. Abbildungen von Flächengruppen in freie Produkte, der ein Resultat von Zieschang über freie Gruppen verallgemeinert (Zieschang, H., 1964). Ähnliche Sätze für freie Produkte mit Amalgam sind mit allgemeineren Kürzungsmethoden in (Lyndon, R.C., 1979), (Zieschang, H., 1981) und (Rosenberger, G., 1980) hergeleitet, die Beweise sind jedoch bedeutend komplizierter, und man erhält keine entsprechend einfachen Aussagen wie im Fall freier Produkte. Letzteres ist der Hauptgrund für unsere Beschränkung auf $Sl_2(\mathbb{Z})$ an Stelle von $Gl_2(\mathbb{Z})$ bei der Klassifikation 4-dimensionaler Seifertscher Faserräume: im Gegensatz zu $PSl_2(\mathbb{Z}) \cong \mathbb{Z}_2 * \mathbb{Z}_3$ ist $PGl_2(\mathbb{Z})$ nur ein freies Produkt mit nichttrivialem Amalgam. Eine Klassifikation sollte auch hier prinzipiell möglich sein, sähe jedoch sicherlich bedeutend komplizierter aus. In § 2 klassifizieren wir Homomorphismen von Flächengruppen in freie Produkte zyklischer Gruppen, insbesondere geben wir

Normalformen an. Im Falle nicht-orientierbarer Flächengruppen verbleiben dabei einige Fälle, die wir nicht unterscheiden können, jedoch ist eine vollständige Klassifikation in dem uns in erster Linie interessierenden Fall $PSl_2(\mathbb{Z}) = \mathbb{Z}_2 * \mathbb{Z}_3$ möglich (Satz 2.13). Homomorphismen von Flächengruppen in endlich zykische, Dieder - bzw. endliche abelsche Gruppen sind von Nielsen (1937), Vogt (1973) bzw. Smith (1967) klassifiziert worden. In § 3 behandeln wir Homomorphismen von Flächengruppen nach $Sl_2(\mathbb{Z})$. Zu jeder Untergruppe $G \subset Sl_2(\mathbb{Z})$ gibt es bis auf Äquivalenz endlich viele Homomorphismen $\phi : F \to Sl_2(\mathbb{Z})$ mit G als Bild, die sich leicht explizit angeben lassen (Normalformen). Wiederum im nichtorientierbaren Fall können wir allerdings einige davon nicht als verschieden nachweisen. Im 4. Paragraphen kommen wir dann zur Isomorphieklassifikation der Fundamentalgruppen Seifertscher Faserräume. Wir beginnen mit einem Satz (4.3), der die Isomorphie der Gruppen auf einen Vergleich der Operationen und der numerischen Grössen w, w_j zurückführt und Methoden und Resultate in (Orlik, P. et al, 1967) verallgemeinert. Zusammen mit den Ergebnissen in § 3 ergibt sich hieraus das Hauptresultat der Arbeit: man kann durch endlich viele elementare Umformungen entscheiden, ob zwei Fundamentalgruppen 0.2 4-dimensionaler transversal orientierbarer Seifertscher Faserräume von hyperbolischem Typ isomorph sind oder nicht, und gegebenenfalls einen Isomorphismus explizit konstruieren. Mit Hilfe dieses Satzes und der Normalformen der Operationen ϕ in § 3 lassen sich bei fest vorgegebener Gruppe $G \subset Sl_2(\mathbb{Z})$ Repräsentanten der Isomorphieklassen der Fundamentalgruppen mit Strukturgruppe G angeben.

1. *QUADRATISCHE GLEICHUNGEN IN FREIEN PRODUKTEN*

Jedes Element x in einem freien Produkt $G = \underset{\nu}{*} G_\nu$ besitzt eine eindeutig bestimmte Normalform $x = a_1 \dots a_n b_n^{-1} \dots b_1^{-1}$ bzw. $x = a_1 \dots a_n c\, b_n^{-1} \dots b_1^{-1}$, wobei aufeinanderfolgende a_i, b_i bzw. c in verschiedenen G_ν liegen, sowie eine Länge $L(x) = 2n$ bzw. $L(x) = 2n+1$, $L(1) := 0$ (vgl. z.B. (Lyndon, R.C. & Schupp, P.E., 1977, Chapter IV.1)). Wir nennen $v(x) := a_1 \dots a_n$ die vordere Hälfte, $h(x) := b_1 \dots b_n$ die hintere Hälfte und, falls vorhanden, $z(x) := c$ das Zentrum von x. Gilt $v(x) = h(x)$, so nennen wir x *konjugiert;* zwei konjugierte Elemente heissen *äquivalent,* falls die vorderen (bzw. hinteren) Hälften übereinstimmen und die Zentren aus demselben Faktor G_ν

sind. Die folgende Definition geht zurück auf Nielsen.

1.1 Definition einer Ordnung für die Elemente in $G = \underset{\nu}{*} G_\nu$.

(a) alphabetische Ordnung $\dot{<}$

Wir wählen Wohlordnungen der freien Faktoren G_ν von G sowie der Elemente jedes einzelnen freien Faktors G_ν.

Für $x, y \in G$ sei $x \dot{<} y$ falls

$L(x) < L(y)$ oder

$L(x) = L(y)$ und x vor y steht bzgl. der lexikographischen Ordnung der in Normalform geschriebenen Elemente x, y.

(b) semi-alphabetische Ordnung $<$

Es sei $x < y$ falls

$L(x) < L(y)$, oder

$L(x) = L(y)$ und $v(x) \dot{<} v(y)$, oder

$L(x) = L(y)$, $v(x) = v(y)$ und $h(x) \dot{<} h(y)$, oder

$L(x) = L(y)$, $v(x) = v(y)$, $h(x) = h(y)$ und $z(x) \dot{<} z(y)$.

Es seien F eine freie Gruppe und $K_1, \dots, K_n$ Elemente eines freien Erzeugendensystems von F. Ein Wort $Q = Q(K_1, \dots, K_n)$ in den K_i heisst *quadratisch*, falls jedes K_i genau zweimal in Q vorkommt (mit Exponenten +1 oder -1). Wir benötigen folgende Übergänge ("Zweiteilungen"), die aus einem quadratischen Wort $Q(K_1, \dots, K_n)$ ein neues quadratisches Wort $\overline{Q}(\overline{K}_1, \dots, \overline{K}_n)$ bilden (vgl. zum folgenden auch (Zieschang, H., 1964) (Zieschang, H. et al, 1980, § 5).

1.2

(a) Es sei $W_1 K_i^\varepsilon W_2$ ein Teilwort von Q, so dass $K_i^{\pm 1}$ in W_1 und W_2 nicht vorkommt. Dann sei $\overline{K}_i^{-\varepsilon} := W_1 K_i^\varepsilon W_2$ und $\overline{K}_k := K_k$ für $k \neq i$. Wir ersetzen in Q die Stelle $W_1 K_i^\varepsilon W_2$ durch $\overline{K}_i^{-\varepsilon}$ und drücken die restlichen K_j in Q durch die $\overline{K}_j$ aus.

(b) Es sei $\overline{K}_i := K_i^{-1}$, $\overline{K}_k := K_k$ sonst. Wir ersetzen K_i^{-1} durch $\overline{K}_i$ in Q und K_i durch $\overline{K}_i^{-1}$.

Im allgemeinen werden wir nach Anwendung der Prozesse 1.2 den Querstrich über Q bzw. den K_i gleich wieder weglassen. Die Prozesse 1.2 entsprechen den Zweiteilungen bei der kombinatorischen Klassifikation der Flächen, vgl. (Zieschang, H. et al, 1980),

(Stillwell, J., 1980). Jedem quadratischen Wert Q kann man auf kanonische Weise ein 2n-Eck in der Ebene zuordnen; den Kanten seien die Elemente $K_i^{\pm 1}$ zugeordnet, so dass der positive Randweg das Wort Q ergibt. Identifizieren wir Kanten, denen dasselbe K_i zugeordnet ist, so erhalten wir eine Fläche. Wir nennen das quadratische Wort Q *flächenhaft*, falls die Eckpunkte des 2n-Ecks dabei zu genau einem Punkt identifiziert werden.

1.3 Satz. Es sei $Q(K_1, \ldots, K_n)$ ein quadratisches Wort in der freien Gruppe F und $\phi : F \to G = \ast_\nu G_\nu$ ein Homomorphismus mit $\phi(q) = 1$. Dann lässt sich explizit eine Folge von Prozessen 1.2 angeben, die Q in ein quadratisches Wort Q' (mit $\phi(Q') = 1$) überführen von folgender Gestalt:

$$Q' = Q'(E_{k,\ell}, A_i, B_i, C_j, D_j, F_m)$$

$$= \prod_{k=1}^{k_0} W_k \prod_{i=1}^{i_0} [A_i, B_i] \prod_{j=1}^{j_0} \{C_j, D_j\} \prod_{m=1}^{m_0} F_m F_m^{-1},$$

wobei $[A_i, B_i] := A_i B_i A_i^{-1} B_i^{-1}$, $\{C_j, D_j\} := C_j D_j C_j^{-1} D_j$.

Dabei gilt $\phi(W_k) = \phi(B_i) = \phi(D_j) = 1$; W_k ist ein flächenhaftes quadratisches Wort in den $E_{k,\ell}$, und für festes k sind die Elemente $\phi(E_{k,\ell})$ äquivalente Konjugierte, d.h. liegen in einer festen zu einem G_ν konjugierten Untergruppe von G. Genau dann ist Q flächenhaft, wenn kein F_m in Q' vorkommt, d.h. $m_0 = 0$.

Beweis. Wir führen den Weweis durch Induktion über die Anzahl n der K_i.

Für $n = 1$ ist nichts zu zeigen.

Beim Induktionsschritt unterscheiden wir mehrere Fälle.

1.4 Q ist nicht flächenhaft. Dann können wir durch Zweiteilungen erreichen, dass in Q eine Stelle FF^{-1} vorkommt, vgl. z.B. (Stillwell, J., 1980, § 1.3.3.). Wir streichen FF^{-1} in Q und erhalten ein quadratisches Wort $\tilde{Q}$, das wir nach Induktionsvoraussetzung auf die Normalform des Satzes bringen können. Die dabei angewandten Zweiteilungen lassen sich in kanonischer Weise auf Q übertragen, vgl. (Zieschang, H., 1964, Hilfssatz 1). Daher können wir ohne Einschränkung annehmen, dass Q durch Einsetzen von FF^{-1} an geeigneter Stelle in ein Wort in

Normalform entsteht. Das entstehende Wort lässt sich dann leicht auf Normalform bringen.

Wir nehmen in folgenden an, dass Q flächenhaft ist.

1.5 Es gibt ein K_i *in* Q *mit* $\phi(K_i) = 1$. Durch Streichen der beiden K_i in Q erhalten wir ein quadratisches Wort $\tilde{Q}$, auf das wir die Induktionsvoraussetzung anwenden; daher sei ohne Einschränkung

$$\tilde{Q} = \underbrace{\Pi W_k \Pi [A_i, B_i] \Pi \{C_j, D_j\}}_{\tilde{Q}_1} \underbrace{\Pi F_m F_m^{-1}}_{\tilde{Q}_2}$$

Wie in 1.4 können wir annehmen, dass Q aus $\tilde{Q}$ durch Einsetzen der beiden K_i entsteht. Werden beide K_i im ersten Teil $\tilde{Q}_1$ von $\tilde{Q}$ eigesetzt, so können wir durch Zweiteilungen erreichen, dass sie nebeneinander stehen, da $\tilde{Q}_1$ flächenhaft ist. Eine Stelle $K_i K_i$ bringen wir durch Zweiteilungen an den Anfang von $\tilde{Q}_1$ (d.h. $K_i K_i$ spielt dann die Rolle eines W_k), eine Stelle $K_i K_i^{-1}$ ans Ende von $\tilde{Q}_1$. In beiden Fällen ergibt sich die Behauptung des Satzes.

Kommt genau ein K_i in $\tilde{Q}_1$ vor, so bringen wir es ans Ende von $\tilde{Q}_1$; daher können wir annehmen, dass beide K_i in $\tilde{Q}_2$ eingesetzt werden. Das dabei entstehende Wort Q_2 ist wiederum flächenhaft. Daher gibt es für Q_2 nur die folgenden Möglichkeiten:

$$F_1 K_i^{\varepsilon} F_1^{-1} K_i^{\pm\varepsilon} \quad \text{bzw.} \quad K_i^{\varepsilon} F_1 K_i^{\pm\varepsilon} F_1^{-1}, \quad \varepsilon = \pm 1.$$

In allen Fällen lässt sich Q nun leicht auf Normalform bringen.

1.6 Es gibt ein (notwendigerweise flächenhaftes) echtes quadratisches Teilwort W_k *von* Q *mit den Eigenschaften wie in Satz 1.3, insbesondere* $\phi(W_k) = 1$.

Das Wort, das wir durch Streichen von W_k erhalten, ist wiederum flächenhaft, daher können wir W_k an den Anfang von Q bringen und auf den Rest die Induktionsvoraussetzung anwenden.

1.7 Wir nennen ein quadratisches Wort $\tilde{Q}(\tilde{K}_1, \ldots, \tilde{K}_n)$ *reduziert*, wenn sich durch Zweiteilungen keines der $\phi(\tilde{K}_i)$ durch ein bezgl. der Ordnung < kleineres Element in $G = \ast_\nu G_\nu$ ersetzen lässt. Wir wenden auf das Wort Q so lange verkürzende Zweiteilungen 1.2 an, bis entweder eine Situation 1.5 oder 1.6 eintritt oder das entstehende Wort reduziert ist. Die Behauptung des Satzes ergibt sich dann aus dem folgenden Lemma.

1.8 Lemma. Es sei $Q(K_1, \ldots, K_n)$ ein reduziertes flächenhaftes quadratisches Wort, für das weder 1.5 noch 1.6 auftritt. Dann gilt $\phi(Q) \neq 1$.

Beweis. In dem Wort $\phi(Q) \in G$ in den $\phi(K_i)$ fassen wir nebeneinanderstehende äquivalente Konjugierte so lange zusammen, bis dies nicht mehr möglich ist. Dabei erhalten wir stets nur nichttriviale Elemente in G: ist das zusammengefasste Teilwort von Q quadratisch, so folgt dies aus 1.6 ; ist es nicht quadratisch, so muss der Wert ungleich Eins sein, da andernfalls ein verkürzender Übergang 1.2 möglich wäre.

Wir erhalten $\phi(Q)$ als Wort $g_1 \cdot \ldots \cdot g_m \in G$ mit folgenden Eigenschaften: $g_i \neq 1$, $g_i g_{i+1} \neq 1$, g_i und g_{i+1} sind nich äquivalente Konjugierte. Wir können jetzt (geeignet modifiziert) die Beweismethode von (Hall, M. jnr., 1953, Lemma 3) anwenden, um $g_i \cdot \ldots \cdot g_m \neq 1$ zu zeigen. □

Damit ist Satz 1.3 bewiesen. □

Bemerkung. Satz 1.3 lässt sich mit geometrischen Methoden beweisen. Es sei der Einfachheit halber Q flächenhaft und $G = G_1 * G_2$. Es sei F die zu Q gehörende Fläche und $\mathcal{G}_1$ und $\mathcal{G}_2$ $K(\pi, 1)$-Komplexe zu G_1 und G_2. Wir verbinden $\mathcal{G}_1$ und $\mathcal{G}_2$ durch ein Intervall $[-1, +1]$ und erhalten einen $K(G, 1)$-Komplex $\mathcal{G}$. Zu ϕ können wir dann eine Abbildung $f : F \to \mathcal{G}$ konstruieren, die wir transversal bzgl. $0 \in [-1, +1]$ machen. Dann besteht $f^{-1}(0)$ aus einfach geschlossenen Kurven auf F, die durch ϕ trivial abgebildet werden, und wir können geeignet eine Induktion anwenden. Wir haben hier einen algebraischen Beweis vorgezogen, da er konstruktiver ist und es direkt gestattet, das quadratische Wort Q durch einfache Prozesse auf Normalform zu bringen.

Um Satz 1.5 weiter anwenden zu können, muss man etwas über die freien Faktoren G wissen. Da wir im folgenden Paragraphen in erster Linie am Fall $G = \mathbb{Z}_2 * \mathbb{Z}_3$ der Modulgruppe interessiert sind, behandeln wir jetzt den Fall freier Produkte zyklischer Gruppen.

1.9 Korollar.

Es sei $S := \langle a_1, b_1, \dots, a_g, b_g \mid \prod_{i=1}^{g} [a_i, b_i] = 1 \rangle$

bzw. $S := \langle v_1, \dots, v_g \mid \prod_{i=1}^{g} v_i^2 = 1 \rangle$

isomorph zur Fundamentalgruppe einer orientierbaren bzw. nichtorientierbaren Fläche vom Geschlecht g und G ein freies Produkt zyklischer Gruppen. Es sei $\phi : S \to G$ ein Homomorphismus. Dann gibt es eine Darstellung

$$S = \langle \bar{v}_k, \bar{u}_\ell, \bar{w}_\ell, \bar{a}_i, \bar{b}_i, \bar{c}_j, \bar{d}_j \mid \prod_{k=1}^{k_0} \bar{v}_k^2 \prod_{\ell=1}^{\ell_0} \bar{u}_\ell^2 \bar{w}_\ell^2 \prod_{i=1}^{i_0} [\bar{a}_i, \bar{b}_i] \prod_{j=1}^{j_0} \{\bar{c}_j, \bar{d}_j\} = 1 \rangle$$

mit folgenden Eigenschaften:

$$\phi(\bar{v}_k^2) = \phi(\bar{u}_\ell^2 \bar{w}_\ell^2) = \phi(\bar{b}_i) = \phi(\bar{d}_j) = 1,$$

$$\phi(\bar{u}_\ell^2) \neq 1, \quad \phi(\bar{u}_\ell) \neq \phi(\bar{w}_\ell)^{-1}$$

(insbesondere erzeugen $\phi(\bar{u}_\ell)$ und $\phi(\bar{w}_\ell)$ eine endliche zyklische Untergruppe von G).

Ist ϕ explizit gegeben, so lassen sich die neuen Erzeugenden von S explizit als Worte in den alten angeben und umgekehrt.

Beweis. Wir wenden Satz 1.3 auf die definierende Relation Q von S an und erhalten ein quadratisches Wort Q' wie in 1.3, wobei $m_0 = 0$ ist, da Q flächenhaft ist. Jedes W_k ist flächenhaft und lässt sich daher durch Zweiteilungen 1.2 auf eine der folgenden Formen bringen (Klassifikation der Flächen):

$\prod_i [x_i, y_i]$ (orientierbarer Fall);

$v^2 \prod_i [x_i, y_i]$ bzw. $u^2 w^2 \prod_i [x_i, y_i]$ (nichtorientierbarer Fall).

Dabei liegen die Bilder unter ϕ der v, u, w, x_i, y_i in einer zyklischen Untergruppe von G. Daher gilt $\phi(v\ ) = 1$ bzw. $\phi(u^2 w^2) = 1$. Ist $\phi(u) = \phi(w)^{-1}$, so ersetzen wir u, v durch $c := u$, $d := uv$ mit $u^2 w^2 = c d c^{-1} d$, $\phi(d) = 1$.

Wir wenden auf jeden Kommutator $[x_i, y_i]$ Prozesse der Form

$$x_i \to x_i y_i^n, \quad y_i \to y_i \quad \text{bzw.} \quad x_i \to x_i, \quad y_i \to y_i x_i^n$$

an, $n \in \mathbb{Z}$. Diese Prozesse überführen $[x_i, y_i]$ in sich, und wir können $\phi(y_i) = 1$ erreichen (euklidischer Algorithmus). Durch Umorden der einzelnen Teile erhalten wir die gewünschte Form der definierenden Relation. Ausserdem erhalten wir die neuen Erzeugenden als Worte in den alten. Durch Anwenden der inversen Prozesse erhalten wir umgekehrt die alten Erzeugenden als Worte in den neuen. □

Im nächsten Paragraphen werden wir nach einigen Vorbereitungen die Darstellung von S bzw. den Homomorphismus ϕ weiter normieren.

2. *ÜBER HOMOMORPHISMEN VON FLÄCHENGRUPPEN IN FREIE PRODUKTE ZYKLISCHER GRUPPEN*

Wir wollen die im Titel genannten Homomorphismen klassifizieren und beschäftigen uns zunächst mit Untergruppen in freien Produkten zyklischer Gruppen.

Im folgenden bezeichne $G = \ast_\nu G_\nu$ stets ein freies Produkt (endlich vieler) zyklischer Gruppen G_ν. Eine Ordnung für G sei wie in 1.1 definiert; dabei sei als kleinstes Element nach 1 in G_ν stets ein erzeugendes Element e gewählt, und die Elemente in G_ν seien nach aufsteigenden Potenzen von e geordnet. Es seien $x_1, \ldots, x_n$ Elemente in G, die eine Untergruppe U erzeugen. Wir benötigen die folgenden Prozesse, die $x_1, \ldots, x_n$ in ein neues Erzeugendensystem von U überführen; die unveränderten Erzeugenden werden nicht angegeben.

2.1

(a) $x_i \to x_i^\alpha$, $\alpha \in \mathbb{Z}$, falls x_i und x_i^α dieselbe Untergruppe von G erzeugen.

(b) falls x_i konjugiert ist (vgl. § 1): $x_i \to x_j^\alpha x_i x_j^{-\alpha}$; dabei ist $\alpha \in \mathbb{Z}$, falls x_j konjugiert ist, andernfalls ist $\alpha \in \{+1, -1\}$.

(c) falls x_i nicht konjugiert ist, oder x_i und x_j äquivalente Konjugierte sind: $x_i \to x_i^\varepsilon x_j^\alpha$ bzw. $x_i \to x_j^\alpha x_i^\varepsilon$, $i \neq j$, $\varepsilon \in \{+1, -1\}$; dabei ist $\alpha \in \mathbb{Z}$, falls x_j konjugiert ist, andernfalls ist $\alpha \in \{+1, -1\}$.

2.2 Reduktion von $x_1, \ldots, x_n$. Wir wenden so lange Prozesse 2.1 an, bis sich die Gesamtlänge der x_i nicht mehr verkürzen lässt und kein x_i durch ein bzgl. der Ordnung $<$ früheres Element ersetzt werden kann. Der Prozess endet bei einem Erzeugendesystem $y_1, \ldots, y_m$ von U, das wir *reduziert* nennen; dabei haben wir eventuell auftretende Einsen weggelassen.

Es sei $Y = \{y_1, \ldots, y_m\}$ ein reduziertes Erzeugendensystem von U . Die Menge $\bar{Y}$ bestehe aus allen Elementen in Y, deren Inversen, sowie allen nichttrivialen Potenzen konjugierter Elemente in Y .

Das folgende Lemma ist eine Verallgemeinerung von (Hall, M., 1953, Lemma 3). Es lässt sich analog durch Induktion über μ beweisen.

2.3 Lemma. Es sei $u_1 \cdot \ldots \cdot u_\mu$ ein Produkt mit folgenden Eigenschaften: $u_i \in \bar{Y}$ (insbesondere $u_i \neq 1$), $u_i u_{i+1} \neq 1$, u_i und u_{i+1} sind nich äquivalente Konjugierte. Dann gilt:

$$u_1 \cdot \ldots \cdot u_\mu > u_1 \cdot \ldots \cdot u_{\mu-1} > \ldots > u_1 u_2 > u_1 \; ;$$

$$u_1 \cdot \ldots \cdot u_\mu > u_2 \cdot \ldots \cdot u_\mu > \ldots > u_{\mu-1} u_\mu > u_\mu \; ;$$

insbesondere ist $u_1 \cdot \ldots \cdot u_\mu \neq 1$. □

Der folgende Satz ist im wesentlichen eine Kombination der Sätze von Kurosh und Grushko-Neumann.

2.4 Satz. Je zwei reduzierte Erzeugendensysteme von U stimmen überein. Zwei beliebige Erzeugendensysteme lassen sich daher durch Prozesse 2.1 ineinander überführen.

Beweis. Es sei $x_1 < x_2 < \ldots < x_n$ ein reduziertes Erzeugendensystem von U . Wegen Lemma 2.3 ist x_1 nach dem Einselement das kleinste Element bzgl. $<$, x_2 ist das kleinste Element, das nicht in der von x_1 erzeugten Untergruppe von U liegt, usw. Daher sind die x_i eindeutig bestimmt. □

Bemerkung. Ist $x_1, \ldots, x_n$ ein reduziertes Erzeugendensystem von U , so gilt wegen Lemma 2.3

$$U = \langle x_1 \rangle * \langle x_2 \rangle * \ldots * \langle x_n \rangle .$$

Insbesondere ist jedes Element endlicher Ordnung in U konjugiert zu einer Potenz eines Konjugierten unter den x_i . Lässt man in 2.1 (b) und

(c) für x_j zusätzlich beliebige Produkte äquivalenter Konjugierter unter den x_k zu, so bleibt 2.3 richtig für freie Produkte beliebiger Gruppen, und man erhält den Satz von Kurosh sowie die Eindeutigkeit der durch ein reduziertes Erzeugendensystem gegebenen Produktdarstellung der Untergruppe. Ausserdem ergibt sich aus 2.3 auch der Satz von Grushko, da die Elemente eines reduzierten Erzeugendensystems eines freien Produktes wegen 2.3 in den freien Faktoren liegen.

2.5 Satz. Es seien reduzierte Erzeugendensysteme $x_1, \ldots, x_n$ bzw. $y_1, \ldots, y_m$ zweier Untergruppen U_1 bzw. U_2 von G gegeben. Dann kann man algorithmisch entscheiden, ob U_1 und U_2 konjugiert sind in G, und gegebenenfalls einen Konjugationsfaktor explizit berechnen.

Beweis. Es ist entscheidbar, ob ein Element $g \in G$ in U_1 (bzw. U_2) liegt. Hierzu produziere man systematisch alle Elemente in U_1 nach aufsteigender Grösse bzgl. der Ordnung $<$; dies ist möglich wegen Lemma 2.3. Irgendwann erreicht man bei diesem Prozess g oder es entstehen nur noch Elemente grösser als g.

Wir zeigen im folgenden, dass nur endlich viele Elemente in G als mögliche Konjugationsfaktoren getestet zu werden brauchen. Daraus ergibt sich die Behauptung des Satzes.

Wir unterscheiden 2 Fälle:

(a) Ein x_i hat endliche Ordnung. Aus $k U_1 k^{-1} = U_2$ folgt $k x_i k^{-1} = m y_j^{\nu} m^{-1}$, $m \in U_2$, d.h. $x_i = (k^{-1}m) y_j^{\nu} (m^{-1}k)$ und $(m^{-1}k) U_1 (m^{-1}k)^{-1} = U_2$. Daher gibt es in der Doppelrestklasse $U_2 k U_1$ einen Konjugationsfaktor mit einer Länge kleiner als $L(x_i)/2$.

(b) Kein x_i hat endliche Ordnung. Dann ist U eine freie Gruppe vom Rang n. In diesem Fall haben die Elemente $x_1, \ldots, x_n$ die Nielsensche Eigenschaft (vgl. z.B. (Lyndon, R.C. & Schupp, P.E., 1977, S. 6)), wie man sofort nachrechnet. Man kann dann das Verfahren im Beweis von Proposition 2.22 in (Lyndon, R.C. & Schupp, P.E., 1977) auf unsere Situation übertragen; es ergibt sich, dass in der Doppelrestklasse $U_2 k U_1$ eines Konjugationsfaktors k ein Konjugationsfaktor mit einer Länge kleiner gleich der Länge des grössten x_i bzw. y_j liegt. Dabei wird angenommen, dass mindestens ein x_i zyklisch reduziert ist, was sich durch eine Konjugation erreichen lässt. □

2.6 Satz. Es sei $x_1, \dots, x_n$ ein reduziertes Erzeugendensystem einer Untergruppe $U \neq \{1\}$ von G. Es sei $N(U)$ der Normalisator von U in G. Dann ist der Index von U in $N(U)$ endlich, und man kann ein Restklassenvertretersystem von U in $N(U)$ explizit berechnen.

Beweis. Es sei $kUk^{-1} = U$, $k \in G$. Nach dem Beweis von Satz 2.5 gibt es in der Restklasse $Uk = kU$ einen Restklassenvertreter mit einer Länge kleiner gleich einer festen Zahl, die nur von $x_1, \dots, x_n$ abhängt. Indem wir alle Elemente bis zu dieser Länge durchtesten, finden wir das gesuchte Restklassenvertretersystem. □

Nach diesen Vorbereitungen können wir uns der Klassifikation der im Titel des Paragraphen angegebenen Homomorphismen zuwenden. Dabei legen wir folgenden Äquivalenzbegriff zugrunde.

2.7 Definition. Es sei S isomorph zur Fundamentalgruppe einer geschlossenen Fläche und G ein freies Produkt zyklischer Gruppen. Zwei Homomorphismen ϕ_1 und ϕ_2 von S nach G heissen *äquivalent*, falls es einen Isomorphismus $\alpha : S \to S$ und ein kommutatives Diagramm wie folgt gibt:

$$\begin{array}{ccc} S & \xrightarrow{\phi_1} & G \\ \alpha \downarrow & & \downarrow i_g \\ S & \xrightarrow{\phi_2} & G \end{array}$$

dabei bezeichnet i_g den inneren Automorphismus mit einem Element $g \in G$. Wir nennen ϕ_1 und ϕ_2 s-äquivalent, falls $g \in \phi_1(S) = \phi_2(S)$ (dann kann man $g = 1$ erreichen).

Es sei $\phi : S \to G$ ein Homomorphismus. Wir wollen die Darstellung von S in 1.9 weiter normieren und behandeln zunächst den einfacheren orientierbaren Fall. Es sei

$$S := \langle a_1, b_2, \dots, a_g, b_g \mid \prod_{i=1}^{g} [a_i, b_i] = 1 \rangle, \text{ mit } \phi(b_i) = 1,$$

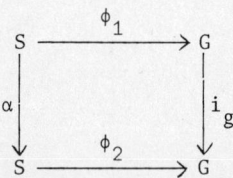

vgl. 1.9. Wir benötigen die folgenden Erzeugendenwechsel von S, die alle Form der definierenden Relation von S unverändert lassen. Die nicht abgeänderten Erzeugenden lassen wir im folgenden weg.

2.8

(a) $a_i \to \bar{a}_i := b_i^{-1} a_i^{-1} b_i^{-1}$,

$b_i \to \bar{b}_i := b_i a_i b_i^{-1} a_i^{-1} b_i^{-1}$;

$\phi(\bar{a}_i) = \phi(a_i)^{-1}$, $\phi(\bar{b}_i) = \phi(b_i) = 1$.

Prozess (a) kann erreicht werden durch folgende Prozesse:

(b) $a_i \to \bar{a}_i := a_i b_i^n$, $b_i \to \bar{b}_i := b_i$ bzw.

$a_i \to \bar{a}_i := a_i$, $b_i \to \bar{b}_i := b_i a_i^n$, $n \in \mathbb{Z}$.

Durch Prozesse (b) kann man erreichen, dass a_i, b_i durch $\bar{a}_i$, $\bar{b}_i$ ersetzt werden mit $\phi(\bar{a}_i) = \phi(a_i)^m$, $\phi(\bar{b}_i) = \phi(b_i) = 1$, falls $\phi(a_i)$ und $\phi(a_i)^m$ dieselbe Untergruppe von G erzeugen.

(c) $a_i \to a_{i+1}$, $b_i \to b_{i+1}$,

$a_{i+1} \to [a_{i+1}, b_{i+1}]^{-1} a_i [a_{i+1}, b_{i+1}]$,

$b_{i+1} \to [a_{i+1}, b_{i+1}]^{-1} b_i [a_{i+1}, b_{i+1}]$.

Durch (c) dann eine beliebige Permutation der Werte $\phi(a_i)$ erreicht werden.

(d) $\bar{a}_1 = a_1 a_2 b_2^{-1}$, $\bar{b}_1 = (b_2 a_2^{-1}) b_1 (a_2 b_2^{-1})$,

$\bar{a}_2 = ((b_2 a_2^{-1}) b_1 (a_2 b_2^{-1}) b_1^{-1}) a_2 ((b_2 a_2^{-1}) b_1^{-1} (a_2 b_2^{-1}))$,

$\bar{b}_2 = b_2 ((b_2 a_2^{-1}) b_1^{-1} (a_2 b_2^{-1}))$;

$\phi(\bar{a}_1) = \phi(a_1)\phi(a_2)$, $\phi(\bar{b}_1) = 1$, $\phi(\bar{a}_2) = \phi(a_2)$, $\phi(\bar{b}_2) = 1$.

Dieser Prozess ist das wesentliche Hilfsmittel in Nielsen (1937) bei der Klassifikation periodischer Abbildungen von Flächen, vgl. auch (Vogt, E., 1973).

Konjugationen der $\phi(a_i)$ können am einfachsten durch folgenden Prozess erreicht werden:

(e) $\bar{a}_1 = a_1 a_2 a_1^{-1}$, $\bar{b}_1 = a_1 b_2 a_1^{-1}$,

$\bar{a}_2 = a_1$, $\bar{b}_2 = [a_2, b_2]^{-1} b_1$;

$$\phi(\bar{a}_1) = \phi(a_1)\,\phi(a_2)\,\phi(a_1)^{-1}\,,\quad \phi(b_1) = \phi(a_1)\,\phi(b_2)\,\phi(a_1)^{-1}\,,$$
$$\phi(\bar{a}_2) = \phi(a_1)\,,\quad \phi(\bar{b}_2) = \phi(b_1)\,.$$

2.9 Satz. Es sei $\phi : S \to G$ ein Homomorphismus, wobei $S = \langle a_i, b_i \mid \prod_{i=1}^{g} [a_i, b_i] = 1 \rangle$ isomorph zur Fundamentalgruppe einer orientierbaren geschlossenen Fläche und G freies Produkt zyklischer Gruppen ist. Dann gibt es eine Darstellung $S = \langle \bar{a}_i, \bar{b}_i \mid \prod_{i=1}^{g} [\bar{a}_i, \bar{b}_i] = 1 \rangle$, so dass $\phi(\bar{b}_i) = 1$ und $\phi(\bar{a}_1), \ldots, \phi(\bar{a}_g)$ das nach Satz 2.4 eindeutig bestimmte reduzierte Erzeugendensystem von $\phi(S) \subset G$ ist (wenn man die Einsen weglässt). Ist ϕ explizit durch die Bilder der Erzeugenden a_i, b_i gegeben, so kann man die neuen Erzeugenden $\bar{a}_i, \bar{b}_i$ explizit als Werte in den a_i, b_i angeben (und umgekehrt).

Beweis. Wir gehen aus von der Darstellung von S in Korollar 1.9 mit $\phi(b_i) = 1$. Auf das Erzeugendensystem $\phi(a_1), \ldots, \phi(a_g)$ von $\phi(S)$ wenden wir Reduktionsprozess 2.2 an. Dabei soll die Form der definierenden Relation von S nicht verändert werden. Dieses liefern gerade die Prozesse 2.8. Als Ergebnis des Reduktionsprozesses erhalten wir die gewünschte Darstellung von S. □

Aus 2.9, 2.5 und 2.6 folgt:

2.10 Korollar. Es seien S, G wie in Satz 2.9 und $\phi_1, \phi_2 : S \to G$ zwei Homomprphismen.

(a) Genau dann sind ϕ_1 und ϕ_2 s-äquivalent bzw. äquivalent, wenn $\phi_1(S) = \phi_2(S)$ bzw. wenn $\phi_1(S)$ und $\phi_2(S)$ konjugiert sind in G.

(b) Sind ϕ_1 und ϕ_2 explizit gegeben, so lässt sich entscheiden, ob ϕ_1 und ϕ_2 s-äquivalent bzw. äquivalent sind, und gegebenenfalls kann man einen Isomorphismus α wie in 2.7 explizit berechnen (sowie den inversen Isomorphismus). □

Der Fall nichtorientierbarer Flächen ist komplizierter, und eine Lösung des Äquivalenzproblems analog zu 2.9 ist uns nicht in allen Fällen gelungen. Wir behandeln daher nur den Fall der Modulgruppe $G = \mathbb{Z}_2 * \mathbb{Z}_3$, unserer Hauptanwendung in den folgenden Paragraphen.

Es sei S nichtorientierbar und $\phi : S \to G$ ein Homomorphismus. Wegen Korollar 1.9 können wir annehmen, dass eine Darstellung

$$S = \langle v_k, u_\ell, w_\ell, a_i, b_i, c_j, d_j \mid \prod_{k=1}^{k_0} v_k^2 \prod_{\ell=1}^{\ell_0} u_\ell^2 w_\ell^2 \prod_{i=1}^{i_0} [a_i, b_i] \prod_{j=1}^{j_0} \{c_j, d_j\} = 1 \rangle$$

vorliegt mit $\phi(v_k^2) = \phi(u_\ell^2 w_\ell^2) = \phi(b_i) = \phi(d_j) = 1$. Wir benötigen zusätzlich zu 2.8 die folgenden Prozesse:

2.11

(a) $\bar{c}_j = c_j d_j c_j^{-1} d_j c_j^{-1}$, $\bar{d}_j = c_j d_j c_j^{-1}$;

$\phi(\bar{c}_j) = \phi(c_j)^{-1}$, $\phi(\bar{d}_j) = \phi(d_j) = 1$.

(b) $\bar{c}_j = c_j d_j^n$, $\bar{d}_j = d_j$.

Der andere Prozess in 2.8 (b) ist nicht übertragbar; dieses ist eine der Hauptschwierigkeiten im nichtorientierbaren Fall.

(c) eine beliebige Permutation der einzelnen Teile v_k^2, $u_\ell^2 w_\ell^2$, $[a_i, b_i]$, $\{c_j, d_j\}$ der definierenden Relation; Vertauschung von u_ℓ und w_ℓ (wir verzichten auf die explizite Angabe der Prozesse).

2.12

(a) $\{c_1, d_1\}\{c_2, d_2\} = \{c, d\}[a, b]$ mit

$c = c_1$, $d = d_1 d_2$, $a = d_2^{-1} d_1^{-1} c_1 c_2^{-1} d_2$,

$b = d_2^{-1} c_2 d_2^{-1} c_2^{-1} d_2$;

$\phi(c) = \phi(c_1)$, $\phi(d) = 1$, $\phi(a) = \phi(c_1)\,\phi(c_2)^{-1}$, $\phi(b) = 1$.

(b) $\{c, d\}[a, b] = \{\bar{c}, \bar{d}\}[\bar{a}, \bar{b}]$ mit

$\bar{c} = cab^{-1}$, $\bar{d} = ba^{-1} dab^{-1}$,

$\bar{a} = ba^{-1} d^{-1} ab^{-1} daba^{-1} dab^{-1}$,

$\bar{b} = bba^{-1} dab^{-1}$;

$\phi(\bar{c}) = \phi(c)\,\phi(a)$, $\phi(\bar{d}) = 1$, $\phi(\bar{a}) = \phi(a)$, $\phi(\bar{b}) = 1$.

Wir bemerken, dass man Konjugationen analog zu 2.8 (e) wieder durch einfachere Prozesse erreichen kann; wir verzichten auf die explizite Angabe.

(c) $v^2[a,b] = \bar{v}^2[\bar{a},\bar{b}]$ mit

$\bar{v} = a^{-1}va\,,\ \bar{a} = a^{-1}v^{-2}av^2a\,,\ \bar{b} = ba^{-1}v^2a\,;$

$\phi(\bar{v}) = \phi(a)^{-1}\phi(v)\phi(a),\ \phi(\bar{a}) = \phi(a),\ \phi(\bar{b}) = 1\,.$

$v^2\{c,d\} = \bar{v}^2\{\bar{c},\bar{d}\}$ mit

$\bar{v} = c^{-1}v^{-1}c\,,\ \bar{c} = c^{-1}v^2cc\,,\ \bar{d} = c^{-1}v^2cd\,;$

$\phi(\bar{v}) = \phi(c)^{-1}\phi(v)\phi(c),\ \phi(\bar{c}) = \phi(c),\ \phi(\bar{d}) = 1\,.$

(d) $v^2\{c,d\} = \bar{v}^2[a,b]$ mit

$\bar{v} = vd\,,\ a = d^{-1}v^{-1}c^{-1}d\,,\ b = d^{-1}cd^{-1}c^{-1}d\,;$

$\phi(\bar{v}) = \phi(v),\ \phi(a) = \phi(v)^{-1}\phi(c)^{-1},\ \phi(b) = 1\,.$

(e) $v_1^2v_2^2 = cdc^{-1}d$ mit $c = v_1\,,\ d = v_1v_2\,.$

2.13 Satz. Es sei $S = \langle v_1,\dots,v_g \mid \prod_{k=1}^{g} v_k^2 = 1\rangle$ isomorph zur Fundamentalgruppe einer geschlossenen nichtorientierbaren Fläche, und $\phi : S \to G = \mathbb{Z}_2 * \mathbb{Z}_3$ ein Homomorphismus.

(a) Es gibt eine Darstellung

$$S = \langle \bar{v}_k, a_i, b_i, c_j, d_j \mid \prod_{k=1}^{k_0}\bar{v}_k^2 \prod_{i=1}^{i_0}[a_i,b_i] \prod_{j=1}^{j_0}\{c_j,d_j\} = 1\rangle$$

mit folgenden Eigenschaften:

(i) $\phi(\bar{v}_k^2) = \phi(b_i) = \phi(d_j) = 1\,;$

(ii) lässt man die Einsen weg, so ist $\phi(\bar{v}_k),\ \phi(a_i),\ \phi(c_j)$ das nach Satz 2.4 eindeutig bestimmte reduzierte Erzeugendensystem von $\phi(S)$;

(iii) es gibt höchstens ein v_k mit $\phi(\bar{v}_k) = 1$; gibt es ein solches oder ein $\phi(c_j)$ ungerader Ordnung (d.h. der Ordnung 1 oder 3), so ist $i_0 = 0$, d.h. es kommen keine a_i, b_i vor.

(b) Eine Darstellung mit den Eigenschaften (i) - (iii) ist (bis auf eine Permutation der Indices k, i bzw. j) eine Invariante der s-Äquivalenzklasse von ϕ; die Zahlen k_0, i_0 und j_0 sind Invarianten der Äquivalenzklasse von ϕ.

(c) Ist ϕ explizit durch die Bilder der Erzeugenden v_k gegeben, so lassen sich die neuen Erzeugenden $\bar{v}_k, a_i, b_i, c_j, d_j$ explizit als Worte in den v_k angeben (und umgekehrt).

Beweis. (a) Wir gehen aus von der in Korollar 1.9 erhaltenen Darstellung von S. Da G ausschliesslich Elemente unendlicher Ordnung sowie der Ordnungen 2 und 3 enthält, ist $\ell_0 = 0$, d.h. es kommen keine $\bar{u}_\ell, \bar{w}_\ell$ vor. Die Elemente $\phi(\bar{v}_k)$, $\phi(\bar{a}_i)$, $\phi(\bar{c}_j)$ bilden ein Erzeugendensystem von $\phi(S)$, auf das wir mittels der Prozesse 2.8, 2.11 und 2.12 das Reduktionsverfahren 2.2 anwenden. Haben zwei v_k den gleichen ϕ-Wert, so ersetzen wir sie mittels 2.12(e) durch einen Kommutator $\{c, d\}$ mit $\phi(d) = 1$. Als Ergebnis erhalten wir eine Darstellung von S mit den Eigenschaften (i), (ii). Gibt es ein $\bar{v}_k$ mit $\phi(\bar{v}_k) = 1$ bzw. ein $\phi(c_j)$ ungerader Ordnung, so ersetzen wir mittels 2.12 (d) bzw. 2.12 (a) die Kommutatoren der Form $[a, b]$ durch Kommutatoren der Form $\{c, d\}$; dabei ist 2.12 (a) m-mal anzuwenden, falls m die Ordnung von $\phi(c_j)$ ist. Damit ist auch (iii) erreicht. (Wir bemerken, dass es im Fall freier Produkte beliebiger zyklischer Gruppen nicht immer möglich ist, ein $\phi(c_j)$ endlicher Ordnung durch eine primitive Potenz zu ersetzen; dieses ist eine der Hauptschwierigkeiten bei der Klassifikation).

(b) Die $\phi(a_i)$ sind genau die Werte unter den $\phi(\bar{v}_k)$, $\phi(a_i)$, $\phi(c_j)$, in deren Urbild in S nur orientierungserhaltende Elemente liegen: es folgt aus Lemma 2.3, dass genau dann ein orientierungsänderndes Element im Kern von ϕ liegt, wenn es ein $\bar{v}_k$ mit $\phi(\bar{v}_k) = 1$ oder ein $\phi(c_j)$ ungerader Ordnung gibt; in beiden Fällen ist $i_0 = 0$. Daher sind die $\phi(a_i)$ durch die s-Äquivalenzklasse von ϕ eindeutig bestimmt und i_0 sogar durch die Äquivalenzklasse.

Die Invarianz der $\phi(\bar{v}_k)$ sieht man folgendermassen. Es gibt in $S_{ab} = S/[S, S]$ genau ein nichttriviales Element x der Ordnung 2, in dessen Urbild in S das Element $\prod_{k=1}^{k_0} \bar{v}_k \prod_{j=1}^{j_0} d_j$ liegt. Wir betrachten den Homomorphismus $\bar{\phi} : S_{ab} \to \phi(S)_{ab}$. Wegen der Bemerkung nach Satz 2.4 ist $\phi(S)$ bzw. $\phi(S)_{ab}$ direktes Produkt bzw. direkte Summe zyklischer Gruppen; die Erzeugenden dieser Gruppen sind die Elemente des reduzierten Erzeugendensystems von $\phi(S)$. Das Bild $\bar{\phi}(x)$ von x in $\phi(S)$ ist durch die s-Äquivalenzklasse von ϕ eindeutig bestimmt. Daher sind auch die Werte $\phi(\bar{v}_1),\ldots,\phi(\bar{v}_{k_0})$ durch die s-Äquivalenzklasse von ϕ eindeutig bestimmt, und k_0 sogar durch die Äquivalenzklasse (man beachte, dass ein triviales $\phi(\bar{v}_k)$ bestimmt, ob das Geschlecht g von S gerade oder ungerade ist).

(c) Wir haben in 1.2 und 2.8, 2.11, 2.12 alle Prozesse explizit angegeben, die zur Darstellung von S führen. □

2.14 Korollar. Es seien S, G wie in Satz 2.13 und $\phi_1, \phi_2 : S \to G$ zwei explizit gegebene Homomorphismen. Dann kann man entscheiden, ob ϕ_1 und ϕ_2 (s-)äquivalent sind, und gegebenenfalls einen Isomorphismus ϕ wie in 2.7 berechnen (sowie den inversen Isomorphismus). □

Eine vollständige Klassifikation der Homomorphismen $\phi : S \to G$ nichtorientierbarer Flächengruppen X in beliebige freie Produkte G zyklischer Gruppen ist uns nur gelungen in einem "stabilen Bereich", in dem das Geschlecht der Flächengruppe S sehr viel grösser ist als der Rang (minimale Anzahl der Erzeugenden) des Bildes $\phi(S)$ von S. Wir wollen darauf hier nicht näher eingehen; genaueres findet sich in (Zimmermann, B, 1982).

3. *QUADRATISCHE GLEICHUNGEN IN $SL_2(\mathbb{Z})$*

Es ist $Sl_2(\mathbb{Z}) = \left\{\begin{pmatrix} a & b \\ c & d\end{pmatrix} : ad - bc = 1;\ a,b,c,d \in \mathbb{Z}\right\} \cong \mathbb{Z}_4 *_{\mathbb{Z}_2} \mathbb{Z}_6$ (vgl. z.B. (Zieschang, H., 1981, Theorem 23.1)). Das Amalgam $\mathbb{Z}_2$ ist das Zentrum von $Sl_2(\mathbb{Z})$ und wird erzeugt von $z := \begin{pmatrix} -1 & 0 \\ 0 & -1\end{pmatrix}$. Mit $x := \begin{pmatrix} 0 & -1 \\ 1 & 0\end{pmatrix}$ bzw. $y = \begin{pmatrix} 0 & 1 \\ -1 & 1\end{pmatrix}$ bezeichnen wir Erzeugende von $\mathbb{Z}_4$ bzw. $\mathbb{Z}_6$, $x^2 = y^3 = z$. Durch Herauskürzen des Zentrums $\mathbb{Z}_2$ erhalten wir einen surjektiven Homomorphismus $p: Sl_2(\mathbb{Z}) \to \mathbb{Z}_2 * \mathbb{Z}_3 \cong PSl_2(\mathbb{Z})$, und $\bar{x} := p(x)$ bzw. $\bar{y} := p(y)$ sind Erzeugende von $\mathbb{Z}_2$ bzw. $\mathbb{Z}_3$.

Es sei S isomorph zur Fundamentalgruppe einer geschlossenen Fläche. Wir wollen Homomorphismen von S nach $Sl_2(\mathbb{Z})$ klassifizieren und legen den folgenden Äquivalenzbegriff analog zu 2.7 zugrunde.

3.1 Definition.

Zwei Homomorphismen ϕ_1, ϕ_2 von S nach $Sl_2(\mathbb{Z})$ heissen *äquivalent*, falls es einen Isomorphismus $\alpha : S \to S$, ein Element

$N \in GL_2(\mathbb{Z}) = \left\{ \begin{pmatrix} a & b \\ c & d \end{pmatrix}, \ ad - bc = \pm 1; a, b, c, d \in \mathbb{Z} \right\}$ und ein kommutatives Diagramm wie folgt gibt:

$$\begin{array}{ccc} S & \xrightarrow{\phi_1} & Sl_2(\mathbb{Z}) \\ {\scriptstyle\alpha}\downarrow & & \downarrow {\scriptstyle \text{Konjugation mit } N} \\ S & \xrightarrow[\phi_2]{} & Sl_2(\mathbb{Z}) \end{array}$$

Wir nennen ϕ_1 und ϕ_2 s-äquivalent, falls $N \in \phi_1(S) = \phi_2(S)$ (dann kann man $N = 1$ erreichen).

Es sei $\phi : S \to Sl_2(\mathbb{Z})$ ein Homomorphismus, $G := \phi(S) \subset Sl_2(\mathbb{Z})$, $\bar{G} := p(G) \subset \mathbb{Z}_2 * \mathbb{Z}_3$. Es sei $\bar{E} = \{ \bar{x}_\alpha, \bar{y}_\beta, \bar{t}_\gamma, \ 0 \le \alpha \le \alpha_0, \ 0 \le \beta \le \beta_0, \ 0 \le \gamma \le \gamma_0 \}$ das nach 2.4 eindeutig bestimmte reduzierte Erzeugendensystem von $\bar{G}$; dabei seien die $\bar{x}_\alpha$ bzw. $\bar{y}_\beta$ konjugiert zu $\bar{x}$ bzw. $\bar{y}$, die $\bar{t}_\gamma$ seien von unendlicher Ordnung. Mit $x_\alpha, y_\beta, t_\gamma$ bezeichnen wir im folgenden fest gewählte Urbilder von $\bar{x}_\alpha, \bar{y}_\beta, \bar{t}_\gamma$ in $Sl_2(\mathbb{Z})$. Die t_γ seien Elemente von G, die x_α bzw. y_β konjugiert zu x bzw. y; hierdurch sind die x_α, y_β und, falls $z \notin G$, auch die t_γ eindeutig bestimmt. Zur einfacheren Formulierung des folgenden Satzes benötigen wir:

3.2 Definition. einer Liftung E von $\bar{E}$ nach G:

$E := \{ x_\alpha, y_\beta, t_\gamma \}$ falls $z \in G$;

$E := \{ y_\beta z, t_\gamma \}$ falls $z \notin G$; im zweiten Fall kommen keine $\bar{x}_\alpha$ vor, und die y_β liegen nicht in G.

Wir behandeln zunächst wider den einfacheren Fall orientierungserhaltender Flächengruppen.

3.3 Satz.

Es sei $S = \langle a_1, b_1, \ldots, a_g, b_g \mid \prod_{i=1}^{g} [a_i, b_i] = 1 \rangle$

isomorph zur Fundamentalgruppe einer orientierbaren geschlossenen Fläche und $\phi : S \to Sl_2(\mathbb{Z})$ ein Homomorphismus, $G = \phi(S)$.

(a) Es gibt eine Darstellung

$$S = < a_1, b_1, \ldots, a_g, b_g \mid \prod_{i=1}^{g} [\tilde{a}_i, \tilde{b}_i] = 1 >$$

mit folgenden Eigenschaften:

(i) lässt man die Einsen weg, so ist $\{\phi(\tilde{a}_1), \ldots, \phi(\tilde{a}_g)\}$ gleich der Liftung E von $\bar{E}$ nach G oder gleich $E \cup \{z\}$; der zweite Fall tritt auf, falls $z \in G$, jedoch kein $\tilde{b}_i$ durch ϕ auf z abgebildet wird (vgl. (ii)) und keine x_α, y_β vorkommen;

(ii) gilt $\phi(\tilde{a}_i) = 1, z, x_\alpha, y_\beta$ oder $y_\beta z$, so ist $\phi(\tilde{b}_i) = 1$; gilt $\phi(\tilde{a}_i) = t_\gamma$, so ist $\phi(\tilde{b}_i) = 1$ oder z.

(b) Eine Darstellung von S mit den Eigenschaften (i), (ii) ist (bis auf eine Permutation der Indices (i) eine Invariante der s-Äquivalenzklasse von ϕ.

(c) Ist ϕ explizit durch die Bilder in $Sl_2(\mathbb{Z})$ der Erzeugenden a_i gegeben, so lassen sich die neuen Erzeugenden explizit als Worte in den a_i angeben, und umgekehrt.

Beweis. Es sei $\bar{\phi} := p \circ \phi : S \to \mathbb{Z}_2 * \mathbb{Z}_3$. Liegt z nicht in G, so folgt die Behauptung aus Satz 2.9. Wir können daher im folgenden $z \in G$ annehmen.

(a) Wiederum wegen Satz 2.9 können wir annehmen, dass es sich bei $\bar{\phi}(a_1), \ldots, \bar{\phi}(a_g)$ (abgesehen von den Einsen) um das reduzierte Erzeugendensystem $\bar{E}$ von $\bar{G} = p(G)$ handelt, und dass $\bar{\phi}(b_i) = 1$ gilt für $i = 1, \ldots, g$.

Es sei $(\bar{\phi}(a_i), \bar{\phi}(b_i)) = (\bar{u}, 1) \in (\mathbb{Z}_2 * \mathbb{Z}_3)^2$, $\bar{u} = \bar{x}_\alpha, \bar{y}_\beta$ oder $\bar{t}_\gamma$. Dann gilt $(\phi(a_i), \phi(b_i)) = (u, 1) (u, z), (uz, 1)$ oder $(uz, z) \in (Sl_2(\mathbb{Z}))^2$ Wir wollen zeigen, dass man in fast allen Fällen $(\phi(a_i), \phi(b_i)) = (u, 1)$ erreichen kann durch Prozesse 2.8, und unterscheiden zwischen den verschiedenen Typen $x_\alpha, y_\beta, t_\gamma$ von Elementen in E. Im folgenden bezeichne $\underset{2.8}{\longleftrightarrow}$ einen Erzeugendenwechsel von S gemäss 2.8, ein Tupel (u, v) steht für $(\phi(a_i), \phi(b_i))$.

(1) $(x_\alpha, 1) \underset{2.8(b)}{\longleftrightarrow} (x_\alpha, x_\alpha^2 = z) \underset{2.8(b)}{\longleftrightarrow} (x_\alpha z, z) \underset{2.8(b)}{\longleftrightarrow} (x_\alpha z, 1)$;

(2) $(y_\beta, 1) \underset{2.8(b)}{\longleftrightarrow} (y_\beta, y_\beta^3 = z) \underset{2.8(b)}{\longleftrightarrow} (y_\beta z, z)$;

(3) $(z,1) \underset{2.8(b)}{\longleftrightarrow} (z,z) \underset{2.8(b)}{\longleftrightarrow} (1,z)$; $(t_\gamma, z) \underset{2.8(b)}{\longleftrightarrow} (t_\gamma z, z)$;

(4) es sei ein Tupel der Form $(x_\alpha, 1)$, $(y_\beta, 1)$ oder $(z, 1)$ im Bild von ϕ vorhanden (oder ein nach (1)-(3) gleichwertiges Tupel):

$(y_\beta, 1) \underset{2.8(d),(a),(c)}{\longleftrightarrow} (y_\beta z, 1)$;

$(t_\gamma, 1) \underset{2.8(d)(a)(c)}{\longleftrightarrow} (t_\gamma z, 1)$:

$(1,1) \underset{2.8(d)(a)(c)}{\longleftrightarrow} (z,1)$.

Ist $(x_\alpha, 1)$, $(y_\beta, 1)$ oder $(z, 1)$ bzw. ein gleichwertiges Tupel nicht im Bild von ϕ vorhanden, so muss ein Tupel der Form (t_γ, z) vorhanden sein, da $z \in G$. In diesem Fall benötigen wir noch den folgenden Prozess:

(P) $a_1 \to a_1 b_1^{-1} b_2^{-1}$, $b_1 \to b_2 b_1 b_2^{-1}$,

$a_2 \to b_2 b_1 b_2^{-1} b_1^{-1} a_2 b_1^{-1} b_2^{-1}$, $b_2 \to b_2 b_1 b_2 b_1^{-1} b_2^{-1}$,

mit $[a_1, b_1][a_2, b_2] \to [a_1, b_1][a_2, b_2]$.

Dann lassen sich die obigen Umformungen ebenfalls erreichen, z.B.

$$(t_{\gamma_1}, 1), (t_{\gamma_2}, z) \underset{2.8(c),(P)}{\longleftrightarrow} (t_{\gamma_1} z, 1)(t_{\gamma_2} z, z) \underset{2.8(d)(a)(c)}{\longleftrightarrow} (t_{\gamma_1} z, 1)(t_{\gamma_2}, z).$$

Mittels (1)-(4) können wir die Tupel $(\phi(a_i), \phi(b_i))$ auf die gewünschte Form (i),(ii) bringen. Für ein t_γ kann dabei das Tupel $(t_\gamma, 1)$ oder (t_γ, z) auftauchen; diese beiden Fälle sind sind s-äquivalent, vgl.(b).

(b) Es bleibt zu zeigen, dass zwei Homomorphismen $\phi_1, \phi_2 : S \to Sl_2(\mathbb{Z})$ mit $\phi_1(S) = \phi_2(S) = G$ und mit den Eigenschaften (i) und (ii) nicht s-äquivalent sind, falls $(\phi_1(a_1), \phi_1(b_1)) = (t_1, z)$ und $(\phi_2(a_1), \phi_2(b_1)) = (t_1, 1)$ gilt.

Wir setzen in $G = \phi_1(S) = \phi_2(S)$ alle t_γ bis auf t_1 gleich Eins, identifizieren alle x_α, $x_\alpha = x_1$, und identifizieren alle y_β mit z, $y_\beta = z$. Wir machen die entstehende Gruppe abelsch und erhalten $\mathbb{Z} \oplus \mathbb{Z}_n$ sowie eine Projektion $\pi : G \to \mathbb{Z} \oplus \mathbb{Z}_n$. Kommt kein x_α vor, so ist $n = 2$ und $\pi(t_1)$ bzw. $\pi(z)$ erzeugen $\mathbb{Z}$ bzw. $\mathbb{Z}_2$; kommt ein x_α vor, so ist $n = 4$ und $\pi(t_1)$ bzw. $\pi(x_1)$ erzeugen $\mathbb{Z}$ bzw. $\mathbb{Z}_4$. Wir wollen zeigen, dass $\tilde{\phi}_1 := \pi \circ \phi_1$ und $\tilde{\phi}_2 := \pi \circ \phi_2$ nicht s-äquivalent sind. Andernfalls existiert ein kommutative Diagramm

$$\begin{array}{ccc} S & & \\ & \searrow^{\tilde{\phi}_1} & \\ \tilde{\alpha}\downarrow & & \mathbb{Z} \oplus \mathbb{Z}_n \\ & \nearrow_{\tilde{\phi}_2} & \\ S & & \end{array}$$

Durch Prozesse 2.8 und (P) können wir wie im Beweis von (a) erreichen, dass $\tilde{\phi}_1$ und $\tilde{\phi}_2$ auf den erzeugenden Paaren (a_i, b_i) von S folgende Werte annehmen:

$$\tilde{\phi}_1 : (\pi(t_1), \pi(z)), (\pi(u), 1), (1, 1), (1, 1) \ldots$$

$$\tilde{\phi}_2 : (\pi(t_1), 1), (\pi(u), 1), (1, 1), (1, 1) \ldots\ldots ;$$

dabei ist $u = x_1$ im Falle $n = 4$ und $u = z$ im Falle $n = 2$.

Wir machen obiges Diagramm abelsch mod n und erhalten

$$\begin{array}{ccc} S_{ab} \bmod n & & \\ & \searrow^{\bar{\phi}_1} & \\ \bar{\alpha}\downarrow & & \mathbb{Z}_n \oplus \mathbb{Z}_n \\ & \nearrow_{\bar{\phi}_2} & \\ S_{ab} \bmod n & & \end{array}$$

Es wird Kern $\bar{\phi}_2$ erzeugt von $\bar{b}_1, \bar{b}_2, \bar{a}_3, \bar{b}_3, \bar{a}_4, \bar{b}_4, \ldots$; dabei bezeichnen wir die abelsch gemachten Erzeugenden mit einem Querstrich. Es gilt:

$$\begin{aligned} \bar{\alpha}(\bar{a}_1) &= \bar{a}_1 v_1, & \bar{\alpha}(\bar{b}_1) &= \bar{a}_2^{n/2} w_1, \\ \bar{\alpha}(\bar{a}_2) &= \bar{a}_2 v_2, & \bar{\alpha}(\bar{b}_2) &= w_2, \\ \bar{\alpha}(\bar{a}_3) &= v_3, & \bar{\alpha}(\bar{b}_3) &= w_3, \\ &\vdots & &\vdots \end{aligned}$$

mit $v_i, w_i \in$ Kern $\overline{\phi_2}$. Daher gehört zu $\bar{\alpha}$ eine (anti)-symplektische Matrix

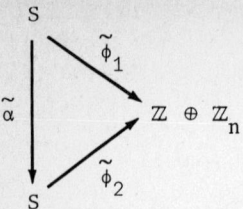

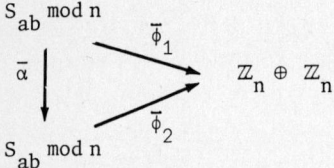

	$\bar{a}_1$	$\bar{b}_1$	$\bar{a}_2$	$\bar{b}_2$	$\bar{a}_3$	$\bar{b}_3$
$\alpha(\bar{a}_1)$:	1	*	0	*	....	
$\alpha(\bar{b}_1)$:	0	*	n/2	*	....	
$\alpha(\bar{a}_2)$:	0	*	1	*	....	
$\alpha(\bar{b}_2)$:	0	*	0	*	....	
$\alpha(\bar{a}_3)$:	.	.	.	.		
	.	.	.	.		
$\alpha(\bar{b}_3)$:	.	.	.	.		

Diese Matrix ist aber offensichtlich nicht (anti-)symplektisch, da das (symplektische) Produkt der 1. und 3. Spalte gleich $n/2 \neq 0$ ist. Aus diesem Widerspruch ergibt sich, dass $\pi \circ \phi_1$ und $\pi \circ \phi_2$ und damit auch ϕ_1 und ϕ_2 nicht s-äquivalent sind. Damit ist der Satz bewiesen. □

Analog zu Korollar 2.14 erhalten wir:

3.4 Korollar.

Es seien S wie in 3.3 und $\phi_1, \phi_2 : S \to Sl_2(\mathbb{Z})$ zwei explizit durch die Bilder der Erzeugenden gegebene Homomorphismen. Dann kann man entscheiden, ob ϕ_1 und ϕ_2 (s-)äquivalent sind, und gegebenenfalls einen Isomorphismus α wie in 3.1 sowie den inversen explizit berechnen. □

Im Falle nichtorientierbarer Flächengruppen S kann man mit analogen Methoden vorgehen und die Homomorphismen $\phi : S \to Sl_2(\mathbb{Z})$ in vielen Fällen bzgl. s-Äquivalenz klassifizieren (ausgehend von Satz 2.13). Z.B. gilt (vgl. (Zimmermann, B., 1982)).

3.5 Satz.

Die Homomorphismen $\phi : S \to Sl_2(\mathbb{Z})$, in deren Bild $\phi(S) \subset Sl_2(\mathbb{Z})$ kein Element konjugiert zu $x = \begin{pmatrix} 0 & -1 \\ 1 & 0 \end{pmatrix}$ liegt, lassen sich analog zu 3.3 bzw. 3.4 bzgl. (s-) Äquivalenz klassifizieren. □

Weiter kann man z.B. leicht die surjektiven Homomorphismen $\phi : S \to Sl_2(\mathbb{Z})$ bzgl. s-Äquivalenz klassifizieren (Zimmerman, B., 1982).

Es verbleiben einige Fälle, in denen wir die s-Äquivalenz zweier Homomorphismen nicht entscheiden konnten. Diese Fälle sind im wesentlichen ähnlich dem folgenden.

Es sei $S = \langle c_1, d_1 c_2, d_2 \mid \{c_1, d_1\}\{c_2, d_2\} = 1$ die nichtorientierbare Flächengruppe vom Geschlecht 4 und eine Untergruppe der Form $\mathbb{Z}_4 *_{\mathbb{Z}_2} \mathbb{Z}_4$ von $Sl_2(\mathbb{Z})$ gegeben. Es seien x_1 und x_2 (konjugiert zu x) Erzeugende der beiden $\mathbb{Z}_4$-Faktoren, $x_1^2 = x_2^2 = z$ (erzeugendes Element des Amalgams $\mathbb{Z}_2$). Durch

$$\phi_1(c_1) = x_1\,,\ \phi_1(c_2) = x_2\,,\ \phi_1(d_1) = \phi_1(d_2) = 1$$

bzw. $$\phi_2(c_1) = x_1\,,\ \phi_2(c_2) = x_2\,,\ \phi_2(d_1) = \phi_2(d_2) = z$$

seien zwei Homomorphismen von S nach $\mathbb{Z}_4 *_{\mathbb{Z}_2} \mathbb{Z}_4 \subset Sl_2(\mathbb{Z})$ definiert. Wir wissen nicht, ob ϕ_1 und ϕ_2 s-äquivalent sind.

4. *ÜBER DIE ISOMORPHIE DER FUNDAMENTALGRUPPEN SEIFERTSCHER FASERRÄUME*

Es seien E_ε, $\varepsilon = 1,2$, die Fundamentalgruppen zweier $(n+2)$-dimensionaler Seifertscher Faserräume M_ε von hyperbolischem Typ. Wir haben exakte Sequenzen

$$1 \to \mathbb{Z}^n \hookrightarrow E_\varepsilon \to F \to 1$$

und Operationen $\phi_\varepsilon : F \to Gl_n(\mathbb{Z})$; wir nennen $G_\varepsilon := \phi_\varepsilon(F)$ die Strukturgruppe von M_ε bzw. E_ε. Da wir E_1 und E_2 auf Isomorphie untersuchen wollen, haben wir bereits $E_1/\mathbb{Z}^n = E_2/\mathbb{Z}^n =: F$ angenommen; F ist eine Fuchssche Gruppe mit einer Darstellung

$$F = \langle a_1, b_1, \ldots, a_g, b_g \text{ bzw. } v_1, \ldots, v_g\,;\ s_1, \ldots, s_q \mid \prod_{i=1}^{g} [a_i, b_i] \prod_{j=1}^{q} s_j = 1$$
$$\text{bzw. } \prod_{i=1}^{g} v_i^2 \prod_{j=1}^{q} s_j = 1\,;\ s_j^{\alpha_j} = 1\,,\ j = 1, \ldots, q \rangle,$$

$$\left(2g - 2 + \sum\left(1 - \frac{1}{\alpha_j}\right)\right) > 0 \text{ bzw.}$$

$$\left(g - 2 + \sum\left(1 - \frac{1}{\alpha_j}\right)\right) > 0, \text{ vgl. (Zieschang, H. et al, 1980).}$$

Wir werden im folgenden stets annehmen, dass die s_j trivial auf $\mathbb{Z}^n$ operieren, $\phi_\varepsilon(s_j) = 1$, d.h., dass die Ausnahmefasern von M_ε ebenfalls Tori sind; für die zu behandelnden Situationen ist dies allgemein genug. Aus der Darstellung von F erhalten wir Darstellungen von E_ε

der Form

$$4.1 \quad E_\varepsilon = \langle \textit{Erzeugende von } \mathbb{Z}^n,$$
$$a_1, b_1, \ldots, a_g, b_g \text{ bzw. } v_1, \ldots, v_g;$$
$$s_1, \ldots, s_q \mid$$
$$\text{Relationen von } \mathbb{Z}^n,$$
$$\prod_{i=1}^{g} [a_i, b_i] \prod_{j=1}^{q} s_j = w_\varepsilon \in \mathbb{Z}^n \text{ bzw.}$$
$$\prod_{i=1}^{g} v_i^2 \prod_{j=1}^{q} s_j = w_\varepsilon \in \mathbb{Z}^n,$$
$$s_j^{\alpha_j} = w_{j,\varepsilon} \in \mathbb{Z}^n, \quad j = 1, \ldots, q,$$
$$a_i x a_i^{-1} = \phi_\varepsilon(a_i)(x), \; b_i x b_i^{-1} = \phi_\varepsilon(b_i)(x) \text{ bzw.}$$
$$v_i x v_i^{-1} = \phi_\varepsilon(v_i)(x),$$
$$s_j x s_j^{-1} = x, \; x \in \mathbb{Z}^n \rangle.$$

Dabei haben wir entsprechende Erzeugende von E_1 und E_2 bzw. von F mit denselben Buchstaben bezeichnet; aus dem Zusammenhang geht jeweils hervor, welche Gruppe gemeint ist.

Umgekehrt ist jede Gruppe mit einer Darstellung 4.1 Fundamentalgruppe eines Seifertschen Faserraumes, falls $\mathrm{ggT}(\alpha_j, w_{j,\varepsilon}) = 1$ gilt für $j = 1,\ldots,q$ (genau dann ist E_ε torsionfrei); für die Isomorphieklassifikation der Gruppen wird diese Bedingung jedoch nicht benötigt.

Der folgende Satz verallgemeinert (und vereinfacht) Resultate und Methoden aus (Orlik, P. et al, 1967) von 3-dimensionalen auf den $(n+2)$-dimensionalen Fall. Er führt die Frage nach der Isomorphie von E_1 und E_2 zurück auf einen Vergleich der Operationen $\phi_\varepsilon : F \to Gl_n(\mathbb{Z})$ und der numerischen Grossen $w_\varepsilon, w_{j,\varepsilon}$. Zuvor benötigen wir noch eine Definition:

4.2 Definition.

Fur $a \in F$ sei $\delta(a) = +1$, falls a orientierungserhaltend ist, andernfalls $\delta(a) = -1$. Dann sei V_ε der von allen Elementen der Form $\delta(a)\,\phi_\varepsilon(a)(x) - x$ erzeugte Untermodul von $\mathbb{Z}^n$, $a \in F$, $x \in \mathbb{Z}^n$;

dabei kann man sich bei den a bzw. x auf die Elemente eines Erzeugendensystems von F bzw. $\mathbb{Z}^n$ beschränken. Es ist V_ε invariant unter der Operation von G_ε.

4.3 Satz.

Es ist E_1 isomorph zu E_2 genau dann, wenn die beiden folgenden Bedingungen (i) und (ii) erfüllt sind:

(i) die Operationen ϕ_1 und ϕ_2 sind äquivalent, d.h. es gibt einen Isomorphismus $\bar{\alpha} : F \to F$, ein Element $N \in Gl_n(\mathbb{Z})$ und ein kommutatives Diagramm

$$\begin{array}{ccc} F & \xrightarrow{\phi_1} & Gl_n(\mathbb{Z}) \\ \bar{\alpha}\downarrow & & \downarrow \text{Konjugation mit } N \text{ (bzw. mit } -N) \\ F & \xrightarrow{\phi_2} & Gl_n(\mathbb{Z}) \end{array}$$

(ii) es gibt Elemente $K_j \in G_2$, $\omega_j \in \{+1, -1\}$ mit folgenden Eigenschaften:

Nach einer geeigneten Permutation der Indices j der $w_{j,2}$ gilt (wobei nur Indices mit dem gleichen α_j permutiert werden dürfen; wir werden im folgenden stets ohne Einschränkung annehmen, dass es sich um die identische Permutation handelt):

$$N(w_{j,1}) \equiv K_j(\omega_j w_{j,2}) \bmod \alpha_j ; \tag{4.4}$$

$$\Big(N(w_1) - w_2 - \sum_{j=1} \frac{N(w_{j,1}) - K_j(\omega_j w_{j,2})}{\alpha_j}\Big) \in V_2 ; \tag{4.5}$$

dabei ist $\omega_j = +1$ bzw. $\omega_j = -1$, falls im Urbild von K_j bzgl. ϕ_2 nur orientierungserhaltende bzw. orientierungsändernde Elemente aus F liegen (in den verbleibenden Fällen liegt ein orientierungsänderndes Element aus F im Kern von ϕ_2).

Beweis. Wir benötigen die folgenden Erzeugendenwechsel von E_1 und E_2 bzw. von F (nicht erwähnte Erzeugende bleiben jeweils fest):

(1) $s_j \to s_j s_{j+1} s_j^{-1}$, $s_{j+1} \to s_j$;

(2) $a_i \to a_{i+1}$, $b_i \to b_{i+1}$,

$a_{i+1} \to [a_{i+1}, b_{i+1}]^{-1} a_i [a_{i+1}, b_{i+1}]$,

$b_{i+1} \to [a_{i+1}, b_{i+1}]^{-1} b_i [a_{i+1}, b_{i+1}]$, bzw.

$v_i \to v_{i+1}$, $v_{i+1} \to v_{i+1}^{-2} v_i v_{i+1}^{2}$,

sowie die inversen Prozesse;

(3) $a_g \to a_g s_1$, $b_g \to s_1^{-1} b_g s_1$,

$s_1 \to (s_1^{-1} b_g) s_1 (b_g^{-1} s_1)$; oder

$a_g \to a_g b_g^{-1} s_1^{-1} b_g$, $b_g \to (b_g^{-1} s_1 b_g) b_g (b_g^{-1} s_1^{-1} b_g)$,

$s_1 \to b_g^{-1} s_1 b_g$;

(4) $a_g \to a_g$, $b_g \to s_1^{-1} b_g$,

$s_1 \to (s_1^{-1} b_g a_g b_g^{-1}) s_1 (b_g a_g^{-1} b_g^{-1} s_1)$; oder

$a_g \to s_1^{-1} a_g s_1$, $b_g \to s_1^{-1} a_g^{-1} s_1 a_g b_g s_1$,

$s_1 \to (s_1^{-1} a_g^{-1}) s_1 (a_g s_1)$;

(5) $v_g \to v_g s_1$, $s_1 \to (s_1^{-1} v_g^{-1}) s_1^{-1} (v_g s_1)$; oder

$v_g \to v_g^2 s_1 v_g^{-1}$, $s_1 \to v_g s_1^{-1} v_g^{-1}$.

Mit Hilfe der Prozesse (1)-(5) kann man erreichen, dass ein Wert $w_{j,\varepsilon}$ in der Darstellung von E_ε durch den Wert $K(\omega w_{j,\varepsilon})$ ersetzt wird für beliebiges $K \in G_\varepsilon$ (und zwar ohne Abänderung von ϕ_ε); dabei ist $\omega = +1$ bzw. bzw. $\omega = -1$, falls im Urbild $\phi_\varepsilon^{-1}(K)$ von K nur orientierungserhaltende bzw. orientierungsändernde Elemente aus F liegen; liegt ein orientierungsänderndes Element im Kern von ϕ_ε, so kann $\omega \in \{+1, -1\}$ beliebig gewählt werden.

(a) Angenommen, 4.3 (i) und (ii) gelten. Es sei $\bar{F}$ die freie Gruppe in den Erzeugenden von F. Nach (Zieschang, V., et al, 1980, Theorem 5.8.2) (verallgemeinerter Satz von Nielsen) wird $\bar{\alpha}$ von einem Automorphismus $\bar{\alpha}'$ von $\bar{F}$ induziert mit

$$\bar{\alpha}'\Big(\prod_{i=1}^{g}[a_i, b_i] \prod_{j=1}^{q} s_j\Big) = \ell \Big(\prod_{i=1}^{g}[a_i, b_i] \prod_{j=1}^{q} s_j\Big)^{\omega\tilde{\omega}} \ell^{-1}$$ (analog für die v_i),

$$\bar{\alpha}'(s_j) = \ell_j\, s_j^{\omega\tilde{\omega}_j}\, \ell_j^{-1}\,, \ \omega = \pm 1, \ell, \ell_j \in \bar{F}\ ;$$

dabei ist $\tilde{\omega}_j = +1$, falls ℓ_j orientierungserhaltend ist (aufgefasst als Element von F), andernfalls $\tilde{\omega}_j = -1$; analog für $\tilde{\omega}$ und ℓ (tatsächlich induziert $\bar{\alpha}'$ eine Permutation der Indices j ; durch Prozesse (1) kann man erreichen, dass es sich um die identische Permutation handelt).

Wir werden im folgenden die Darstellungen von E_1 bzw. E_2 durch einfache Prozesse abändern, bis sie übereinstimmen; die geänderten Grössen bezeichnen wir danach wieder mit denselben Symbolen. Wir schliessen an $\bar{\alpha}$ den inneren Automorphismus mit ℓ^{-1} an und können daher ohne Einschränkung $\ell = 1$ annehmen; bei diesem Prozess ändert sich N , die Bedingungen 4.4 und 4.5 sind mit geeigneten K_j jedoch weiterhin erfüllt. Wir ersetzen in der Darstellung von E_1 die Grössen w_1, $w_{j,1}$ und ϕ_1 durch $N(w_1)$, $N(w_{j,1})$ und $N\phi_1 N^{-1}$ (entspricht einem Basiswechsel von $\mathbb{Z}^n$ für E_1) und können daher ohne Einschränkung $N = 1$ annehmen. Durch Erzeugendenwechsel für E_2 mittels der Prozesse (1)-(5) ersetzen wir $w_{j,2}$ durch $K_j(\omega_j w_{j,2})$ und können daher $K_j = 1$ und $\omega_j = 1$ annehmen. Durch Abändern von $\bar{\alpha}$ mittels (1)-(5) können wir $\phi_2(\ell_j) = 1$ und $\tilde{\omega}_j = 1$ erreichen. In der Darstellung von E_1 ersetzen wir s_j durch die neue Erzeugende $s_j \cdot \left(\frac{w_{j,2} - w_{j,1}}{\alpha_j}\right)$. Wegen 4.4 stimmen für die neuen Darstellungen von E_1 und E_2 die Grössen $w_{j,1}$ und $w_{j,2}$ dann überein. Wegen 4.5 gilt $w_1 - w_2 \in V_2$. Durch Ersetzen von a_i, b_i bzw. v_i in der Darstellung von E_2 durch geeignete neue Erzeugende $a_i x_i$, $b_i y_i$ bzw. $v_i x_i$ mit $x_i, y_i \in \mathbb{Z}^n$ kann man w_2 in w_1 abändern (Beweis durch Induktion über g ; man drücke $w_1 - w_2$ durch ein geeignetes endliches Erzeugendensystem von V_2 aus, vgl. 4.2). Nun stimmen die Werte $w_{j,\varepsilon}$ bzw. w_ε für $\varepsilon = 1$ und 2 überein, und es ergibt sich leicht, dass der Automorphismus $\bar{\alpha} : F \to F$ von einem Automorphismus $\alpha : E_1 \to E_2$ mit $\alpha \mid \mathbb{Z}^n = \omega \cdot \mathrm{id}_{\mathbb{Z}^n}$ induziert wird.

(b) Es seien jetzt umgekehrt E_1 und E_2 isomorph durch einen Isomorphismus $\phi : E_1 \to E_2$. Dann gibt es ein kommutatives Diagramm

$$\begin{array}{ccccccccc} 1 & \longrightarrow & \mathbb{Z}^n & \hookrightarrow & E_1 & \longrightarrow & F & \longrightarrow & 1 \\ & & \downarrow{\scriptstyle N\in Gl_n(\mathbb{Z})} & & \downarrow{\scriptstyle \alpha} & & \downarrow{\scriptstyle \bar\alpha} & & \\ 1 & \longrightarrow & \mathbb{Z}^n & \hookrightarrow & E_2 & \longrightarrow & F & \longrightarrow & 1 \end{array};$$

hieraus erhält man leicht das Diagramm in 4.3 (i).

Es wird $\bar\alpha$ von einem Automorphismus $\bar\alpha'$ der freien Gruppe $\bar F$ mit den Eigenschaften wie in Teil (a) des Beweises induziert. Es folgt

$$\alpha(s_j) = \ell_j s_j^{\omega\tilde\omega_j} \ell_j^{-1} . \lambda_j , \quad \lambda_j \in \mathbb{Z}^n ;$$

$$N(w_{j,1}) = \alpha(w_{j,1}) = \alpha(s_j^{\alpha_j}) = \alpha(s_j)^{\alpha_j} = (\ell_j s_j^{\omega\tilde\omega_j} \ell_j^{-1} \lambda_j)^{\alpha_j}$$

$$= \ell_j s_j^{\omega\tilde\omega_j \alpha_j} \ell_j^{-1} \lambda_j^{\alpha_j} = \phi_2(\ell_j)(\omega\tilde\omega_j w_{j,2}) + \alpha_j . \lambda_j \quad \text{(additiv geschrieben)},$$

$$\text{d.h. } \lambda_j = \frac{N(w_{j,1}) - K_j(\omega\tilde\omega_j w_{j,2})}{\alpha_j} \in \mathbb{Z}^n , \quad K_j := \phi_2(\ell_j) ;$$

dies ist Gleichung 4.4. Gleichung 4.5 erhalten wir folgendermassen:

$$N(w_1) = \alpha(w_1) = \alpha\Big(\prod_{i=1}^{g} [a_i, b_i] \prod_{j=1}^{q} s_j\Big) = \prod_{i=1}^{g} [\alpha(a_i), \alpha(b_i)] \prod_{j=1}^{q} \alpha(s_j) =$$

(man schreibe jedes der Elemente $\alpha(a_i)$, $\alpha(b_i)$, $\alpha(s_i)$ (bzw. $\alpha(v_i)$) als Produkt der a_i, b_i, s_j mit einem Element aus $\mathbb{Z}^n$ und bringe die Elemente aus $\mathbb{Z}^n$ dann auf die rechte Seite des Ausdrucks)

$$= \ell \Big(\prod_{i=1}^{g} [a_i, b_i] \prod_{j=1}^{q} s_j\Big)^{\omega\tilde\omega} \ell^{-1} . \lambda_1 \ldots \lambda_q . x \quad (\text{wobei } x \in V_2)$$

$$= \phi_2 (\ell)(\omega\tilde\omega w_2) + \lambda_1 + \ldots + \lambda_q + x .$$

Es folgt

$$N(w_1) - \phi_2(\ell)(\omega\tilde\omega w_2) - \lambda_1 - \ldots - \lambda_q \in V_2 .$$

Wir multiplizieren den Ausdruck mit ω und können daher $\omega = +1$ annehmen (dabei ändert sich N in ωN). Wegen $\phi_2(\ell)(\tilde\omega w_2) - w_2 \in V_2$ ergibt sich nun 4.5 .

Damit ist der Satz bewiesen. □

Bemerkungen. (a) In 4.3 (i) und (ii) kann man N durch $\omega K N$ ersetzen mit $K \in G_2$; dabei sei $\omega = -1$, falls nur orientierungsändernde Elemente aus F durch ϕ_2 auf K abgebildet werden, andernfalls $\omega = +1$. Die Bedingungen sind mit geeigneten K_j weiterhin erfüllt.

(b) Der wesentliche Beweisschritt ist in beiden Richtungen jeweils die Anwendung des verallgemeinerten Satzes von Nielsen. Sollen statt Isomorphieklassen Äquivalenzklassen von Erweiterungen von $\mathbb{Z}^n$ durch F zu einer fest vorgegebenen Operation $\phi : F \to Gl_n(\mathbb{Z})$ klassifiziert werden, so wird dieser Satz nicht benötigt und der Beweis vereinfacht sich. Dabei heissen 2 Erweiterungen E_1, E_2 äquivalent, falls ein kommutatives Diagramm folgender Form existiert:

$$\begin{array}{ccccccccc} 1 & \longrightarrow & \mathbb{Z}^n & \hookrightarrow & E_1 & \longrightarrow & F & \longrightarrow & 1 \\ & & \downarrow{\scriptstyle id_{\mathbb{Z}^n}} & & \downarrow{\scriptstyle \alpha} & & \downarrow{\scriptstyle id_F} & & \\ 1 & \longrightarrow & \mathbb{Z}^n & \hookrightarrow & E_2 & \longrightarrow & F & \longrightarrow & 1 \end{array} .$$

Die Äquivalenzklassen von Erweiterungen werden klassifiziert durch $H^2_\phi(F, \mathbb{Z}^n)$, vgl. z.B. (MacLane, S., 1963). Es sei F der Einfachheit halber eine Flächengruppe (d.h. $q = 0$). Dann folgt aus 4.3 $H^2_\phi(F, \mathbb{Z}^n) \cong \mathbb{Z}^n / V$, wobei V wie in 4.2 definiert sei. Dieses kann man auch folgendermassen bestätigen: F ist eine 2-dimensionale Poincaré-Dualitätsgruppe (Bieri, R., 1976); daher gilt

$$H^2_\phi(F, \mathbb{Z}^n) \cong H_0(F, \mathbb{Z}^n \otimes \overline{\mathbb{Z}}) = (\mathbb{Z}^n \otimes \overline{\mathbb{Z}}) \otimes_F \mathbb{Z} = \mathbb{Z}^n / V .$$

Dabei operieren orientierungserhaltende bzw. -ändernde Elemente aus F als id bzw. -id auf $\overline{\mathbb{Z}} \cong \mathbb{Z}$.

(c) Besonders einfach wird Satz 4.3 im Falle trivialer Operationen von F auf $\mathbb{Z}^n$. Es sei F orientierbar. Dann ist $V_1 = V_2 = \{0\}$ und es sei

$$e_\varepsilon := \left(w_\varepsilon - \sum_{j=1}^{q} (w_{j,\varepsilon} / \alpha_j) \right) \cdot \alpha_1 \cdots \alpha_q \in \mathbb{Z}^n .$$

Es sei $\tilde{e}_\varepsilon$ der grösste gemeinsame Teiler der Komponenten von e_ε in $\mathbb{Z}^n$; $\tilde{e}_\varepsilon$ entspricht der Euler-Zahl für klassische 3-dimensionale Seifertsche Faserräume, vgl. (Neumann, W.D. & Raymond, F., 1978)(Macbeath, A.M.). Aus Satz 4.3 folgt

4.6 Korollar.

$E_1 \cong E_2 \Longrightarrow \tilde{e}_1 = \tilde{e}_2$, d.h. die "Euler-Zahl" ist eine Invariante der Isomorphieklasse. □

(d) Im Falle trivialer Operationen von F auf $\mathbb{Z}^n$ ist es einfacher, statt die Gruppen auf Isomorphie direkt die Seifertschen Faserräume auf fasertreue Homöomorphie hin zu untersuchen. Hierfür wird dann der verallgemeinerte Satz von Nielsen nicht benötigt, vgl. (Seifert, H., 1933)(Orlik, P., & Raymond, F., 1974). Ausserdem kann man die Grössen $w_{j,\varepsilon}$ und w_ε vollständig normieren durch Reduktion mod α_j (bis auf einen Basiswechsel in $\mathbb{Z}^n$, der in Satz 4.3 durch N gegeben ist). Andererseits ergibt sich im Falle trivialer Operation aus Satz 4.3 sofort, dass homöomorphe Räume auch fasertreu homöomorph sind: nach Normierung (und Basiswechsel von $\mathbb{Z}^n$ für einen der beiden Räume) stimmen die Invarianten $w_{j,\varepsilon}$ und w_ε wegen 4.3 überein; daher sind die Räume auch fasertreu homöomorph (Seifert, H., 1933)(Orlik, P. & Raymond, F., 1974).

(e) Für nichttriviale Operationen der s_j auf $\mathbb{Z}^n$ findet man leicht notwendige Bedingungen für die Isomorphie von E_1 und E_2 analog zur 4.4 und 4.5 ; es ist uns jedoch nicht gelungen zu zeigen, dass diese Bedingungen auch hinreichend sind, da sich die Konjugationsfaktoren ℓ_j im Beweis von 4.3 nicht unabhängig voneinander behandeln lassen (analog zu den Prozessen (1)-(5)).

Wir kommen jetzt zum zentralen Satz dieser Arbeit.

4.7 Satz.

Es seien M_1 und M_2 zwei 4-dimensionale Seifertsche Faserräume von hyperbolischem Typ. Es seien Darstellungen der Form 4.1 der Fundamentalgruppen E_1 und E_2 von M_1 und M_2 explizit gegeben, und es liege eine der beiden folgenden Situationen vor:

(i) M_1 ist orientierbar mit orientierbarer Zerlegungsfläche (insbesondere ist $G_1 \subset Sl_2(\mathbb{Z})$);

(ii) M_1 ist transversal orientierbar (d.h. $G_1 \subset Sl_2(\mathbb{Z})$); ausserdem sei entweder $G_1 = Sl_2(\mathbb{Z})$ oder es liege kein Element konjugiert zu $x = \begin{pmatrix} 0 & -1 \\ 1 & 0 \end{pmatrix}$ in G_1 (allgemeiner liege eine Situation vor, für die man die s-Äquivalenz der Operationen entscheiden kann).

Dann lässt sich algorithmisch entscheiden, ob E_1 und E_2 isomorph, d.h. M_1 und M_2 fasertreu homöomorph sind, und gegebenenfalls kann man einen Isomorphismus explizit konstruieren.

Beweis. Da die Strukturgruppe G_1 in (i) und (ii) eine Untergruppe von $Sl_2(\mathbb{Z})$ ist, können wir annehmen, dass die s_j in den Darstellungen von E_1 und E_2 jeweils trivial auf $\mathbb{Z}^2$ operieren, d.h. dass die Ausnahmefasern von M_1 und M_2 ebenfalls Tori sind (im Falle $Gl_2(\mathbb{Z})$ können als Ausnahmefasern zusätzlich Kleinsche Flaschen auftreten).

Wir wenden Satz 4.3 an und geben in 2 Schritten einen Algorithmus an, der die Bedingungen 4.3 (i) und (ii) nachprüft. Dabei nehmen wir an dass die Faktorgruppen $E_1/\mathbb{Z}^2$ und $E_2/\mathbb{Z}^2$ bereits isomorph sind, $F := E_1/\mathbb{Z}^2 = E_2/\mathbb{Z}^2$, d.h. dass die Werte g bzw. $\alpha_1, \ldots, \alpha_q$ für E_1 und E_2 übereinstimmen (notwendige Bedingung für die Isomorphie von E_1 und E_2).

1. Schritt: Vergleich der Operationen ϕ_1 und $\phi_2 : F \to Sl_2(\mathbb{Z})$. Wegen 2.5 konnen wir entscheiden, ob $G_1 = \phi_1(F)$ und $G_2 := \phi_2(F)$ konjugiert sind in $Gl_2(\mathbb{Z})$ (notwendige Bedingung für die Isomorphie von E_1 und E_2), und gegebenenfalls einen Konjugationsfaktor explizit berechnen. Wir können im folgenden daher ohne Einschränkung $G_1 = G_2$ annehmen, $G := G_1 = G_2$.

Es sei $N(G)$ der Normalisator von G in $Gl_2(\mathbb{Z})$. Wegen 2.6 können wir ein Restklassenvertretersystem $R'(G)$ von G in $N(G)$ explizit berechnen. Es sei $R(G) := Gl_2(\mathbb{Z})$, falls $G = \{id\}$ oder $\{id, -id\}$, andernfalls sei $R(G) := \{\pm x,\ x \in R'(G)\}$ ($R(G)$ ist im zweiten Fall endlich).

Wir entscheiden nach 3.3 bzw. 3.5, ob ein kommutatives Diagramm

$$\begin{array}{ccc} F & \xrightarrow{\ \phi_1\ } & Sl_2(\mathbb{Z}) \\ {\scriptstyle \bar{\alpha}}\big\downarrow & & \big\downarrow {\scriptstyle \text{Konjugation mit } N \in R(G)} \\ F & \xrightarrow{\ \phi_2\ } & Sl_2(\mathbb{Z}) \end{array}$$

existiert, d.h. ob $N\phi_1 N^{-1}$ und ϕ_2 s-äquivalent sind. Dabei lassen wir N der Reihe nach alle Elemente in $R(G)$ durchlaufen. Es sei $\tilde{R}(G) \subset R(G)$ die Menge derjenigen $N \in R(G)$, für die ein Diagramm wie oben existiert; im Falle $G = \{id\}$ oder $\{id, -id\}$ ist $\tilde{R}(G)$ leer oder gleich $Gl_2(\mathbb{Z})$. (Dass F im allgemeinen keine Flächengruppe wie in § 3 ist, spielt keine Rolle, da die s_j trivial auf $\mathbb{Z}^2$ operieren). Es sei im folgenden $R(G) \neq \emptyset$ (notwendige Bedingung für die Isomorphie von E_1 und E_2, vgl. Bemerkung (a) nach Satz 4.3). Wir bemerken, dass man die Prozesse, die auf die Darstellung von F bei der Normierung von ϕ_1 bzw. ϕ_2 angewendet werden, in kanonischer Weise auf die Darstellungen von E_1 bzw. E_2 übertragen kann; die numerischen Invarianten in diesen Darstellungen ändern sich dabei nicht.

2. *Schritt:* Vergleich der numerischen Invarianten von E_1 und E_2. Wir unterscheiden mehrere Fälle.

(a) $G \neq \{id\}, \{id, -id\}$ und V_2 hat endlichen Index in $\mathbb{Z}^2$. Dann gibt es eine Zahl c mit $c \cdot \mathbb{Z}^2 \subset V_2$. Wir haben nach Satz 4.3 zu entscheiden, ob es Elemente $N \in \tilde{R}(G)$ und $K_j \in G$ gibt, die das Gleichungssystem 4.4, 4.5 erfüllen. Es sei m das kleinste gemeinsame Vielfache von $\alpha_1, \ldots, \alpha_q$ und c. Wir versuchen, das Gleichungssystem mod m zu lösen, und zwar der Reihe nach für die endlich vielen Elemente $N \in \tilde{R}(G)$. Da $Sl_2(\mathbb{Z}_m)$ endlich ist, sind alle möglichen Tupel $(\bar{K}_1, \ldots, \bar{K}_q)$ in $(G \bmod m)^q \subset Sl_2(\mathbb{Z}_m)^q$ berechenbar, die 4.5 und darüber hinaus 4.4 mod α_j erfüllen; zur Bestimmung der ω_j hat man dabei das Bild in G mod m der orientierungserhaltenden Elemente in F zu berechnen (für ϕ_2). Dann wählt man $(K_1, \ldots, K_q) \in G^q \subset Sl_2(\mathbb{Z})^q$ mit $(\bar{K}_1, \ldots, \bar{K}_q) = (K_1, \ldots, K_q) \bmod m$ (unter Berücksichtigung der ω_j). Für diese K_j ist das Gleichungssystem 4.4, 4.5 dann erfüllt.

(b) $G = \{id\}$ oder $\{id, -id\}$ und $V_2 = 2 \cdot \mathbb{Z}^2$. Es ist $\tilde{R}(G) = Gl_2(\mathbb{Z})$. Wir setzen für K_j jeweils id oder - id ein und lösen wie unter (a) die Gleichungen 4.5 bzw. 4.4 mod m bzw. mod α_j, wobei jetzt $N \in Gl_2(\mathbb{Z})$ gesucht ist.

(c) $G = \{id\}$ oder $\{id, -id\}$ und $V_2 = \{0\}$. Es ist $\tilde{R}(G) = Gl_2(\mathbb{Z})$. Wir haben ein Gleichungssystem der Form

$$N(w_{j,1}) \equiv w_{j,2} \bmod \alpha_j, \quad N\underbrace{\left(w_1 - \sum_{j=1}^{q} \frac{w_{j,1}}{\alpha_j}\right)}_{=:\, e_1} = \underbrace{w_2 - \sum_{j=1}^{q} \frac{w_{j,2}}{\alpha_j}}_{=:\, e_2}$$

zu lösen. Durch einen Basiswechsel für $\mathbb{Z}^2$ in der Darstellung von E_1 können wir $e_1 = \binom{x}{0} \in \mathbb{Z}^2$ erreichen, $x \in \mathbb{Z}$. Angenommen es gilt $N_0(e_1) = e_2$, $N_0 \in Gl_2(\mathbb{Z})$. Dann kommen für N genau die Elemente $N_0\begin{pmatrix} 1 & y \\ 0 & \pm 1 \end{pmatrix}$ in Frage, $y \in \mathbb{Z}$. Für die Gleichungen 4.4 können wir dann mod $(\alpha_1 \cdot \ldots \cdot \alpha_q)$ reduzieren und so feststellen, ob eine gemeinsame Lösung N für 4.4 and 4.5 existiert.

Der noch verbleibende Fall ist

(d) $V_2 = \mathbb{Z} \subset \mathbb{Z}^2$, insbesondere $G \neq \{id\}, \{id, -id\}$ und $\tilde{R}(G)$ endlich. Es sei $g = \begin{pmatrix} a & b \\ c & d \end{pmatrix} \in G$. Wird ein orientierungserhaltendes Element aus F durch ϕ_2 auf g abgebildet, so liegen $\binom{a-c}{c} = g\binom{1}{0} - \binom{1}{0}$ und $\binom{b}{d-1} = g\binom{0}{1} - \binom{0}{1}$ in V_2; im orientierungsändernden Fall liegen $\binom{a+1}{c} = g\binom{1}{0} + \binom{1}{0}$ und $\binom{b}{d+1} = g\binom{0}{1} + \binom{0}{1}$ in V_2. Da V_2 eindimensional ist, folgt

$$\det\begin{pmatrix} a-1 & b \\ c & d-1 \end{pmatrix} = 0 \quad \text{bzw.} \quad \det\begin{pmatrix} a+1 & b \\ c & d+1 \end{pmatrix} = 0 .$$

Es folgt $|\mathrm{Spur}(g)| = |a+d| = 2$. Daher besteht das Bild von G in $PSl_2(\mathbb{Z})$ ausschliesslich aus parabolischen Elementen und ist unendlich zyklisch. Durch einen Basiswechsel für $\mathbb{Z}^2$ können wir erreichen, dass G von Elementen der Form $\begin{pmatrix} \pm 1 & x \\ 0 & \pm 1 \end{pmatrix}$ oder von $\begin{pmatrix} \pm 1 & x \\ 0 & \pm 1 \end{pmatrix}, \begin{pmatrix} -1 & 0 \\ 0 & -1 \end{pmatrix}$ erzeugt wird, $0 < x \in \mathbb{Z}$. Da V_2 eindimensional ist, gilt $V_2 = \langle \binom{x}{0} \rangle$. Der Normalisator $N(G)$ von G in $GL_2(\mathbb{Z})$ wird erzeugt von $\begin{pmatrix} 1 & 1 \\ 0 & 1 \end{pmatrix}$, $\begin{pmatrix} -1 & 0 \\ 0 & -1 \end{pmatrix}, \begin{pmatrix} 1 & 0 \\ 0 & -1 \end{pmatrix}$, daher ist ein Restklassenvertretersystem von G in $N(G)$ in der Menge $R := \left\{ \begin{pmatrix} \varepsilon_1 & y \\ 0 & \varepsilon_2 \end{pmatrix}, \varepsilon_1, \varepsilon_2 \in \{+1, -1\}, -x < y < x \right\}$ enthalten. Wir setzen für N in 4.4 und 4.5 der Reihe nach alle Elemente in R ein und versuchen, das Gleichungssystem mit $K_j = \begin{pmatrix} \pm 1 & k_j \\ 0 & \pm 1 \end{pmatrix}$ zu lösen, $k_j \in \mathbb{Z}$. Ist Gleichung 4.5 in der zweiten Komponente von $\mathbb{Z}^2$ erfüllt, so rechnen wir wieder mod m, $m = \mathrm{kgV}(\alpha_1, \ldots, \alpha_q, x)$, und können wie unter (a) feststellen, ob eine Lösung existiert.

Mit Hilfe der Schritte 1 und 2 lässt sich entscheiden, ob E_1 und E_2 isomorph sind. Da alle Prozesse, die die Darstellungen von

E_1 und E_2 im Falle von Isomorphie ineinander überführen, explizit angegeben wurden, lässt sich gegebenenfalls ein Isomorphismus zwischen E_1 und E_2 explizit konstruieren. □

Die Beweismethoden von Satz 4.7 gestatten es, zu fest vorgegebener Gruppe $G \subset Sl_2(\mathbb{Z})$ alle 4-dimensionalen Seifertschen Faserräume von hyperbolischem Typ mit Strukturgruppe G zu klassifizieren, z.B. durch Angabe eines vollständigen Repräsentantensystems für die Isomorphieklassen der Fundamentalgruppen. Dabei gibt es in den Fällen (a) und (b) des Beweises von Satz 4.7 endlich viele, in den Fällen (c) und (d) unendlich viele Isomorphieklassen für festes G und F. Liegen Ausnahmefasern vor, d.h. besitzt F nichttriviale Torsion, so ist das Verfahren wegen des Auftretens der K_j allerdings langwierig. Ist der Normalisator $N(G)$ von G in $Gl_2(\mathbb{Z})$ bekannt und liegen keine Ausnahmefasern vor, so lassen sich die Isomorphieklassen verhältnismässig einfach aufzählen.

In höheren Dimensionen gilt:

4.8 Satz.

Es sei $\mathcal{E}_{n+2}$ die Klasse der Fundamentalgruppen (orientierbarer) (n+2)-dimensionaler Seifertscher Faserräume (mit orientierbarer Zerlegungsfläche), $n > 3$. Dann ist das Isomorphieproblem für die Klasse $\mathcal{E}_{n+2}$ unlösbar, d.h. es gibt keinen Algorithmus, der zu entscheiden gestattet, ob zwei beliebige Gruppen in $\mathcal{E}_n$ isomorph sind oder nicht.

Beweis. In (Miller III, C.F., 1971, Chapter III.C) wird eine endlich erzeugte Untergruppe G von $Sl_n(\mathbb{Z})$, $n > 3$, konstruiert, sowie eine Klasse $\mathcal{K}$ endlicher Teilmengen von G, für die sich algorithmisch nicht entscheiden lässt, ob die Elemente einer beliebig vorgegebenen Menge der Klasse die Gruppe G erzeugen oder nicht. Ausserdem ergibt sich aus der Konstruktion, dass, falls die Elemente G nicht erzeugen, die von ihnen erzeugte Untergruppe nicht isomorph, insbesondere auch nicht konjugiert ist zu G. [Diesen Hinweis verdanke ich C.F. Miller III.] Daher lässt sich nicht entscheiden, ob eine beliebige Menge der Klasse eine zu G konjugierte Untergruppe erzeugt.

Es sei F_{2g} die orientierbare Flächengruppe vom Geschlecht $2g$, $F_{2g} = \langle a_1, b_1, \ldots, a_{2g}, b_{2g} \mid \prod_{i=1}^{2g} [a_i, b_i] = 1 \rangle$. Ist M eine

Menge in K mit g oder weniger Elementen, so sei $\phi : F_{2g} \to \langle M \rangle \subset Sl_n(\mathbb{Z})$ (die von M erzeugte Untergruppe von G) ein surjektiver Homomorphismus mit $\phi(b_i) = 1$ für $1 \le i \le 2g$, $\phi(a_i) = 1$ für $g+1 \le i \le 2g$, und $E_{M,g}$ das semidirekte Produkt von $\mathbb{Z}^n$ mit F_{2g}. Der Isomorphietyp von $E_{M,g}$ hängt nicht von der speziellen Wahl von ϕ ab (man wende die Prozesse 2.8 (c) und (d) an). Ausserdem sind zwei Gruppen $E_{M_1,g}$ und $E_{M_2,g}$ für M_1, M_2 in K genau dann isomorph, wenn die von M_1 und M_2 erzeugten Untergruppen konjugiert sind in $Gl_n(\mathbb{Z})$. Daher lässt sich allgemein nicht entscheiden, ob eine beliebige Gruppe $E_{M,g}$ zu einer Gruppe $E_{M_0,g}$ isomorph ist, wobei M_0 ein festes Erzeugendensystem von G ist. □

Wegen 4.8 kann es also einen Satz analog zu 4.7 für $(n+2)$-dimensionale Seifertsche Faserräume mit $n > 3$ nicht geben. Man kann aber z.B. versuchen, die $(n+2)$-dimensionalen Seifertschen Faserräume mit fest vorgegebener Strukturgruppe G oder fest vorgegebener Strukturabbildung $\phi : F \to Sl_n(\mathbb{Z})$ bzw. $Gl_n(\mathbb{Z})$ zu klassifizieren. Die Schwierigkeit liegt hierbei in einem Vergleich der Operationen ϕ und $N\phi N^{-1}$ für Elemente N in Normalisator $N(G)$ von G (falls der Normalisator sich überhaupt bestimmen lässt). Für den Fall, dass $N(G) = G$ gilt, lassen sich die Seifertschen Faserräume zu fester Strukturabbildung $\phi : F \to Gl_n(\mathbb{Z})$ mit $\phi(F) = G$ analog zum Beweis von 4.7 klassifizieren durch geeignete numerische Betrachtungen.

Wir kommen jetzt noch einmal auf den 4-dimensionalen Fall zurück. Es sei E die Fundamentalgruppe eines 4-dimensionalen Seifertschen Faserraumes M von hyperbolischem Typ, $G \subset Gl_2(\mathbb{Z})$ sei die Strukturgruppe, V sei definiert wie in 4.2. Es sei $G_0 := G \cap Sl_2(\mathbb{Z})$ und $\bar{G}_0$ das Bild von G_0 in $PSl_2(\mathbb{Z})$.

Der Beweis von Satz 4.7 legt eine Einteilung in die folgenden 3 Typen nahe:

4.9 Definition.

Die Strukturgruppe G von E bzw. M heisst

(a) elliptisch, falls G endlich ist, d.h. $\bar{G}_0$ nur aus elliptischen Elementen besteht;

(b) parabolisch, falls $\bar{G}_0$ nichttrivial ist und ausser der Identität nur parabolische Element enthält;

(c) hyperbolisch, falls $\bar{G}_0$ ein hyperbolisches Element enthält.

Der folgende Satz fasst einige einfache Eigenschaften von M bzw. E zusammen.

4.10 Satz.

(a) G ist elliptisch genau dann, wenn eine endliche Überlagerung von M eine effektive Operation des 2-dimensionalen Torus $S^1 \times S^1$ besitzt.

(b) G ist parabolisch genau dann, wenn eine endliche Überlagerung von M eine effektive S^1-Operation, jedoch keine effektive $S^1 \times S^1$-Operation besitzt.

(c) Ist G hyperbolisch, so ist E Untergruppe von endlichem Index in dem semidirekten Produkt von $\mathbb{Z}^2$ mit $F = E/\mathbb{Z}^2$. Zu fest vorgegebener Operation $\phi : F \to Gl_2(\mathbb{Z})$ gibt es nur endlich viele nichtisomorphe Gruppen E.

Beweis. (a) Ist G endlich, so gibt es eine Untergruppe E_1 von endlichem Index in E mit trivialer Strukturgruppe G_1. Dann liegt $\mathbb{Z}^2$ zentral in E_1. Dies ist äquivalent zur Existenz einer effektiven $(S^1 \times S^1)$-Operation auf der Überlagerung M_1 zu E_1, vgl. (Conner, P.E. & Raymond, F., 1977 § 9),(Conner, P.E. & Raymond, F., 1969).

(b) Genau dann bestitzt eine endliche Uberlagerung M_1 von M mit Strukturgruppe $G_1 \subset Sl_2(\mathbb{Z})$ eine S^1-, aber keine $(S^1 \times S^1)$-Operation, wenn das Zentrum von $E_1 = \pi_1 M_1$ isomorph zu $\mathbb{Z} \subset \mathbb{Z}^2$ ist (Conner, P.E. & Raymond, F., 1969 & 1977). Dies ist genau dann der Fall, wenn die Elemente in G_1 nur Eigenwerte $+1$ besitzen und G_1 nichttrivial ist.

(c) Enthält $\bar{G}_0$ ein hyperbolisches Element, so ist V von endlichem Index in $\mathbb{Z}^2$. Wegen 4.3 gibt es dann zu fester Operation $\phi : F \to Gl_2(\mathbb{Z})$ nur endlich viele nicht-isomorphe Gruppen E, insbesondere ist $H^2_\phi(F, \mathbb{Z}^2)$ endlich. Es beschreibe $x \in H^2_\phi(F, \mathbb{Z}^2)$ die Erweiterung $1 \to \mathbb{Z}^2 \hookrightarrow E \to F \to 1$. Es sei $\phi : \mathbb{Z}^2 \to \mathbb{Z}^2$ die Multiplikation mit m für ein $m > 0$ mit $m \cdot x = 0$. Für die induzierte Abbildung $\alpha^* : H^2_\phi(F, \mathbb{Z}^2) \to H^2_\phi(F, \mathbb{Z}^2)$ gilt dann $\alpha^*(x) = m \cdot x = 0$, d.h. $\alpha^*(x)$ beschreibt das semidirekte Produkt $\tilde{E}$ von $\mathbb{Z}^2$ mit F, und E liegt von endlichem Index m in $\tilde{E}$. □

LITERATURVERZEICHNIS

Bieri, R. (1976). Homological Dimension of Discrete Groups. Queen Mary College Math. Notes.

Conner, P.E. & Raymond, F. (1970). Actions of compact Lie groups on aspherical manifolds. In Topology of Manifolds (Proc. Inst. Univ. of Georgia, Athens, Ga., 1969), pp.227-264. Chicago, Illinois, Markham.

Conner, P.E. & Raymond, F. (1977). Deforming homotopy equivalences to homeomorphisms in aspherical manifolds. Bull. Amer. Math. Soc. 83, 36-85.

Hall, M. Jr., (1953). Subgroups of free products. Pacific J. Math. 3, 115-120.

Holmann, H. (1964). Seifertsche Faserräume. Math. Ann. 157, 138-166.

Lyndon, R.C. (1979). Quadratic equations in free products with amalgamation. Houston J. Math. 4, 91-103.

Lyndon, R.C. & Schupp, P.E. (1977). Combinatorial Group Theory. Berlin-Heidelberg-New York, Springer.

Macbeath, A.M. (Preprint) The fundamental groups of the three dimensional Brieskorn manifolds.

Miller III, C.F. (1971). On Group-Theoretic Decision Problems and their Classification. Annals of Math. Studies 68, Princeton University Press.

MacLane, S. (1963). Homology. Berlin-Heidelberg-New York, Springer.

Nielsen, J. (1937). Die Struktur periodischer Abbildungen von Flächen. Det. Kgl. Dansk Videnskaternes Selskab. Mat.-fys. Meddelerer 15, 1-77.

Neumann, W.D. & Raymond, F. (1978). Seifert manifolds, plumbing, μ-invariant and orientation reversing maps. Proc. Conf. at Santa Barbara, pp.163-196. Lecture Notes in Math. 664, Berlin, Springer.

Orlik, P. (1970). Seifert Manifolds. Lecture Notes in Math. 291. Berlin-Heidelberg-New York, Springer.

Orlik, P. & Raymond, F. (1970),(1974). Actions of the torus on 4-manifolds-I,II. Trans. Amer. Math. Soc. 152, 531-559. Topology 13, 89-112.

Orlik, P., Vogt, E. & Zieschang, H. (1967). Zur Topologie gefaserter dreidimensionaler Mannigfaltigkeiten. Topology 6, 49-64.

Rosenberger, G. (1980). Gleichungen in freien Produkten mit Amalgam. Math. Z. 173, 1-12.

Seifert, H. (1933). Zur Topologie dreidimensionaler gefaserter Räume. Acta Math. 60, 147-238.

Serre, J.-P. (1980). Trees. Berlin-Heidelberg-New York, Springer.

Smith, P.A. (1967). Abelian actions on 2-manifolds. Mich. Math. J. 14, 257-275.

Stillwell, J. (1980). Classical Topology and Combinatorial Group Theory. New York-Heidelberg-Berlin, Springer.

Thornton, M.C. (1967). Singularly fibered manifolds. Illinois J. Math. 11, 189-201.

Thornton, M.C. (1968). On closing manifolds fibered over surfaces. Proc. Amer. Math. Soc. 19, 890-894.

Vogt, E. (1973). Stable foliations of 4-manifolds by closed surfaces. Inventiones math. 22, 321-348.

Vogt, E. (1977). Foliations of codimension 2 with all leaves compact on closed 3-, 4- and 5-manifolds. Math. Z. 157, 201-223.

Waldhausen, F. (1967). Eine Klasse von 3-dimensionalen Mannigfaltigkeiten I, II. Invent. Math. 3, 308-333, 4, 87-117.

Zieschang, H. (1964). Alternierende Produkte in Freien Gruppen. Abh. math. Sem. Univ. Hamburg 27, 13-31.

Zieschang, H. (1969). On toric fiberings over surfaces (russisch). Mat. Zametki 5, 569-576; englische Übersetzung: Math. Notes 5, 341-345.

Zieschang, H. (1981). Finite Groups of Mapping Classes of Surfaces. Lecture Notes in Math. 875, Berlin, Springer.

Zieschang, H. (Preprint). On decompositions of discrete groups of motions of the plane.

Zieschang, H., Vogt, E., Coldeway, H.-D. (1980). Surfaces and Planar Discontinuous Groups. Lecture Notes in Math. 835, Berlin-Heidelberg-New York, Springer.

Zimmermann, B. (1982). Über höherdimensionale Seifertsche Faserräume. Habilitationsschrift, Bochum.

CONFERENCE PROBLEM LIST

1. Let $\bar{\mu}$ and μ^* denote the Milnor numbers of a link in R^3. Is it true that for all indices $\bar{\mu}(i_1 i_2 \ldots i_k) = \mu^* (i_1 i_2 \ldots i_k)$? The definition of the Milnor numbers is too complicated to be given here; see J. Milnor 'Isotopy of Links', Algebraic Geometry and Topology, a symposium in honour of S. Lefshetz. Princeton U. Press, 1957.

2. Is the unknotting number additive over knot composition? (C. Kearton)

3. Let $G(k) = \pi_1(S^3 - k)$ where k is a knot. Define $G^n(k) = \langle g^n | g \in G(k) \rangle$. So $G^n(k)$ is the normal subgroup of $G(k)$ generated by all nth powers. Is it true that G/G^n is finite for most knots? If n is prime, is it true that $G/G^n \cong \mathbb{Z}_n$.
(K. Murasugi)

4. Define a *meridian* of a knot or link k to be a simple closed curve spanning a disc which meets k transversely in a single point. Let $m(k)$ the *meridian number* of k denote the minimum number of meridians needed to generate the group of k. Is it true that $m(k) = b(k)$ the bridge number of k? For 2-bridge knots this question has been answered in the affirmative by M. Boileau. He, M. Rost and H. Zieschang have verified the truth of this conjecture for a large class of knots and links. (Torus and Montesinos knots and links.)

If $G(k) = \{x_1, \ldots, x_m | r_1, \ldots, r_m\}$ is a Wirtinger presentation with $m = m(k)$ is it true that $\langle x_1 \rangle \cap \langle x_2, \ldots, x_m \rangle = \{1\}$? A consequence of this question if it should turn out to be true is that $m(k_1 \# k_2) = m(k_1) + m(k_2) - 1$.
(T. Maeda)

5. Consider the integer lattice in R^3 and the closed curves k formed from lattice edges.

Conjecture: The probability of getting a genuine knot with n edges tends to 1 as n tends to infinity.

6. Let $h(M)$ denote the Heegaard genus of the 3-manifold M, that is the smallest genus of an orientable closed surface which divides M into two handlebodies. Let $r(M)$ denote the rank of $\pi_1 M$. Then $r(M) \leq h(M)$. Equality holds for all Seifert fibre spaces except those of type $S = S(0; e_o; 1/2, \ldots, 1/2, \beta_m/2\lambda + 1)$ $(\lambda > 0$, m even, $m \geq 4$, $e_o \nequiv 1/2(2\lambda + 1))$. For $m = 4$ it is known that

$$2 = r(S(0;e_o;1/2,1/2,1/2,\beta_m/2\lambda+1)) < h(S(0;e_o;1/2,1/2,1/2,\beta_m/2\lambda+1)) = 3.$$

In general it is

$$m - 2 = r(S) \leq h(M) \leq m - 1 .$$

([Boileau-Zieschang, C.R. or Invent. Math.])

Problems: a) Determine $h(S)$ for $m > 4$,

b) Find irreducible manifolds M with $h(M) - r(M) > 1$.

7. Let M be a 3-manifold with a Riemannian metric and with $\pi_2(M) = 0$. Let F be a closed surface, not S^2 or P^2 and let $f : F \to M$ be a smooth inversion with least area in its homotopy class and with $f_* : \pi_1(F) \to \pi_1(M)$ injective. If f is homotopic to an embedding and has trivial normal bundle then Freedman, Han & Scott, Invent. Math. (1983) have shown that either f is an embedding or is the double cover of a one-sided embedded surface. What can be said when the normal bundle of f is non-trivial? (P. Scott)

8. Let F be a closed compact surface of genus γ in R^3 with p local maxima and q local minima. Can F be isotoped so that the number of maxima and minima are both equal to k where $k \leq \min\{p, q\}$?

This is true for $\gamma \leq 1$ and also for $p = 1$. (H. Morton)

9. A *link homotopy* of the disjoint union $X \cup Y$ in Z is an arbitrary homotopy in which the images of X and Y are distinct. Milnor studied this equivalence for links in S^3.

Is there a non-trivial link $S^2 \cup S^2$ in R^4 which is not homotopic to zero? Note that $S^p \cup S^1$ link in R^{p+2}. (W. Massey communicated by D. Rolfsen)

Addendum R. Fenn and D. Rolfsen have shown that such a link exists if both components are allowed to have singularities.

10. Any orientable closed compact 3-manifold can be constructed from its surgery data $(\underline{V}, \underline{J})$. That is, $\underline{V} = V_1 \cup \ldots \cup V_k$ is a disjoint union of solid tori in S^3 and $\underline{J} = J_1 \cup \ldots \cup J_k$ where the J_i are essential simple closed curves in ∂V_i which span discs after the surgery.

Write $M(\underline{V}, \underline{J})$ for the resulting manifold. Let $G = \pi_1(S^3 - \underline{V})$ and let H be the normal closure in G of the classes $[J_1], \ldots, [J_n]$. Note that $\pi_1(M) = G/H$. Let $\overline{G} = G/[G, H]$ and $\overline{H} = H/[G, H]$.

Theorem. (Fukuhara : *If* $M(\underline{V}, \underline{J}) \cong M(\underline{V}', J')$ *then* $(\overline{G}, \overline{H}) \times \mathbb{Z}^m \cong (\overline{G}', \overline{H}') \times \mathbb{Z}^n$ *for some integers* m, n.

Are there non-homeomorphic M_1 and M_2 which have the same π_1 but are distinguished by their stable classes of pairs $(\overline{G}, \overline{H})$? (D. Rolfsen)